The Structure of Amorphous Materials using Molecular Dynamics

Online at: https://doi.org/10.1088/978-0-7503-2436-6

The Structure of Amorphous Materials using Molecular Dynamics

Carlo Massobrio
Université de Strasbourg, CNRS, Institut de Physique et Chimie des Matériaux de Strasbourg, UMR 7504, Strasbourg F-67034, France

IOP Publishing, Bristol, UK

ISBN 978-0-7503-2436-6 (ebook)
ISBN 978-0-7503-2434-2 (print)
ISBN 978-0-7503-2437-3 (myPrint)
ISBN 978-0-7503-2435-9 (mobi)

DOI 10.1088/978-0-7503-2436-6

Version: 20221101

IOP ebooks

British Library Cataloguing-in-Publication Data: A catalogue record for this book is available from the British Library.

Published by IOP Publishing, wholly owned by The Institute of Physics, London

IOP Publishing, No.2 The Distillery, Glassfields, Avon Street, Bristol, BS2 0GR, UK

US Office: IOP Publishing, Inc., 190 North Independence Mall West, Suite 601, Philadelphia, PA 19106, USA

Contents

Preface

The book is focussed on modeling of disordered materials (mostly, but not only, amorphous) at the atomic scale by using a theoretical tool known (in the areas of physics and chemistry of condensed matter) as molecular dynamics (MD). MD allows following the atomic trajectories of a collection of atoms by accounting for the interatomic forces and it is particularly well suited when the structural arrangements are unknown or hard to grasp experimentally. Due to the extended time scales inherent in the relaxation of disordered materials and the intrinsic time limitations of molecular dynamics, the quest of predictive power when modeling disordered materials might appear hopeless or at very best bound to remain an academic exercise. The main intention of this book is to demonstrate that this is not the case, by providing revealing examples of structural studies carried out by employing molecular dynamics, over a time span of more than 30 years, by the author and his co-workers (but not only) on a variety of systems. The realistic character of this approach is underlined by the extensive use made of the first-principles version of MD (FPMD), in which the interatomic interactions are obtained by accounting explicitly for the electronic structure. Application of FPMD to disordered materials (amorphous in particular) has been characterized by a conflict between the high computational cost of the method (due to the explicit account of bonding through density functional theory (DFT)) and the length of the temporal trajectories needed to obtain statistically meaningful results. In this book it will be shown how to address this issue for specific cases, by providing examples of outstanding agreement between properties calculated via FPMD and experimental data. To achieve these results, two challenges have been met. First, one has to adopt a theoretical scheme capable of describing realistically a variety of bonding situations including those combining fundamental chemical characters (for instance, iono-covalent). Second, the computer production of the disordered phases has to be made compatible with the space and time limitations. Regarding these issues, the book shows in detail how they have been addressed and how technical stumbling blocks have been circumvented by adapting as much as possible the FPMD description to the specific systems under consideration. Systems treated encompass (for the classical MD part), Lennard-Jones and metallic amorphous alloys, while chalcogenide systems are mostly (but not exclusively) targeted by the FPMD studies. This choice reflects the route I followed throughout my engagement in the field but it does not exclude results obtained by other researchers (largely quoted throughout the book) especially when they are useful to complement and enrich understanding. In this respect, focusing on the realization of one team over the years is proposed as a valuable strategy to follow the evolution of MD techniques applied to disordered materials. Indeed, via my research efforts, I have witnessed a period characterised by an impressive increase in conceptual advances and computational power, thereby allowing for substantial adapting of methods and extension in the number of systems prone to be modeled. Attention is paid to limit the biographical flavor to what is strictly necessary to define the context of the scientific challenge, even though several autobiographical notes are present to enrich the presentation and dwell on personal experience. Therefore, this book is not intended to celebrate the scientific accomplishments of one person. Instead, it takes advantage of what a scientist (and his co-workers) have been able to achieve to (hopefully) enhance awareness of the enormous capabilities of atomic-scale calculations to increase knowledge among computational material scientists interested in glasses and amorphous systems.

Acknowledgement

The simplest way to deal with a section devoted to acknowledgements is to include all of the colleagues I have interacted with over the past 35 years or so (even 40 if I go back to the early days of graduate school at the University of Rome). It might sound a bit trivial and corny but, with a few exceptions not worthy of mention, I have learned something from all of my co-workers regardless of their age and status. There is no point in compiling an explicit list since they can be easily found when looking up the quoted papers I co-wrote with them. Concerning this book, I have to admit I worked on it almost exclusively by myself since I wanted it to have some self-biographical touch. I am convinced it would have been impossible to achieve this goal if I had decided to share this venture with some partner. In spite of these considerations, a great help with figures and some part of specific chapters has come from Evelyne Martin, also quite effective in ensuring a critical reading of the entire book. The section devoted to the CPMD code is a result of fruitful exchanges with my teammate Mauro Boero. Guido Ori, Assil Bouzid and Sébastien Le Roux have also helped in this context.

Coming back to acknowledge scientists that greatly contributed to shape up my background and evolution as a scientist, and by reiterating that all co-workers have been special to some extent, I have selected five outstanding colleagues that I would like to put in my short list of people I owe more than some gratitude. In chronological order, Giovanni Ciccotti is the first. I met him in 1982 when I was a student of Physics at the University of Rome. He was a quite nice story teller and he was proposing something new and stimulating: molecular dynamics. I started punching cards under his direction and I published my first papers with him. He sent me to the USA to work with W G Hoover and he was the first to believe that some talent was hiding behind a quite meticulous good student. We worked together for a short period of time and I must have met him not more than 10 times over the past 30 years. But he played a big role and, as such, he has to be recognized.

I enjoyed working with Vassilis Pontikis on disordered systems of interest in metallurgy. That period marked a growth of interest in issues more closely related to materials science, amorphization by solid state reaction being the first. Vassilis was very tactful, extremely helpful and he acted as an experimentalist fully knowledgeable in computational methods. He had great, modern views on the role of simulation in materials, much in advance with respect to those times. The part of this book devoted to classical molecular dynamics applications is the results of our collaboration during my early days in France.

I began working on first-principles molecular dynamics in 1993 with two great teachers, Alfredo Pasquarello and Roberto Car, from whom I learned a lot, not only in terms of notions but also, and most likely more, in terms of approach to work, ambition and eagerness to strive for high quality achievements. Car–Parrinello methodology was already well established but not to a very large extent in the area of amorphous and glassy systems. By selecting $GeSe_2$ the idea was to look for analogies and differences with SiO_2. In reality, at least for me, it was the beginning of a long story that continued ever since, by including more and more systems worthy of interest of that family. Alfredo is the most rigorous and demanding scientist I have ever met. I learned from him never to give up when searching the solution of a problem and to never settle for a less prestigious result just to have a paper published quicker.

I am grateful to my teammate and co-worker Mauro Boero for the role he plays in bridging the gap between materials science and computing. Mauro is behind the CPMD code, installed on almost any high performance computer of this … galaxy (!) as a result of his hard work over the past 20 years and more. Limiting my acknowledgement to this simple sentence would be quite unfair. Mauro is most of all a great scientist, active in many areas of computer science (physics, chemistry and biology). However, he has a defect slightly undermining his potential: his immeasurable modesty. In any event I owe him a lot for the many hours spent, for the benefit of his closest co-workers to which I have the chance to belong, programming, debuggging, adapting and technically improving our main working tool.

Finally, I would like to thank Caroline Mitchell of IOP Publishing for inviting me to write a book on my research activities and for her continued support during a writing period that turned out to be longer than expected. She has been very patient and encouraging, by providing useful advice during our periodic on-line meetings with a reassuring attitude.

Author biography

Carlo Massobrio

I graduated in physics from the University of Rome in 1983 with a diploma thesis on the thermal conductivity of a Lennard-Jones fluid by non-equilibrium molecular dynamics (MD), supervised by G Ciccotti. At that time, the Italian education system was offering, at the end of the undergraduate studies (the so-called 'laurea'), only a very few PhD scholarships, not particularly attractive from a financial point of view. Hired by W G Hoover as post-doctoral fellow in 1984, I spent a year at the University of California, Davis, and pursued molecular dynamics work in connection with non-equilibrium properties and statistical mechanics approaches. After a second year in the USA at Clemson University, I moved to France to work as a postdoc on the structure of amorphous systems. After this period spent at the Physics Metallurgy laboratory of CEA (Commissariatl'énergie atomique et aux énergies), MD studies on realistic systems (metallic alloys, binary oxides) were pursued in CNRS (Centre National de la Recherche Scientifique) laboratories located in Meudon and Vitry in 1988. In 1989 I was hired by the Department of Chemistry of CNRS on one of the first positions explicitly created to study materials at the atomic scale. The goal was to develop and apply molecular dynamics to materials chemistry, nanostructures and, more generally, materials science. The underlying scientific mission was to drive the transition from qualitative to quantitative, by including in my field of expertize first-principles techniques to account for chemical bonding. In 1991, I had the opportunity, while keeping my permanent position within CNRS, to move to the Ecole Polytechnique Fédérale de Lausanne, EPFL (Suisse) to work in the area of clusters and nanosystems on an extended sabbatical leave that lasted until 1995. That period was marked by the collaboration with Alfredo Pasquarello and Roberto Car that resulted in a strong training in first-principles molecular dynamics, employed first to study the structural and photoelectron properties of clusters and then, beginning in 1995, to focus on the structure of disordered materials. While keeping the interaction with my EPFL co-workers, I came back to France and to CNRS in 1996 by moving to Strasbourg at IPCMS, my laboratory of affiliation ever since.

I like defining myself as a computational material scientist with an academic background of physics, complemented and enriched by a fruitful exposure to solid state chemistry and nanoscience, nurtured by a wealth of interactions with colleagues in Strasbourg (some fine experimentalists), Stuttgart (I have been a regular visitor of the Michele Parrinello group at MPI for about three years), Bath (the neutron scattering team of P S Salmon) and several others laboratories in France and abroad. While most of my production is on disordered materials, I do my best to feel comfortable with any potential application of first-principles molecular dynamics, without refraining from using the classical version for particular situations (as it was done for atoms and cluster deposition on metallic surfaces). Relevant results have been obtained on hybrid organic–inorganic materials and their magnetic properties as well as on doped heterofullerenes. On the methodological point of view, my philosophy is based on the idea that quantitative predictions have to be preferred, regardless of their cost or the time needed to obtain valuable results. For this reason, a non-negligible part of my research work is devoted to testing models and the underlying theories, like the role played by the exchange–correlation functional within density functional theory or the inclusion of dispersion forces.

I have always attached a great deal of importance to networking and sharing information with colleagues at the national and international levels. A series of workshops on disordered materials were launched to stimulate interactions among computational materials scientists active in the areas of glassy and amorphous systems. The last has taken place in Sao Carlos (Brazil) in 2017 (the 2020 issue scheduled in Corning was canceled because of the Covid pandemic). Finally, my research activity did not prevent me from playing a relevant role in management and administration (deputy director of IPCMS, 2013–6 and head of an International Joint Laboratory France–Korea, 2015–8).

IOP Publishing

The Structure of Amorphous Materials using Molecular Dynamics

Carlo Massobrio

Chapter 1

Introduction

This chapter sets the scene for the variety of concepts, ideas and results that will be presented hereafter, by providing a short historical overview and some motivations underlying the description of amorphous systems by molecular dynamics. One important point has to do with the notion of *direct experience* that has guided the presentation both in terms of methodology and achievements. An overwhelming part of the content of this book comes from the direct involvement of the author in the area of simulation of glasses and, as such, a natural boundary between what can or what cannot be found in there is easily recognizable. As a result, this book has some undeniable (limited) self-biographical tones that should by no means undermine its general character and purposes. This introduction is completed with a presentation of the content of different chapters, so as to allow the reader choosing the best strategy to go through it, either in an ordered manner or by picking up useful information from one specific chapter/section.

1.1 Why this book?

The purpose of this book is to describe how molecular dynamics simulations can be instrumental to understand amorphous materials by providing, as a first and essential ingredient, a detailed knowledge of their atomic structure. At first sight, one may wonder why and how the simplicity of the above sentence should entail an elaborated level of description in terms of scientific facts as is typical of a book targeting an academic environment and of a monograph intended to be a 'detailed treatment on a specific theme' bound to be useful to researchers and practitioners in the specific area of disordered matter. Such questions can be addressed by focusing on two main aspects invoked from the outset: methodology and applications, and how they meet to create a very active field of research. Molecular dynamics (MD) is a methodology based on the solution of the Newtonian equation of motion for a collection of atoms over time trajectories that in principle are only limited in length by the available computational power.

doi:10.1088/978-0-7503-2436-6ch1

To capture the essence of MD we can think of the correspondence invoked in elementary statistical mechanics between the microscopic character of the equations of motion and the macroscopic character of the thermodynamic variables obtained therewith as ensemble averages [1][1]. As an undergraduate student one is quite customarily taught that, for any practical purpose, there is not a realistic counterpart on the microscopic side to the thermodynamic (or, equivalently, macroscopic and measurable) variables, since the corresponding expressions depend on an extremely high number of degrees of freedom (the atoms, when the electronic structure is not explicitly accounted for). This is because following in time these variables on time trajectories is too long to be tractable on any computer. In practice, this statement was meant to exclude *a priori* any viable connection between the evolution of a collection of atoms at a given temperature and the resulting measurable properties, leaving this correspondence virtual or in any case not persuable in practice. Historically, the only exceptions were those based on analytical theories attempting to solve the integrals involved in the expression of the ensemble averages for highly idealized systems with simplified descriptions of bonding (hard spheres, rare gases) [2][2].

In the early 1950s, condensed matter theoreticians began facing the dilemma of linking the microscopic (atomic scale) world to what is observable macroscopically via a tool capable of accounting for individual atoms moving under the action of interatomic forces, no matter how complicated analytically these could be [3][3]. For this they needed a devoted instrument capable of working under their direction at will. This is where computers came into play, since the creation of trajectories produced by the atomic movements requires the implementation of an automated iterative process totally out of reach of even the most ingenious and motivated individual. Merging theoretical expressions (the Newton equations of motion) and a technical tool (computers capable of dealing with thousands of operations within a reasonable time) marks the birth of MD as a methodology to bridge the gap between statistical mechanics (the microscopic world) and thermodynamics (the macroscopic world).

Having established the genesis of the methodology referred to in this book, and referring to chapter 3 for a more detailed description of the essence of MD approaches, we can turn to what is meant by applications, by keeping in mind that the keyword ‘amorphous materials’ is intended to be an operational label for

[1] The book by Hoover is quite instructive since it presents concepts of statistical mechanics exemplified and developed by using computers. In particular, the link between atomic-scale and macroscopic descriptions is highlighted and made accessible to the non-specialist.

[2] The book by Hansen and McDonald, first appeared in the 1970s, is a masterpiece of statistical mechanics of disordered matter, making available both the foundations and developments of analytical treatments, with an attentive eye to the progresses made in this area by computer simulation.

[3] If there is one scientist that should be quoted and labeled as ‘inventor’ of what can be considered as a new discipline the name of Berni Alder should be invoked for the novelty and the impact of his ideas connecting computers and statistical mechanics. In this context, and in addition to reference [3], a joint historical and scientific perspective is provided in the recent book by G Battimelli, G Ciccotti and P Greco *Computer Meets Theoretical Physics, The New Frontier of Molecular Simulation* [4].

the systems that will be considered. Throughout this book, the term 'amorphous' will be employed in its most general meaning, standing for a system topologically disordered that has a crystalline counterpart but differs from the liquid state due to vanishing diffusion coefficients and reduced mobility even on time scales macroscopically short (nanoseconds) but indeed quite extended for a computer counterpart. This simple minded definition has the only ambition of being easy to grasp while encompassing several sub-categories of disordered systems more accurately defined in the literature, where the differences between glasses, amorphous and noncrystalline systems has been made explicit and refined very accurately over the years. In practice, the actual difference between glassy and amorphous materials lies in the existence of a glass transition in the first case, thereby introducing a well precise physical concept in the definition of a disordered structure [5][4]. While these conceptual issues are of undoubtful interest, we are convinced the purpose and the focus of this book are better served by employing the generic term of *amorphous* materials (and, whenever sensible, *glassy* materials) for any collection of atoms bound to represent a bulk system devoid of crystalline order, essentially not evolving in time on even the most extended computer time scales (say, a few ns) and yet sensitive to substantial increase of temperature to the point of being representative also of a liquid state. To be more explicit, all systems considered hereafter as *amorphous* feature diffusion coefficients that cannot be calculated with a meaningful error bar even by the most extended molecular dynamics run. This point can be exemplified by selecting a value for a diffusion coefficient typical of a supercooled liquid state (10^{-7} cm^2 s^{-1}), a distance expressing significant atomic diffusion out of the shell of nearest neighbors (say, two interatomic distances between atoms not necessarily of the same kind, ~8 Å) and looking for the time interval needed to achieve the long time diffusive behavior through the relationships

$$\langle r_\alpha^2(t) \rangle = \frac{1}{N_\alpha} \left\langle \sum_{i=1}^{N_\alpha} |\mathbf{r}_{i\alpha}(t) - \mathbf{r}_{i\alpha}(0)|^2 \right\rangle \tag{1.1}$$

and

$$D_\alpha = \lim_{t\to\infty} \frac{\langle r_\alpha^2(t) \rangle}{6t}. \tag{1.2}$$

The outcome in terms of temporal trajectory (roughly 10^{-8} s) allows reaching the conclusion that dynamical evolutions of amorphous systems are hard to describe by atomic-based simulations, knowing that trajectories are discretized on timesteps not larger than 10^{-15} s and the diffusion coefficients at room temperature are orders of magnitudes smaller than 10^{-7} cm^2 s^{-1}. Clearly, running more than several millions

[4] The quoted paper by Zanotto and Mauro focuses on the very definition of the glassy state of matter. The most clear-cut appears to be the one stating that 'a glass is a non-equilibrium, noncrystalline condensed state of matter that exhibits a glass transition'. We refer to that highly pedagogical piece of work for further thoughts on this topic.

of timesteps for systems containing a number of atoms representative of a bulk system is a prohibitive task especially when the description of the interatomic forces includes the electronic structure and cannot be set as a reasonable goal of modeling, regardless of the available computational power and of any technical or conceptual effort aimed at reducing the costs.

Does this finding mean that there is no point in combining molecular dynamics (our methodology) and amorphous systems (our application), such effort belonging to the category of academic exercises? What follows provides evidence that this is indeed not the case. To substantiate this assertion, I have decided[5] to rely on an extended set of results spanning over more than 30 years and witnessing the evolution of MD as applied to the amorphous system. The idea is to follow a valuable track of accomplishments to understand limitations and progresses of this methodology and the success (together with some failures) of its application. To this end, one needs to adopt a guideline resulting from direct experience in MD calculations reflecting the conceptual advances and the computational power at the time when they were realized.

1.1.1 The guideline: relying on direct experience

In this book I have made the choice of pursuing this endeavor by relying on results issued by my own work performed with a variety of colleagues and in different scientific institutions since 1987. Such strategy has advantages (largely predominant) and a few, undeniable disadvantages that are not expected to undermine the overall quality or impact of the whole project but have to be considered and counteracted very carefully.

- Let's consider the possible **disadvantages** first by playing a sort of devil's advocate for the sake of truthfulness. Selecting extensively out of the research work done by a single individual and his team might appear somewhat self-centered and quite narrow-minded since even the most reputable scientist cannot have the pretension of representing an entire area of research. In addition, a drawback can be an excessive biographical tone unduly embedding the full presentation, as if the main intention was to report on some success story more than describing achievements of a larger community.
- On the side of the **advantages**, it is worthwhile to quote the true intention of describing results obtained via a direct involvement spanning the entire process of realization (from the motivation to the main findings, via the technical implementation). Also, the scope is more to focus on how to make things happen (how to obtain and describe an amorphous system via MD, why first-principles MD has a higher predictive power, etc) than on describing what happened in the field via a comprehensive account of the main results obtained over the past, say, 30 years. Without being a user manual of MD for

[5] Throughout this monograph, first-person singular or plural will be employed with a preference, not systematic, for singular in case of reference to a close biographical experience or to a quite subjective point of view (advice, recommendation) while plural will be adopted when presenting the published scientific results.

disordered matter, the content presented shows how to achieve a robust structural description of amorphous system via examples extracted from specific cases but easily applicable to any other one.

- Finally, such highly oriented focus does not have to be mistaken for the neglect of what else is available in the literature or, worse, an overestimate of the impact of our achievements. In addition to the fact that relevant results obtained by other scientists will be reported and referred to whenever needed, I would like to ask the reader to view our choice of presentation as a proof of modesty and awareness of being both unsuited and unwilling to provide a supposedly exhaustive view of MD applications in the area of disordered (amorphous) system. As an author, I consider this task much less interesting and stimulating than the account of a longstanding direct experience in the field.

To summarize, what would we like to offer to the interested reader of this book? For those more interested in methodology, roadmaps of how to set up calculations to obtain information at the atomic scale on the structure of amorphous systems. For those more keen to learn about applications, several examples of work done on specific systems for which the link with the methodological part remains inseparable. For both kinds of readers, a view on how a set achievements on disordered matter have been obtained to come to conclusions readily comparable with experiments results or complementing them.

1.1.2 Inside each chapter

Having established that this book takes advantage of work done on specific systems to extract general concepts, methodologically and scientific information for well defined concrete cases, it is time to invoke the field of research considered, the techniques employed and specific cases treated to fulfill the primary goal of employing molecular dynamics to better understand amorphous materials.

- After a few considerations on glass science and the need to obtain direct information on the structure, **chapter 2** will take us through a set of different quantities that can be obtained by MD, as a function of the degrees of freedom, to understand more about structural organization in direct and reciprocal space, with a special focus on strategies to link atomic structure and bonding features. This is where useful tools can be found and described, so as to invoke and employ them when presenting the results on specific systems. More specifically, there will be first the account of a set of atomic-scale tools based on coordinates, quite easy to grasp because of their intrinsic simplicity (pair correlation functions and coordination numbers, for instance, readily accessible from standard analyses of temporal trajectories). Moreover, special attention will be devoted to an approach targeting the degree of localization of the valence electrons, so as to infer the different nature of bonding in systems (like disordered chalcogenides or, more generally, disordered network-forming materials) featuring combined (iono-covalent, for instance) bonding characters. Since most of the quantities detailed and employed as numerical probes of the structure are inspired by experimental counterparts, **chapter 2** contains also a section devoted to a presentation of the experimental framework mostly invoked when describing disordered network via first-principles MD approaches. As made explicit in section 1.1.1, this choice reflects the general philosophy of this monograph not planned to include a specific part where experiments on amorphous systems are described comprehensively. Overall, **chapter 2** provides information on what properties to calculate when the focus is on structure at the atomic scale thereby avoiding the repetition of expressions and formulas each time they are applied for the various systems under consideration. It will contain an experimental flavor since, for disordered chalcogenides, our reference measurements have been issued from the same team, to which we owe valuable criticism and collaboration.

- **Chapter 3** goes through the foundations of classical MD (based on interatomic potentials) and first-principles MD (based on a potential energy derived from the electronic structure). The intention of this chapter is to describe how MD, quite well known as a simulation method, has been put to good use to meet some requirements specific to amorphous structures. The chapter sketches first MD as a general methodology, to move on to some notions (including explicit analytical expressions) relative to n-body potentials (employed for metals) and to potential for systems crudely approximated within early classical MD by point charges. This section anticipates considerations on glassy chalcogenides extensively developed in the chapters devoted to results and applications of the methodology.
- **Chapter 4** builds a practical roadmap based on the concept of **chapter 3** to carry out first-principles molecular dynamics (FPMD) on amorphous materials, by dwelling on explicit examples targeting disordered chalcogenides as a playground. The idea was to present a framework ready to be exploited and yet not limited to a list of instructions. Yet, an effort was made to avoid too many technicalities by keeping in mind that the purpose is more informing than teaching. Examples will be given in this context to establish a clear connection between theory and how calculations develops in reality.
- **Chapter 5** is based on two examples of application of classical MD to the study of amorphous systems. These examples are extracted by results obtained on a somewhat academic system (a Lennard-Jones monoatomic glass) and a more realistic model devised to simulate amorphization by solid-state reaction in $NiZr_2$. Besides the intention of following an historical path of achievements, I draw it to the attention of the reader as endeavors quite representative of attempts to extract quantitative information from simulation despite the limits of available computers hampering, at that time, both the account of the electronic structure in combination with MD and the extent, within classical MD, of space and time scales. It turned out that there

was a lot to be learned from a monoatomic Lennard-Jones glass in terms of structural stability and even more from a sensible model of a metallic alloy for which, in addition, one was able to apply the constant pressure simulation technique proposed a few years before.

- Beginning with **chapter 6** disordered chalcogenides (to a large extent those belonging to the Ge_xSe_{1-x} family) are the main object of investigation. Liquid and glassy $GeSe_2$ are treated in great detail by stressing the link between the total energy description (different exchange–correlation functionals) and the structural properties. $GeSe_2$ has served as a reference system for introducing the search of the atomic-scale origins of the first sharp diffraction peak (FSDP) in the total and partial structure factors, an investigation to which a great deal of energy and resources have been devoted over the years. Two additional kind of information can be found in **chapter 6**. The notion of electron localization is invoked to describe chemical bonding in Ge–Se systems. Also, the account of the concentration–concentration and charge–charge structure factors proved valuable to classify disordered network-forming materials on the basis of their topological and chemical order.
- In **chapter 7** we address the issue of the behavior under pressure of two chalcogenide glasses, $GeSe_2$ and $GeSe_4$. Two distinct phase transformations are highlighted, in close connection with experiments. As a prerequisite, the control of pressure in FPMD calculations is discussed with a targeted calculations.
- With **chapter 8** we continue extending our detailed studies of network structures not only within the Ge_xSe_{1-x} family but also to systems useful for comparative purposes like $SiSe_4$. Also, a connection is established with phenomenological theories relating the concentration $x = 0.2$ to a transition between different topologies described in terms of constraints and self-organization. This allows discussing in atomic-scale terms the notion of structural variability and accessing a structure far from stoichiometry composition on the side of Se rich composition, where Se homopolar contacts play a predominant role. Glassy Ge_2Se_3 features the opposite pattern (Ge contacts with all Se occupied in tetrahedra), thereby making possible an instructive comparative study of structure and electronic features for three compositions.
- **Chapter 9** contains results for glassy GeS_2, $GeSe_9$ and a comparison between GeS_4 and $GeSe_4$. For these systems we have taken advantage of larger sizes and well-tested schemes to improve upon previous results and provide analyses correlating the atomic structure to the electronic one, via one of our most employed (and advertised) tools based on the Wannier functions orbital description. Also, we can monitor the differences existing between different sizes by providing useful information on hypothesized (or actual) effects on the structure of periodic boundary conditions. The case of $GeSe_9$

exemplifies the need of larger systems when the concentration of one species is small, as is the case in ternary systems considered in a later chapter.

- **Chapter 10** is about the inclusion of dispersion forces (van der Waals, vdW) in the total energy expression employed to implement FPMD. The treatment of these forces has long been an issue of debate and, in the context of electronic structure calculations, and active field of research over the past 30 years or so. Due to its invoked sensitivity to dispersion forces and to its importance as material prone to applications, glassy $GeTe_4$ gave the opportunity to study the impact of different schemes of dispersion forces on the structure within a three step research accomplishment involving a conspicuous amount of FPMD calculations.
- The consideration of ternary materials containing at least one species in small proportions (around 10%) has been driven by the advent of phase change materials employed for fast and effective data storage (intrinsically unstable against crystallization). On the side of applications, ternary materials possessing extended infrared windows are also worth investigating. In both cases, amorphous phases play an important role and have been studied by FPMD approaches, as detailed in **chapter 11.**
- The final chapter of the book, **chapter 12** will contain a set of considerations addressing the future of molecular dynamics modeling of disordered materials, by attempting to sketch what it is useful to make progresses in this area, including extension of calculations to new properties (as thermal ones) and the impact of machine learning computational schemes. It is made of three sections, bridging past, present and future. The first section is devoted to additional calculations on vibrational properties as an example of what is feasible beyond atomic structure characterization. The second contains results showing how FPMD can give access to thermal properties. Finally, after having established some 'golden rules' to ensure significant progresses to be made in the area of modeling of disordered materials, a few examples are provided of very recent calculations on systems technologically impactful.

References

[1] Hoover W G 1991 *Computational Statistical Mechanics* (Amsterdam: Elsevier)
[2] Hansen J P and McDonald I R 2006 *Theory of Simple Liquids* 3rd edn (Amsterdam: Elsevier)
[3] Battimelli G and Ciccotti G 2018 Berni Alder and the pioneering times of molecular simulation *Eur. Phys. J.* H **43** 303–35
[4] Battimelli G, Ciccotti G and Greco P 2020 *Computer Meets Theoretical Physics, The New Frontier of Molecular Simulation* (Berlin: Springer)
[5] Zanotto E D and Mauro J C 2017 *J. Non-Cryst. Solids* **471** 490–5

Chapter 2

Amorphous materials via atomic-scale modeling

This chapter establishes a link between atomic-scale modelling (in particular, molecular dynamics, MD) and the description of disordered glassy structures. The main idea is to define the quantities that will be used to describe glasses by taking as the starting point the knowledge of the atomic positions, velocities, forces and, whenever useful, the electronic structure. We pay particular attention to the connection between the structure factors and the pair correlation functions that will be extensively employed when presenting results on specific systems. Among the others, we also introduce the coordination numbers and their multiple significances to describe the atomic local environment. Electronic structure tools are described to correlate the atomic structure to bonding properties. Some selected examples are given to highlight how these quantities have been employed and what one can learn from them. Neutron scattering techniques are also considered as the main experimental counterpart to the first-principles MD (FPMD) calculations presented in this monograph. Overall, this chapter is the needed prerequisite to understand how an atomic scale quantitative approach can contribute to gain a better knowledge of a glassy structure.

2.1 The inspiring role of Glass Science

The scientific domain of glasses, or, more generally (as assumed and rationalized in the introduction) of amorphous materials features a large variety of fundamental issues and practical applications. Glasses are ubiquitous in real life and employed in everyday human activities on a daily basis. In my case, a great source of inspiration and learning over the past 15 years or so has been the magnificent talks given at conferences like the Glass and Optical Materials Division (GOMD) meetings of the American Ceramic Society and the International Congress of Glass (ICG). These are the right places to be especially for practitioners approaching the field from

doi:10.1088/978-0-7503-2436-6ch2

statistical mechanics (treating glasses as disordered objects of unknown and yet quite intriguing topology) while those driven by the interest in applications will found there a lot of useful information.

It appears that glasses can be studied from the fundamental and applied points of view, since they are intimately connected to each other, contributing to shaping a truly interdisciplinary area. As clearly established by Steve W Martin (Iowa State University) at his talk for ICG 2016 in Shanghai *Glass Science and Engineering has made innumerable contributions to the quality, enjoyment, safety, productivity, artistry, and general well being of human life. Lighting, windows, optical fibers, food containers, cook and table ware, architectural components, flat panel displays, art glass and medical devices are all examples of the tremendous impact that glass has made upon life as we now enjoy it*[1]. I would like to extend these considerations devoted to amorphous materials as systems prone to computational material sciences endeavors by following, as a guideline, the classification of topics commonly employed in the international meetings. This will be done at the beginning of the next section. But first, a few words to clarify my presentation strategy and some underlying concepts related to glass science are necessary.

In what follows there is no intention to neglect notions, classifications and critical reviews presented in several excellent textbooks and monographs available in the area[2]. My aim is involving the reader through my own motivation of describing materials and properties quantitatively without being aware of all practical applications from the outset. Most students and young researchers approaching the field of glasses (especially from simulations) are proposed an initial system to start with, it could be through simulations with specific codes or measurements by using apparatuses nowadays well controlled and performing. While the motivation for their study is (hopefully) clearly made explicit by some supervisor or older colleague, there is no reason to believe that a deep knowledge of the whole discipline is a prerequisite to succeed more than being fully concentrated on the given assignment. My experience taught me that very fundamental and somewhat elementary questions remain to be answered for amorphous systems (structure being the first). As a consequence, one can legitimately concentrate on a single issue and contribute to the global advancement of the field while expanding little by little his/her own understanding out of the initial range of competences. *Therefore, in the context of the information I would like to convey with this book, I am convinced a comprehensive knowledge of issues related to glasses is neither a prerequisite nor a goal. Accordingly, I made the choice of not going through an exhaustive presentation of all areas of glass research, easily available elsewhere. However, a few ideas are needed to capture how scientists work in this area and what are the main targets of their approaches.*

[1] I am indebted to Prof. S W Martin for allowing me to quote part of the abstract of that talk.

[2] For a list of monographs on amorphous materials and their application, I suggest to take a look at https://www.sanfoundry.com/best-reference-books-amorphous-materials-application/.

2.2 From experiments to modelling: toward a connection with atomic-scale tools

Under the general definition of Glass Science, quite often employed as a general title for main symposia at GOMD for instance, one can look at glasses as classified according to their chemical composition and/or their properties. In very general terms, one can encounter silicate and non-silicate glasses, for instance chalcogenides (that will be the main focus of this book) belong to this second category. Alternatively one can think of oxides and non-oxide glasses, the overlap between these two definitions marking the importance of silica as a reference system for any comparative study of glassy materials. In material science, the applications of glasses are not only numerous, but largely prone to be exploited technologically, as in the far infrared range and for optical and sensing devices. Structural characterization (NMR spectroscopy, Raman, x-ray, neutrons) is currently employed in the search of information on the specific structural motifs and on how they are interconnected in space. Glasses can play the role of host matrices for ionic conduction and their dissolution properties have a great impact in corrosion phenomena. Also, glass surfaces have become an emerging field because of their role in determining wear and friction. Mechanical properties of glasses are of interest due to their influence into the durability and resistance of materials. Last but not least one can mention, as a paradox, the huge interest driven by a class of poor glass formers (i.e. unstable against crystallization on macroscopic scale) known as phase change materials, for which the rapid switching from the crystalline to the amorphous state and viceversa ensures the practical realization of memory devices as those employed in current cell phones or stick memories.

As a whole, Glass Science can be considered as a discipline that shares some similarities with other condensed matter branches but it features in addition an everlasting and enduring request for structural determination as a prerequisite. Another strategy to classify this multifaceted domain of knowledge is to focus on the *amorphous state*. This allows rationalizing fundamental issues that are crucial to optimize the production and the control of real glasses. Among these, worth mentioning is the search of clues unraveling the glass transition and glass relaxation phenomena, strongly connected to the concepts of ideal glass formers, with the related classification in terms of strong and fragile based on the glass viscosity. The extent of structural order beyond the first shell of neighbors, the only one thought to be reminiscent of the parent crystalline state is also a matter of intense investigation as a key element to better understand the *amorphous state*. Most network-forming materials are indeed structurally organized as if each single atom could behave as a viewpoint over the position of neighbors extending well beyond the shortest bonding distance [1]. The so-called intermediate or medium-range order shown via reciprocal space probes has been at the center of a wealth of attempts, more or less successful, to link its appearance to atomic-scale features. Incidentally, it was also the motivation for my renewed interest in amorphous materials back in 1995, after a first experience of calculation on monoatomic Lennard-Jones and amorphous metallic systems (1987–91) reported in chapter 5 of this book.

In the rest of this chapter we aim at establishing a connection between a scientific discipline on its own and an atomic-scale computer based tool. This can be done via the description of quantities that are both prone to be measured experimentally (macroscopically) and to be calculated at the atomic scale (microscopically). From the standpoint of modelling and especially from the viewpoint of MD, glasses are a stimulating playground for challenging the capacities and the appropriateness of an approach relying on three ingredients: atomic coordinates, interatomic forces and equations of motion. In addition, the notion of disorder is particularly appealing since it legitimates the idea of following atomic movements as a method to explore the configurational space as a function of temperature. It appears that MD users are attracted by glasses because of the inherent structural disorder so hard to grasp analytically without resorting to generic models and crude approximations. When invoking disorder, one has to realize that its extent (unlike any realization of order) deserves some precision and, whenever possible, a quantification. For this reason, it is customary to establish a difference between *distinct degrees of order*, as if the boundary between order and disorder could feature some smoothness. Also, introducing different kinds of disorder makes sense since the identity of the degrees of freedom is not necessarily univocal. For instance, in a multicomponent system, atoms can be considered all identical when disregarding their chemical nature, a global topological disorder being found when the atomic positions are not reproduced regularly in space according to a given periodicity. This situation corresponds to topological disorder, in which the atomic positions are not correlated, since by knowing the position of a given atom there is no way to predict the position of other atoms farther apart than the closest interatomic distances (short range topological order). By introducing the chemical identity of the species, one can also consider situations in which the regular arrangement of atomic positions bearing the same chemical identity can be altered, regardless of the fact that atoms are fully ordered topologically. This corresponds to chemical disorder with respect to a given chemical arrangement known for being thermodynamically stable at a given temperature. While chemical disorder does not imply topological disorder, the opposite holds true, since a regular repetition in space of atoms bearing the same chemical nature cannot occur in the absence of structural (topological) order.

2.3 Accessing properties: direct and reciprocal space

Most amorphous materials can be identified as a collection of atoms exhibiting short range topological order at most, this definition being by all means not exhaustive and somehow highly simplified. The above rationale deserves to be completed by introducing the concept of correlation, closely related to any definition of disorder. Atoms are correlated when they appear to be dependent on each other, the most obvious link being the interatomic bonding that is customarily defined especially in covalent systems. Correlated variables (as atomic positions or velocities, for instance) can give rise to properties having specific signatures on well defined scales, allowing for information to be extracted on the structural and dynamical properties of a system. Having established that disorder and correlation are concepts accessible to MD as a method to obtain macroscopic information on it, the question arises on the amount of information available from experiments. For this, one has to address the simple question: where are the atoms? which is **the prerequisite** to any investigation on amorphous materials. A convenient connection between experiments and MD modelling can be found by taking as a primordial quantity the pair correlation function $g(r)$, most likely the simplest and most used quantity calculated and compared to the experimental counterpart for a disordered material. For practical purposes, there are two ways of considering $g(r)$, both implying statistical averages over time trajectories to attain the desired statistical accuracy. The first consists in linking it to a measurable quantity as a partial structure factor through the Fourier transform relationship

$$g_{\alpha\beta}(r) - 1 = \frac{1}{2\pi^2 \rho r} \int_0^\infty dk\, k \left[S_{\alpha\beta}(k) - 1 \right] \sin(kr) \tag{2.1}$$

where ρ is the atomic number density and $S_{\alpha\beta}(k)$ is a partial structure factor. The above relationship implies that we are dealing with a multicomponent system containing species α and β and that the partial structure factors $S_{\alpha\beta}(k)$ are readily available. Since $S_{\alpha\beta}(k)$ are not the easiest quantities to have access to experimentally[3] one is often limited to using the above relationship in terms of total quantities in which the contribution of the different species is not disentangled

$$g_{\mathrm{T}}(r) - 1 = \frac{1}{2\pi^2 \rho r} \int_0^\infty dk\, k\, [S_T(k) - 1] \sin(kr). \tag{2.2}$$

The corresponding relationships between partial and total quantities, for the case of choice of neutron scattering experiments (see section 2.6), read as follows:

$$S_{\mathrm{T}}(k) - 1 \equiv \sum_{\alpha=1}^{n} \sum_{\beta=1}^{n} \frac{c_\alpha c_\beta b_\alpha b_\beta}{\langle b \rangle^2} \left[S_{\alpha\beta}(k) - 1 \right] \tag{2.3}$$

$$g_{\mathrm{T}}(r) - 1 \equiv \sum_{\alpha=1}^{n} \sum_{\beta=1}^{n} \frac{c_\alpha c_\beta b_\alpha b_\beta}{\langle b \rangle^2} \left[g_{\alpha\beta}(r) - 1 \right] \tag{2.4}$$

where n is the number of different chemical species, c_α and b_α are the atomic fraction and coherent neutron scattering length of chemical species α, $\langle b \rangle = c_\alpha b_\alpha + c_\beta b_\beta$ is the mean coherent neutron scattering length.

So far we have considered partial structure factors that are defined with respect to the atomic species. Their most common expression, normalized to 1 at large wavevectors and employed above, is due to Faber and Ziman (FZ) [2, 3]. However, there is another family of partial structure factors that focus on the

[3] We shall address some issues relative to experimental measurements (mostly by neutron diffraction) in section 2.6. Hereafter, when invoking 'experiments' we refer almost exclusively to the technique (neutron diffraction with or without isotopic substitution) we have mostly dealt with via the interaction with the team of P S Salmon in Bath (UK).

notion of number and concentration, by providing information that goes beyond the nature of the chemical species. These are known as the Bhatia–Thornton (BT) [4, 5] partial structure factors, for which the following relationships hold:

$$S_{NN}(k) = c_\alpha c_\alpha S_{\alpha\alpha}(k) + 2c_\alpha c_\beta S_{\alpha\beta}(k) + c_\beta c_\beta S_{\beta\beta}(k) \tag{2.5}$$

$$S_{CC}(k) = c_\alpha c_\beta\{1 + c_\alpha c_\beta[(S_{\alpha\alpha}(k) - S_{\alpha\beta}(k)) + (S_{\beta\beta}(k) - S_{\alpha\beta}(k))]\} \tag{2.6}$$

$$S_{NC}(k) = c_\alpha c_\beta\{c_\alpha(S_{\alpha\alpha}(k) - S_{\alpha\beta}(k)) - c_\beta(S_{\beta\beta}(k) - S_{\alpha\beta}(k))\}. \tag{2.7}$$

For further purposes it is of interest to rewrite the total neutron structure factor given in equation (2.3) in terms of $S_{NN}(k)$, $S_N(k)$ and $S_{CC}(k)$ as follows

$$S_T(k) = S_{NN}(k) + A\,[S_{CC}(k)/c_\alpha\, c_\beta\; - 1] + B\, S_{NC}(k), \tag{2.8}$$

where $A = c_\alpha c_\beta \Delta b^2/\langle b\rangle^2$, $B = 2\Delta b/\langle b\rangle$ and $\Delta b = b_\alpha - b_\beta$.

It is instructive to observe under which conditions $S_{NN}(k)$ in equation (2.8) can be exploited to gather information on the total neutron structure factor. Let us consider the case of disordered $GeSe_2$ systems, to which we shall refer quite extensively in chapter 6. Here the values of $b_{Ge} = 8.185$ fm and $b_{Se} = 7.97$ fm are very close, leading to $A = 1.6 \times 10^{-4}$ and $B = 0.053$. Due to the limited variation of $S_N(k)$ and $S_{CC}(k)$,[4] this means that $S_{NN}(k)$ is a very good approximation of the total neutron structure factor $S_T(k)$.

What is the meaning of a peak appearing in $S_{CC}(k)$? Equation (2.6) can be rationalized by invoking the sensitivity of a given atom to the chemical nature of its environment for the range of distance related to the specific value of k. Peaks are absent when the atoms α or β show no preference for homopolar or heteropolar bonding as if there were no energy cost in changing the atomic identities. The meaning of $S_{NC}(k)$ is even more subtle since it involves the relative weights of the quantities $c_\alpha(S_{\alpha\alpha}(k) - S_{\alpha\beta}(k))$ and $c_\beta(S_{\beta\beta}(k) - S_{\alpha\beta}(k))$, as described in [5] for the corresponding pair correlation functions in real space. Even though $S_{NC}(k)$ has been largely used (for chalcogenides and within FPMD) to complete the panel of quantities to be compared with experiments, I have to admit to not having taken much advantage of it to assess the overall performances of theory in those cases[5].

Equations (2.1) and (2.2) appeal to experimentalists working on diffraction experiments and extracting information from the reciprocal space. However, the information on direct space stems from integrals covering a range in principle limited by the finite diffractometer window (see section 2.6). Within the philosophy of MD for which there are no limits to the identification of atomic positions, pair correlation functions are the most natural quantities to be obtained *directly* from a

[4] $|S_{NC}(k)| < 0.2$, $S_{CC}(k) < 0.5$ for liquid $GeSe_2$ [6], similar considerations being applicable to the case of glassy $GeSe_2$ [7], see also sections 6.2.2 and 6.2.3.

[5] Throughout the chapters devoted to the presentation of result we shall employ either the FZ expressions for the structure factors (when no label is indicated) or the BT ones, explicitly labeled.

set of coordinates knowing that experiments can produce only an *indirect* result via Fourier transformation from reciprocal space. The pair correlation function provides a measure of the departure from a uniform distribution of atoms as expressed by the average density of the system $= \rho = N_{at}/V$ (N_{at} = number of atoms, V volume). By focusing on each atom in a sequence, one has to build up an histogram collecting the occurrences of neighbors in selected shells around distances conveniently discretized. The ratio $N(\Delta(r))/N_{at}$ between the number of atoms in a given shell and N_{at} has to be first normalized to the volume of the shell under consideration to avoid any bias due to increasing values of r. We can call this quantity $g'(r)$

$$g'_{\alpha\beta}(r) = \frac{N(\Delta(r))}{N_{at} 4\pi r^2 \Delta(r)}. \tag{2.9}$$

Equation (2.9) gives an idea of the presence or absence of neighbors at given distances with respect to a uniform distribution. However, it is appropriate to further normalize $g'(r)$ to the density $N_{at}/V = \rho$ to obtain a reference value 1 attained at large distances, where one does not expect any variation from a uniform distribution of atoms:

$$g_{\alpha\beta}(r) = \frac{N(\Delta(r))}{N_{at} 4\pi r^2 \Delta(r)\rho}. \tag{2.10}$$

In addition to the wealth of examples of pair correlation functions available in this monograph, it is useful to provide an example aimed at showing what kind of information can be extracted from a set of partial $g_{\alpha\beta}(r)$ and which others cannot, or, better, cannot be straightforwardly associated to different coordination environments. We are considering the case, treated by FPMD, of glassy GeS_2 intended as a bulk and as a surface, that is two systems having in common the same chemical nature but differing by the topological one (bulk vs surface, in red in figure 2.1). The two sets of $g_{\alpha\beta}(r)$ share the most representative features of the networks (the three peak pattern of $g_{GeGe}(r)$, the very intense main peak of $g_{GeS}(r)$ standing for predominant tetrahedral order and the presence of homopolar bonds in both $g_{GeGe}(r)$ and $g_{SS}(r)$) and yet there is little marking the difference between the bulk and the surface [8]. This shows that the pair correlation functions are quite useful and unreplaceable to describe the atomic structure of a glass but one should be aware of their limits and not stick to them as the only tool of structural investigation.

The methodology underlying equation (2.10) to obtain the pair correlation function is by far the most reliable and accurate within MD, since it is fully independent on any variable in reciprocal space and it can achieve a high statistical accuracy for sufficiently extended time trajectories. As a matter of fact, MD users are accustomed to obtain information on real and reciprocal space in a way (from real to reciprocal or both directly in real and reciprocal space) that is quite different from the one available to experimentalists (from reciprocal to real).

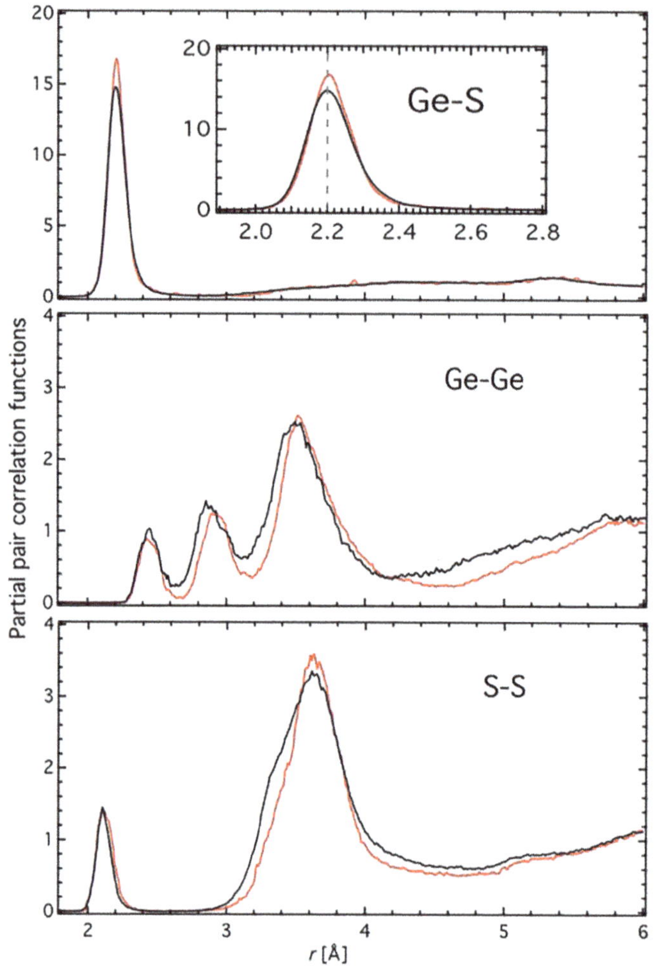

Figure 2.1. An example of pair correlation functions focusing on the comparison between the data obtained via FPMD for a bulk and a surface system of glassy GeS_2. The results for the surface system are given in red. Reprinted figure with permission from [8]. Copyright (2014) by the American Physical Society.

The direct calculation of the factor structure, given below in its most general form

$$S_{\alpha\beta}(k) = \frac{1}{N_{\rm at}} \sum_{i=1}^{N_\alpha} \sum_{j=1}^{N_\beta} \mathrm{e}^{-i\mathbf{k}\cdot(\mathbf{r}_{i\alpha}-\mathbf{r}_{j\beta})} \tag{2.11}$$

is also accessible to MD, thereby allowing for viable alternatives to schemes in which direct and reciprocal space are inevitably linked at the computational level. The limitations of equation (2.11) are those related to insufficient sampling for small wavevectors, a shortcomings often circumvented by integrating from direct space due to smoothening of the statistical errors. Explicit examples exemplifying these issues will be provided in the chapters devoted to FPMD applications.

To summarize on the use of the pair correlation functions within MD, in connection with the existence of experimental data, three different routes can be followed, all of them requiring the use of periodic boundary conditions, universally accepted as the simulation framework to be employed when the purpose is to mimic an essentially infinite disordered system with an otherwise limited number of particles [9, 10].

- **A:** Calculate $g_{\alpha\beta}(r)$ according to equation (2.10) on a range of distances extended enough to cover a number of neighbors well beyond interatomic distances. This will allow a meaningful representation of the structural units present in the systems and that can be interconnected as it is the case in network-forming glasses and, in general, in disordered systems featuring non-trivial structural order. Equation (2.10) can be used by considering or not the chemical identity of the atoms, leading to a global $g(r)$ (equal to equation (2.4) if neutron scattering parameters are included) or the partial $g_{\alpha\beta}(r)$. This gives access to the corresponding partial structures from integration as follows

$$S_{\alpha\beta}(k) - 1 = \frac{4\pi\rho}{k}\int_0^\infty \mathrm{d}r\, r\left[g_{\alpha\beta}(r) - 1\right]\sin(kr) \tag{2.12}$$

  where ∞ has to be intended as the largest range compatible with periodic boundary conditions. This procedure avoids calculating directly the structure factors and relies entirely on the statistical accuracy of the pair correlation function. Incidentally, it can be employed to monitor the impact of different ranges of distances on the pattern taken by the structure factors for small values of k. An example was provided in [6] for liquid $GeSe_2$ (figure 2.2). In that work, it was shown that equations (2.11) and (2.12) do exhibit a different convergence with the size of the periodic cell, that was artificially extended periodically by allowing for larger distances to be probed. More than 20 years after taking those conclusions, we can ensure that the calculations of the structure factors carried out via equation (2.12) have become more effective, due the availability of larger sizes ($N_{\mathrm{at}} = 120$ was the size of the periodic system in [6]).
- **B:** Calculate $g_{\alpha\beta}(r)$ according to equation (2.10) and calculate $S_{\alpha\beta}(k)$ according to equation (2.11). This choice provides independent evaluation in real and reciprocal space and, in principle, it is the one better for exploiting the full capabilities of MD in terms of direct access to atomic-scale properties. However, the partial structure factors could be somewhat noisy in the region of low wavevectors (say, smaller than 2 Å^{-1}, giving information on the largest distances in real space) due to limited sampling for the largest accessible distance. This feature is severe when the number of atoms belonging to given spaces is relatively small (less than 20–30 atoms or so), thereby affecting the corresponding partial quantity. For amorphous system this can be improved, despite the little dynamical evolution at low temperatures, by taking statistical averages on a large number of configurations. Due to the availability of increasingly larger system sizes, allowing for integrals of pair correlation

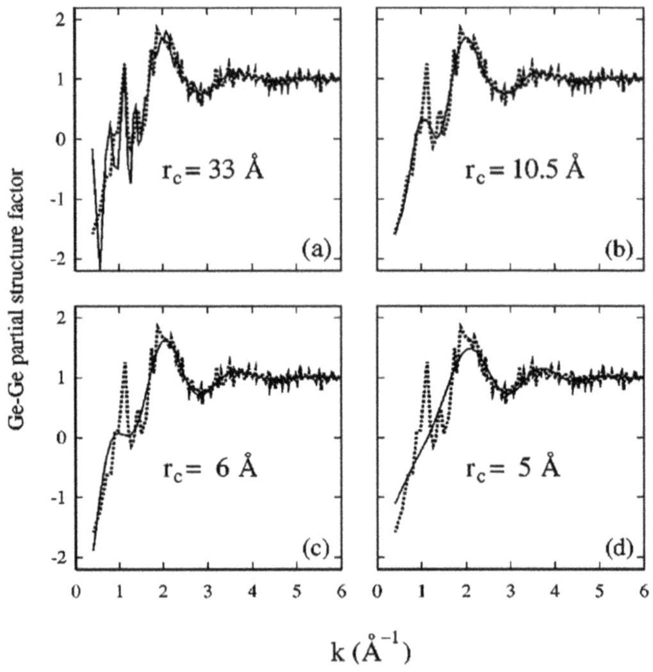

Figure 2.2. Partial structure factor $S_{\mathrm{GeGe}}(k)$ of liquid $GeSe_2$ (solid line) via Fourier integration of the calculated partial pair correlation function $g_{\mathrm{GeGe}}(r)$ for the indicated integration ranges (equation (2.12)). The dotted line corresponds to the partial structure factor directly calculated in k space via equation (2.11). Reprinted figure with permission from [6]. Copyright (2001) by the American Physical Society.

functions to be taken on adequate space intervals, MD users tend nowadays to disregard the direct calculation of the partial structure factor in reciprocal space even though this remains an additional piece of information not to be neglected *a priori*. For instance, this allows averaging structure factors obtained with the two methods to improve statistical accuracy.

- **C:** Calculate $S_{\alpha\beta}(k)$ as in equation (2.11) and $g_{\alpha\beta}(r)$ as in equation (2.1). This option is the one closest in spirit to measurements, since it is based on the upfront calculation of the structure factor, while the pair correlation function is the result of an integration in reciprocal space. In this case, to ensure the highest consistency between theory and experiments, one can even restrict the integration range to that covered by the range of the diffractometer, that is

$$g_{\alpha\beta}(r) - 1 = \frac{1}{2\pi^2 \rho \, r} \int_0^{k^{\max}} \mathrm{d}k \, k \left[S_{\alpha\beta}(k) - 1 \right] \sin(kr). \qquad (2.13)$$

It turns out that the spurious contribution at very small, unphysical values in real space, as well as any difference between $g_{\alpha\beta}(r)$ as given in equation (2.1) or in equation (2.10) vanishes when increasing the upper bound of integration $k^{\max}$ in (2.13). An example is given in section 8.2, figures 8.2 and 8.3.

Over the years, we have employed the three strategies detailed above in a variety of situation for different systems, with a special focus for available comparison between FPMD calculation and experiments on partial structure factors. **A**, **B** and **C** are all worth using and connecting to each other in the search of all available clues of the atomic structure. Based on our own experience of continued efforts to compare our results with experimental data in order to confer to MD calculations a quantitative predictive power (at least for structural determination), we would like to point out the necessity of a comparison based on quantities accessible, as best as possible, on both experiments and calculations. However, there is caveat that should be carefully considered. Not all spectroscopic or diffraction probes are routinely capable of revealing the chemical identity of atoms, being insensitive to the nature of the system components. Among the available methodologies, neutron diffraction with isotopic substitution can overcome this difficulty and it has retained the attention of computational materials scientists as a powerful and reliable way to describe glasses and liquids. This is due to its capacity of attributing specific features in the partial structure factors and the derived pair correlation function to topological features of the network. More on the essence of this experimental technique and the strong interplay developed over the years with our FPMD investigations will be detailed in section 2.6.

2.4 Describing the network topology

2.4.1 Coordination numbers and units

The description of a disordered structure cannot escape the search of the number and the nature of neighbors surrounding a given atom. While this feature is well known for a crystal, it remains a goal by itself for any rationalization of a non-ordered spatial arrangement. The notion of coordination number is equivalent to the counting of atoms that are considered bonded, calculated as an average on the species of interest for a multicomponent system. Once established the relevant pair correlation function $g_{\alpha\beta}$ involving pairs of atoms α and β, it is customary to obtain the coordination number n_α^β by integrating $g_{\alpha\beta}$ up to the first minimum. In more general terms, for any set of distances r_1 and r_2 the following relationship holds:

$$n_\alpha^\beta = 4\pi\ \rho c_\beta \int_{r_1}^{r_2} \mathrm{d}r\ r^2 \left[g_{\alpha\beta}(r)\right]. \tag{2.14}$$

The total coordination numbers for atoms α and β (under the hypothesis of a two-component systems, the generalization to a multiple component system being straightforward) are given by $n_\alpha = n_\alpha^\alpha + n_\alpha^\beta$ and $n_\beta = n_\beta^\beta + n_\beta^\alpha$, where $n_\alpha^\beta / c_\beta = n_\beta^\alpha / c_\alpha$ and the average total coordination number n_{ave} (also referred to as $\bar{n}$ and n_{tot}) is equal to $c_\alpha\ n_\alpha + c_\beta\ n_\beta$. These numbers are useful but they say little about the various local arrangements of the atoms, that can be differently coordinated and bond, in principle, with any neighbor regardless of its chemical identity, especially when the composition does not follow stoichiometry or temperature favors the occurrence of structural defects that do not exist at room

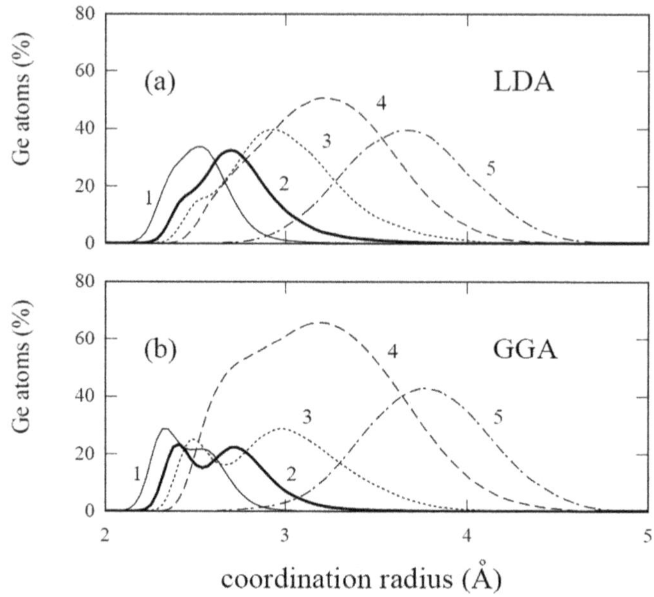

Figure 2.3. Behavior of the coordination units $n_{\alpha}(l)$ in liquid $GeSe_2$ for two different choices of the total energy functional (see section 3.6.5 and [11]). Reprinted figure with permission from [11]. Copyright (1999) by the American Chemical Society.

temperature. Therefore, one can obtain a more complete description of the network by resorting to the individual α–l structural units where an atom of species α (say, A or X) is l-fold coordinated to other atoms. Within this notation, A–AX_3 represents an A atom that is connected to 1 other A atom and 3 X atoms while A–X_4 corresponds to an A atom that is connected to 4 X atoms. Typically, along a MD trajectory, the weight of a specific unit, $n_{\alpha}(l)$, is found by taking the ratio between its average number over the time and the total number of atoms of type α. The significance of $n_{\alpha}(l)$ is more profound than the mere calculation of the coordination numbers, since it is carried out for the entire range of values of the interatomic distances (accessible to the periodic box). This allows accessing the coordination environment of species α as a function of l by choosing whatever r might appear suitable, the preference being given to those values corresponding to stable bonds with the neighbors. A revealing example of the relevance of this analysis is given in figure 2.3, where the focus is on liquid $GeSe_2$ as calculated with two different total energy functionals (for details, see section 3.6.5) leading to different numbers of Ge fourfold coordinated. Only the description based on the generalized gradient approximation (GGA) of the exchange–correlation energy is able to promote the formation of a tetrahedral network, coexisting with the presence of homopolar Ge–Ge bonds visible at short distances. It should be understood that focusing on $n_{\alpha}(l)$ is like 'taking a picture' of what happens inside a sphere of a given radius around a given atom and then counting how many atoms are within it and who they are. Looking at figure 2.3 one might be puzzled by the decreasing trends of $n_{\alpha}(l)$ for increasing r, since the number of atoms inside a sphere is not bound to decrease when the sphere becomes larger. However, any specific unit for small l will becomes less representative of the coordination environment for larger interatomic distances, since the number of neighbors inside the sphere is bound to increase, at the expense of those $n_\alpha(l)$ describing the actual, nearest neighbor bonding environment.

2.4.2 Bond-angle distributions and local order parameter

By knowing how triads of atoms organize themselves in space we can obtain information on the geometries of the structural units inherent in a disordered system and, in particular, in a disordered network where the bonds are highly directional. This is the reason underlying the presence of bond-angle distribution in most reports devoted to atomic-scale descriptions of glasses characterized by an expected predominant structural motif. Such units are connected in space according to well defined angular preferences that result from both packing (steric) and chemical bonding considerations. Bond-angle distributions are quite useful to highlight such structural organization, being straightforward to have access to within, for instance, a code devoted to the calculation of pair correlation functions. Briefly, by focusing attention on a given atom of coordinates **A** and on two neighbors of coordinates **B** and **C** (to be selected among those closer than a suitable cutoff distance, as the minimum of the corresponding pair correlation function), the cosine of the $\langle \mathbf{BAC} \rangle$ angle is given by the ratio

$$\cos \langle \mathbf{BAC} \rangle = \frac{\mathbf{AB} \cdot \mathbf{AC}}{AB \times AC} \tag{2.15}$$

where $\mathbf{AB}\cdot\mathbf{AC}$ is the scalar product between **AB** and **AC** and $AB \times AC$ is the product between the vectors moduli. The angles of all triads thereby identified along the temporal trajectories can be collected within a histogram to provide the so-called bond-angle distributions that will feature peaks, maxima and minima of increasing intensity with lowering the temperature. It is instructive to consider an example of bond-angle distribution $\theta_{\alpha\beta\gamma}$ for a case that will be treated extensively in section 6.2.3, namely glassy $GeSe_2$. For that system, as for other chalcogenides of the Ge_xSe_{1-x} family, there exists a clear relationship between the topology of intertetrahedral and intratetrahedral connections and the peaks appearing in θ_{GeSeGe} and θ_{SeGeSe}. This is visible in figure 2.4.

In θ_{GeSeGe} two main peaks at 83° and 102° can be found. The first peak is due to *edge-sharing tetrahedra*, sharing two Se atoms. One can also relate these subunits to the presence of fourfold rings. The second peak comes from *corner-sharing tetrahedra*, which have in common a single Se atom. These configurations, shown in figure 2.5, are commonly encountered in investigations of disordered network-forming materials. The maximum of θ_{SeGeSe} at 109° stems from the predominant tetrahedral character of the network. The slight not symmetric shape of θ_{SeGeSe} is rationalized in detail in [7].

The notion of local order parameter **q** [12, 13] is quite useful to monitor the onset of a phase transformation, as those that will be detailed in section 7.2 (glassy $GeSe_2$

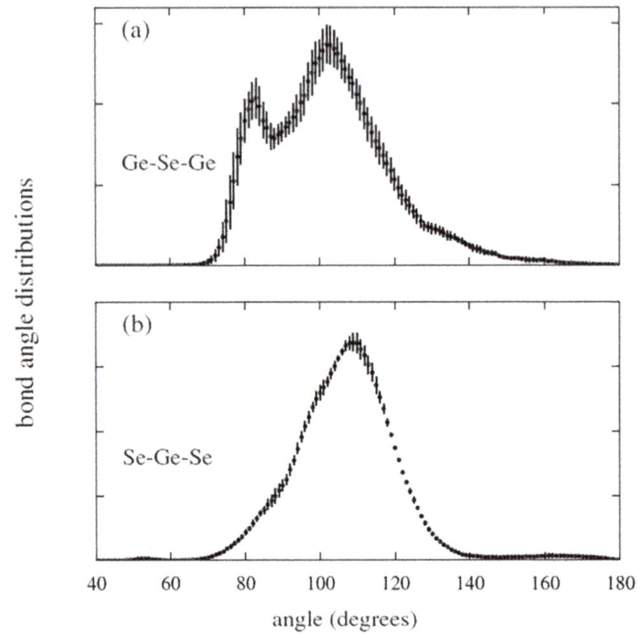

Figure 2.4. Bond-angle distributions θ_{GeSeGe} and θ_{SeGeSe} of glassy $GeSe_2$. Reprinted figure with permission from [7]. Copyright (2008) by the American Physical Society.

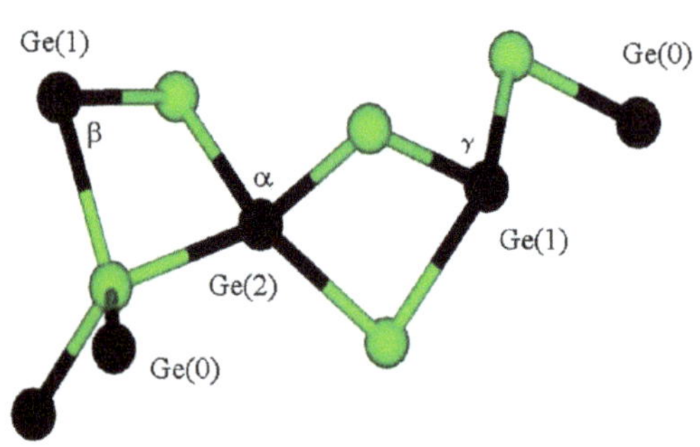

Figure 2.5. Structural subunits of glassy $GeSe_2$ showing representative Se–Ge–Se triads and the angles involved. Ge atoms are black and Se atoms are green. Atoms are connected by bonds when their distance is smaller than 3 Å. Ge(0), Ge(1), Ge(2) are Ge atoms lying in 0, 1, and 2 fourfold rings, respectively. Ge(0) are Ge atoms in corner-sharing configurations, Ge(1) and Ge(2) are Ge atoms in edge-sharing configurations. Reprinted figure with permission from [7]. Copyright (2008) by the American Physical Society.

under pressure), section 10.2 (glassy $GeTe_4$) and in the chapter devoted to ternary chalcogenide systems (chapter 11). At the opposite sites within the [0–1] range one has $\mathbf{q} = 1$ that refers to the ideal tetrahedral network, while $\mathbf{q} = 0$ corresponds to sixfold coordinated octahedral sites.

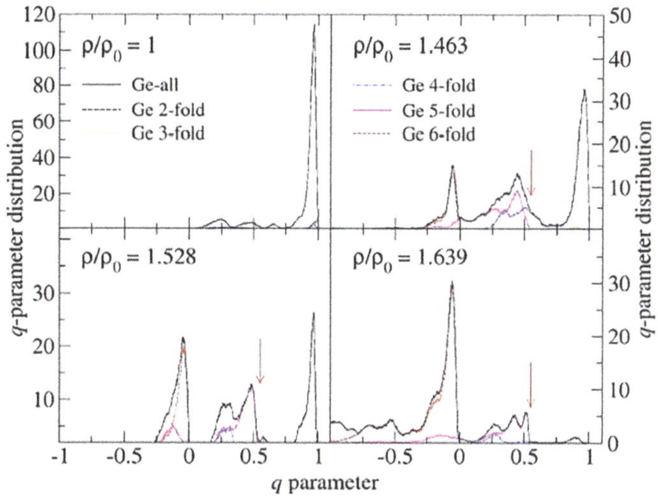

Figure 2.6. Example of behavior of **q** as a property useful to characterize the topology of a glass. Density dependence (expressed relative to the density at ambient pressure) of **q** for glassy $GeSe_2$. At each density, the distribution for all Ge atoms is broken down into its contributions from n-fold coordinated Ge atoms ($n = 3, 4, 5$, or 6). The vertical (red) arrows mark the **q** value for trigonal bipyramidal units (see reference [14]). Reprinted figure with permission from [14]. Copyright (2014) by the American Physical Society.

$$q = 1 - \frac{3}{8}\sum_{k>j}\left[\frac{1}{3} + \cos\theta_{jik}\right]^2 \tag{2.16}$$

where the angle θ_{jik} is the one defined with respect to a central atom i and its neighboring atoms j and k. A revealing example of the usefulness of **q** to describe the structural organization of a glass is provided in [14], where the different arrangements taken by glassy $GeSe_2$, spanning tetrahedra, octahedra and bipyramids (regular and distorted) are highlighted as shown in figure 2.6.

2.4.3 Making sense out of diffusion in glasses via MD

Among the properties characterizing a disordered system one has to keep in mind the diffusion coefficient as revealing the probe of the essential difference (the extent of mobility) separating a liquid from a glass. This is exactly why we turned to this quantity via equations (1.1) and (1.2) in section 1.1, to highlight a physical boundary that is easy to grasp when introducing the amorphous state and the implications of its nature for MD. The calculation of a diffusion coefficient is a simple way to follow the onset of configurational arrest following a period of quench from the liquid or to make sure that a system has diffused enough before undergoing rapid quench. This situation is commonly encountered in simulations of a glassy state or in any simulation aimed at producing a disordered system starting from an unrealistic initial set of position, knowing that in this latter case what counts is to achieve significant diffusion to loose memory of the initial configuration. In figure 2.7 we

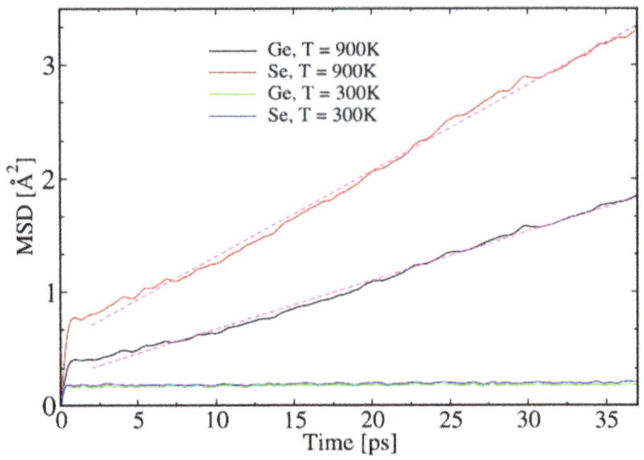

Figure 2.7. Mean square displacement of Ge and Se atoms in $GeSe_4$ at two different temperatures in the liquid and glassy state (see [15]). Note that for this system the diffusion coefficient (slope of the mean square displacement) at $T = 900$ K is larger for Se atoms since many of them evolve in chain-like configurations while all Ge atoms are arranged in tetrahedra. Reprinted figure with permission from [15]. Copyright (2015) by AIP Publishing.

report an example of behavior in the liquid and in the glassy phase for $GeSe_4$ (taken from [15]), showing clearly the different patterns exhibited by the mean square displacements.

2.5 Correlating structural and electronic properties

2.5.1 Electronic density of states

The idea of seeking information on the electronic structure by exploiting the availability of configurations evolving in time is at the very heart of FPMD, thereby pointing out what can be considered the best asset of the methodology when compared to the classical counterpart. Despite the high degree of complexity of electronic structure calculations, the concept of electronic density of states (EDOS) can be made intuitively simple by invoking the existence, in any quantum mechanical modelling of the interaction between an assembly of particles, of the eigenvalues of the total energy (quantity to which we devote some thoughts in section 3.6 when describing the FPMD approach). By arranging these numbers as in a distribution (allowing to highlight the intervals of most (or least) populated values (as in a histogram)) gives a first insight into the basic bonding properties of the system, especially for the region around the highest energy, corresponding to what is commonly called the highest occupied molecular orbital (HOMO). The most crucial detail of the EDOS, roughly separating materials in two classes, namely insulators/semiconductors and metals, is the energy gap between the HOMO and the LUMO (lowest occupied molecular orbital) or, in the solid-state physics terminology, the energy gap between valence and conduction states. In this context, the notion of Fermi energy is quite often invoked as the one midway between HOMO and LUMO

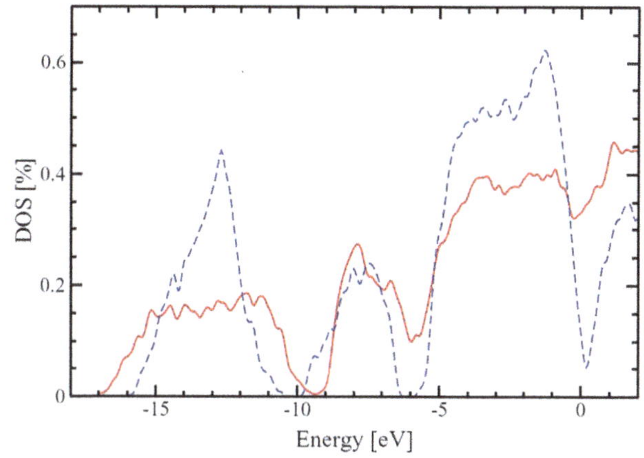

Figure 2.8. An example of comparison between two EDOS calculated for liquid Ge_2Se_3 (solid red curve) and liquid $GeSe_2$ (broken blue curve) [16]. A Gaussian broadening of 0.1 eV has been employed. Reprinted figure with permission from [16]. Copyright (2011) by the American Physical Society.

or, in a simplistic manner, as the highest available energy for the fermionic (electronic) system. In this monograph, EDOS are shown in two specific situations.

- For a given system, we are interested in comparing different theoretical schemes, all of them built into density functional theory but differing by the choice made for the exchange–correlation functional. As it will be detailed with specific examples, we have pointed out that there is a profound correlation, at least for disordered chalcogenides, between the treatment of this part of the total energy functional and the atomic-scale properties (see chapters 3 and 6).
- When comparing different systems, the existence of gaps or the extent of pseudogaps (a visible, strong depletion around the Fermi energy) is a fingerprint of the nature of bonding and it can help with classifying different levels of ionicity/covalency or metallicity, as these two classes of systems are very much identifiable. We have employed this guideline (for instance, see section 8.3.1), by keeping in mind that this kind of correspondence is qualitative and needs to be substantiated, if possible, by other tools like the maximally localized Wannier function, to which we devote the next section. An example of this kind is given in figure 2.8[6]. Liquid Ge_2Se_3 has a more pronounced metallic character than liquid $GeSe_2$, as it should be since it lies on the Ge rich side of the concentration range, approaching pure liquid Ge which is a liquid metal system.

2.5.2 Maximally localized Wannier functions

The idea of describing the nature of the electronic structure via a set of localized orbitals goes back to the early days of materials theory [17] and has fostered more

[6] Throughout this monograph the Fermi energy of the EDOS is set at the zero value for the energy scale.

recent developments able to convey information on the correlation between atomic structure and bonding properties [18, 19]. From the intuitive point of view, one can consider it more convenient to handle a basis set that can be associated to some local feature of bonding, regardless of the fact that we are using plane-waves and periodic (Bloch, as customarily defined in solid-state physics) eigenstates to describe the electronic structure of our system[7]. In what follows, we shall sketch the basic ideas underlying the Wannier functions methodology, by keeping some elementary analytical developments as concise and pedagogical as possible, knowing that the derivation of the entire formalism is fully available in more appropriate references [19, 20]. In principles, the n eigenstates one deals within a periodic system can be written as

$$\psi_{n\mathbf{k}}(\mathbf{r}) = u_{n\mathbf{k}}(\mathbf{r})e^{i\mathbf{k}\cdot\mathbf{r}} \tag{2.17}$$

where $\mathbf{k}$ vectors are in the Brillouin zone and, for convenience, we shall forget about the fact that most often in disordered systems any dependence of $\mathbf{k}$ of the eigenstates is limited to the Γ point ($\mathbf{k} = 0$) only. Let us suppose one seeks to obtain a very localized equivalent of $\psi_{n\mathbf{k}}(\mathbf{r})$ by considering an extended distribution of $\mathbf{k}$ values. This can be achieved via a definition that leads to a first expression for the Wannier function W_0:

$$W_0(\mathbf{r}) = \frac{V}{(2\pi)^3}\int d\mathbf{k}\psi_{n\mathbf{k}}(\mathbf{r}) \tag{2.18}$$

with the integral taken over the Brillouin zone. Equation (2.18) can be extended to include not only the origin but also other $\mathbf{R}$ locations (lattice vectors) in the periodic cell by including $e^{i\mathbf{k}\cdot\mathbf{R}}$ in equation (2.18), so as to generate an orthonormal set of n eigenfunctions linked to the original $\psi_{n\mathbf{k}}(\mathbf{r})$ set by the relationships

$$W_{n\mathbf{R}}(\mathbf{r} - \mathbf{R}) = \frac{V}{(2\pi)^3}\int d\mathbf{k}\psi_{n\mathbf{k}}(\mathbf{r})e^{-i\mathbf{k}\cdot\mathbf{R}} \tag{2.19}$$

$$\psi_{n\mathbf{k}}(\mathbf{r}) = \sum_{\mathbf{R}} W_{n\mathbf{R}}(\mathbf{r} - \mathbf{R})e^{i\mathbf{k}\cdot\mathbf{R}}. \tag{2.20}$$

Equation (2.20) means that the set of $\psi_{n\mathbf{k}}(\mathbf{r})$ can be taken as superposition of the Wannier functions $W_{n\mathbf{R}}(\mathbf{r} - \mathbf{R})$. This implies that both sets are equally valid to describe the electronic structure of the system. However, $W_{n\mathbf{R}}(\mathbf{r} - \mathbf{R})$ are intrinsically localized and potentially quite useful when focusing on the bonding character. We reiterate that this statement holds regardless of the delocalized nature of the plane-waves basis set employed to express their counterpart $\psi_{n\mathbf{k}}(\mathbf{r})$. The tricky point to be tackled and overcome before resorting to the Wannier functions for any practical purpose (for instance, to understand how much covalent or ionic character exists in a given system or, as we shall see shortly in the case of a nanostructure, to appreciate

[7] Our disordered systems are treated in periodic cells with no loss of generality, provided certain conditions are met, as a minimal periodic size, as further discussed in this monograph.

the spatial extent of LUMO and HOMO orbitals) has to do with their indeterminacy. This stems from the existence of an arbitrary phase factor in the definition of $\psi_{n\mathbf{k}}(\mathbf{r})$ calling for some additional requirement to obtain the best suited Wannier functions in terms of localization. This is where the work by Marzari and Vanderbilt [20] comes into play, through the introduction of the minimal spread of the Wannier functions, that results not only in the definition of a unique set of localized orbitals (the maximally localized Wannier functions, MLWF), but also provides the analytical expression for their centers, the so-called Wannier functions centers (WFCs). These theoretical developments have been largely detailed in the literature to which we refer the interested reader [19]. As an alternative, we aim at discussing two specific examples that elucidate the role played by this formalism to understand features of the electronic structure in a prototypical nanostructure (a doped fullerene) but also the topology of disordered systems, that is our main target here. It is worth stressing that the WFCs play the role of a pseudo-atomic systems associated to the real one and carrying information on bonding since, for instance, their lying exactly in between two atoms (of the same kind) is a clear fingerprint of covalency.

2.5.2.1 Example 1: maximally localized Wannier functions for doped fullerenes
Fullerenes have been a fascinating area of research ever since their discovery, the enhancement of their chemical reactivity being one of the first target due to their intrinsic electrophile character. At the turn of the 21st century I had the opportunity to get involved in FPMD calculation of substitutionally doped fullerenes (mostly C_{60-x} Si_x with C atoms replaced by Si atoms) by inferring, as the ultimate goal, the maximum number of Si atoms allowed to replace C atoms without causing the disruption of the cage. Regardless of this specific result, it is useful to take advantage of this kind of achievements to show the usefulness of MLWF concepts to understand the changes occurring in the bonding properties as a result of doping. These examples are quite instructive to become familiar with the notion of (maximally localized) Wannier functions and the corresponding centers. As a prerequisite, one should keep in mind that Si and C belong to the same group of the periodic table and yet their chemical nature is not the same. Carbon can form single, double and triple homo- and heteropolar bonds while silicon prefers single sp^3 bonds. For these reasons, inserting Si in a C sp^2 network is by far not trivial and deserves to be studied via quantitative theoretical methodologies. On an intuitive basis one can expect the presence of Si to be reflected by some local electronic properties, since the presence of a dopant is expected to alter the shape of the electronic density of state especially for the higher energy of the valence band. In what follows, we shall refer mostly to a paper devoted to the silicon doped fullerenes $C_{59}Si$ and $C_{58}Si_2$, where one and two carbon atoms of the fullerene network are replaced by silicon atoms [21]. It turns out that the presence of a Si atom has the effect of inducing a distortion in the fullerence cage very much localized at the Si site. Changes in the electronic structure can be first described in terms of charge transfer from the Si atom to the neighboring C atoms. Concerning the electron localization properties, the situation is better explained by relying on figure 2.9 that shows the

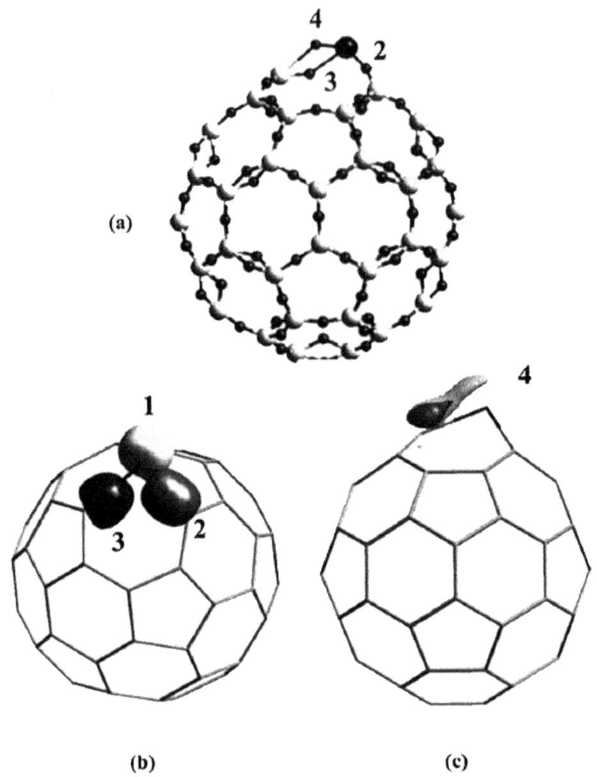

Figure 2.9. (a) Focus on the WFCs (small dark balls) for the doped $C_{59}Si$ fullerene. The Si atom corresponds to large black balls, while the C atoms to light gray balls; (b) isosurfaces (0.02 e/a.u.^{3}) of the Wannier functions having as centers to WFC1, WFC2, and WFC3; (c) isosurfaces of the Wannier function WFC4 for the isosurface densities at 0.02 e/a.u.^{3} (light gray, transparent) and 0.025 e/a.u.^{3} (dark gray). Numbers from 1 to 4 are given to the WFCs around the Si atom. Reprinted figure with permission from [21]. Copyright (1999) by AIP Publishing.

existence of a distinct region of localization in the vicinity of the Si atoms, which coexists with the localization regions in between the Si and C atoms typical of covalent bonding.

In figure 2.9(a) we visualize the WFCs for $C_{59}Si$ around Si. These are defined as WFC1, WFC2, WFC3, and WFC4. Three of them belong to the Si-C bonds and are displaced in the direction of the C atoms, namely WFC1, WFC2, and WFC3. This is a demonstration of the polar character of Si-C connections. WFC4 is closer to the Si atom along the Si-C 6-6′ (hexagon–hexagon) bond that takes the form of a deformed Si-C double bond and goes along with these latter two WFCs (WFC3 and WFC4). The Wannier functions on the two Si-C 5-6′ (pentagon-hexagon) bonds, with centers WFC1 and WFC2, follow the pattern of Wannier functions for the C-C single bonds, with the notable difference of being displaced towards carbon (figure 2.9(b)). The same occurs for WFC3. The remaining Wannier function associated with WFC4 is less localized than the WFC1, WFC2, and WFC3 counterparts. In fact, if one takes the same isosurface value of the density (0.02) as in figure 2.9(b), it extends from above Si down to the C atom. For a higher value, it is localized close to the C atom of the Si-C 6-6′ bond (figure 2.9(c)). This example of behavior of the Wannier functions and related centers exemplifies their importance in the understanding of bonding features, as the degree of polarity and the specific extension of a localization, by showing clearly how single and double bonds are modified by the presence of the dopant. As such, it calls for further applications exploiting the same concepts in the area of disordered systems, for which ionic and covalent character are often difficult to disentangle.

2.5.2.2 Example 2: MLWF applications to amorphous silicon and to a glassy surface of chalcogenide

The first example of application of the Wannier analysis based on the concept of MLWF to disordered system can be traced back to a work on amorphous silicon [22] carried out within the FPMD strategy of reference [23]. This accomplishment clearly marks a transition between traditional methods of counting the nearest neighbors and alternative strategies where the WFCs come into play. The analysis of the structure is based on the combined use of the pair correlation functions $g_{\mathrm{SiSi}}(r)$ and $g_{\mathrm{Si-WFC}}(r)$, the second one correlating the positions of the atoms to the positions of the Wannier centers. Two different methods of counting the neighbors are adopted. In the first, the first minimum of $g_{\mathrm{SiSi}}(r)$ is chosen to define the range of the coordination numbers. In the second, two atoms are considered bonded when they share the same WFC within the position of the first minimum of $g_{\mathrm{Si-WFC}}(r)$. While both analysis give a large majority of Si fourfold coordinated, the amount and kind of miscoordinations are different. In particular, twofold and threefold Si atoms are totally overlooked by the traditional counting based on the minimum of $g_{\mathrm{SiSi}}(r)$, this methodology highlighting the presence of fivefold coordinations only. We have employed atom–WFC correlations for several disordered chalcogenide systems. An instructive example is provided here by the case of the surface of glassy GeS_2 [24], a system that will not be treated in more detail in this monograph but it stands as a highly valuable application of FPMD calculations to a special case (a surface network) of disordered chalcogenide.

In figure 2.10 the pair correlation function $g_{\mathrm{S-WFC}}(r)$ (obtained by considering the S atoms and the Wannier centers as two families of degrees of freedom) features two peaks, the first centered at 0.435 Å and a second peak at 0.875 Å. The first peak reflects the distance between S atoms and the Wannier centers $\mathrm{WF_{lp}}$ associated with lone pairs (lone pairs being two valence electrons not implied in any bond, that is to say not shared with any other atom). The second peak is due to the correlation between S atoms and the Wannier centers $\mathrm{WF_b}$ indicative of bond formation since lying in between two atoms. The S–$\mathrm{WF_b}$ distance, 0.875 Å, obtained for the surface of glassy GeS_2, is slightly larger than the value obtained for the bulk (0.86 Å) and lower than the value obtained for Se–$\mathrm{WF_b}$ in glassy $GeSe_2$ for the bulk. Therefore, the Ge–S bond has a higher ionic character than the Ge–Se one since the Wannier center is closer to the S ionic site. Also, when focusing on bulk and surface glassy GeS_2, the same argument holds when comparing the above S–$\mathrm{WF_b}$ distance: there is

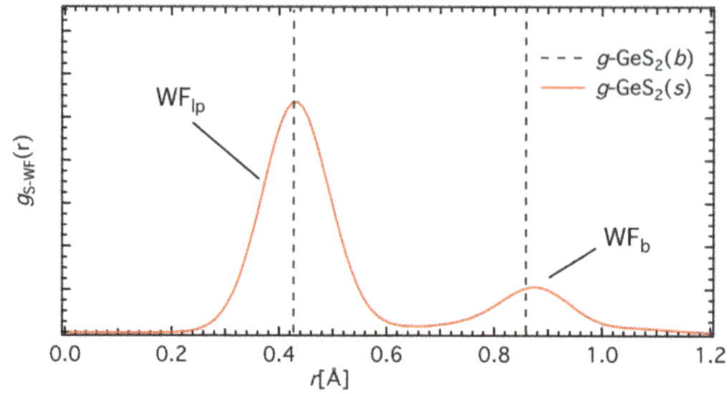

Figure 2.10. Pair correlation function $g_{S-WFC}(r)$ of the glassy surface GeS_2, termed (s) (red line). The dashed lines show the position of the peaks for glassy GeS_2 (bulk system, termed (b)). WF_{lp} indicates the peaks arising from the correlations between S atoms and the WFC related to lone pairs. WF_b indicates the peak due to correlations between S atoms and the WFC due to chemical bonds. Reprinted figure with permission from [24]. Copyright (2014) by the American Physical Society.

a slightly lower ionic character in the surface when compared to the bulk. This kind of analysis is quite useful to extract information on the bonding properties and, as such, it will be exploited in the chapters devoted to the applications of FPMD to glassy chalcogenides. Among the others, an example can be found in section 9.4.3.

2.6 Neutron scattering as experimental counterpart to MD

Before moving to the theoretical framework employed to make the connection between MD and structural characterization of amorphous materials, some more considerations are in order to become accustomed with to link calculations to experiments[8].

In this book, the large majority of the comparisons carried out with experimental data are based on the results obtained via NDIS (neutron diffraction via isotopic substitution) [25][9] by the team of Prof. P S Salmon in Bath (UK). Historically, my FPMD involvement with disordered materials is marked by the encounter with the literature on disordered chalcogenides like $GeSe_2$ and the multiple attempts to characterize them as prototypes of intermediate range order exhibiting strong analogies and differences with other tetrahedral networks like SiO_2 [26–37]. The availability of partial structure factors for liquid $GeSe_2$ made explicit in the seminal NDIS paper [38], together with the intriguing open issues raised on the interplay between the first sharp diffraction peak (FSDP) [39] and the underlying order, were a strong motivation to pursue working on methodologies accounting for chemical bonding via an explicit treatment of the electronic structure. When these efforts

[8] In what follows, I shall refer exclusively to the part of this monograph devoted to disordered chalcogenides, for which the connection theory-experiment has been a real guide for the work described in this monograph.
[9] For the sake of simplicity, in the framework of the NDIS methodology, we shall refer mostly to one comprehensive article (reference [25]) knowing that the basic information useful here can also be found in many other valuable papers quoted in this section.

started (mid 1990s) first-principles studies of disordered network-forming glasses and liquids were at their infancy and little was known on the subtleties hidden in their structural organization. A twofold motivation for studying glassy chalcogenides appeared very soon, resulting from the willingness to better characterize them at the atomic scale and as a playground for testing and improving theoretical approaches clearly challenged by the very special nature of bonding exhibited by these systems, escaping any classification in terms of ionic or covalent. By differing the explicit account of some revealing cases to specific chapters, it is worth underlining that much progress has been made by ourselves due to the continuous and accurate feedback with an experimental team combining high quality measurements with an open-minded attitude toward theory and the capabilities or weaknesses. In this section, I will raise a few points about neutron diffraction experiments and the specificities of NDIS measurements. These concepts will be reiterated and referred to also in some of the following chapters (those devoted to FPMD applications). This entire content is the result of my view of an area not covered by my own expertize. Nevertheless, it has to be dealt with neatly to ensure optimal understanding of glass structural properties. In short, I will draw attention to a few essential points with no intention to be exhaustive. From the viewpoint of a computational material scientist, the purpose is to elucidate how measurements can help quantitative predictions to grow and be constantly improved.

- In the area of disordered materials, the key quantity to start with is the measure by neutron diffraction of the total neutron structure factor

$$F(k) = \sum_{\alpha}\sum_{\beta} c_\alpha c_\beta b_\alpha b_\beta \left[S_{\alpha\beta}(k) - 1 \right] \tag{2.21}$$

where we have employed an expression more readable than equation (2.3) but having strictly the same physical meaning. In equation (2.21) k is the magnitude of the scattering vector, $S_{\alpha\beta}(k)$ is a partial structure factor for chemical species α and β, b_α, c_α are the coherent neutron scattering length and the atomic fraction of chemical species α, respectively. Within the NDIS methodology [25] a system made of n chemical species can be described by $m = n(n + 1)/2$ independent partial structure factors, this means that one needs m samples of different isotopic composition to obtain the full set of $S_{\alpha\beta}(k)$. This calls for a rewriting of equation (2.21) in order to allow for the presence of the corresponding scattering lengths.

$$F_i(k) = \sum_{\alpha}\sum_{\beta} c_\alpha c_\beta b_{\alpha i} b_{\beta i} \left[S_{\alpha\beta}(k) - 1 \right] \tag{2.22}$$

where in a matrix form for two species w and z one has

$$\begin{bmatrix} F_1(k) \\ F_2(k) \\ F_3(k) \end{bmatrix} = \begin{bmatrix} c_w^2 b_{w1}^2 & c_z^2 b_{z1}^2 & 2c_w c_z b_{w1} b_{z1} \\ c_w^2 b_{w2}^2 & c_z^2 b_{z2}^2 & 2c_w c_z b_{w2} b_{z2} \\ c_w^2 b_{w3}^2 & c_z^2 b_{z3}^2 & 2c_w c_z b_{w3} b_{z3} \end{bmatrix} \begin{bmatrix} S_{ww}(k) - 1 \\ S_{zz}(k) - 1 \\ S_{wz}(k) - 1 \end{bmatrix} \tag{2.23}$$

$$= \begin{bmatrix} a_{11} & a_{12} & a_{13} \\ a_{21} & a_{22} & a_{23} \\ a_{31} & a_{32} & a_{33} \end{bmatrix} \begin{bmatrix} S_{ww}(k) - 1 \\ S_{zz}(k) - 1 \\ S_{wz}(k) - 1 \end{bmatrix} \tag{2.24}$$

leading to the concise matrix expression

$$[F(k)] = [A][S(k) - 1] \tag{2.25}$$

that can be solved by inversion to obtain the partial structure factors $S_{ww}(k)$, $S_{zz}(k)$, $S_{wz}(k)$

$$[S(k) - 1] = [A^{-1}][F(k)]. \tag{2.26}$$

Technical points relating to the occurrence of a NDIS matrix close to singular can be found in [25].

- The availability of the partial structure factors makes possible the connection with real space via the application of equation (2.1) leading to the set of $g_{\alpha\beta}(r)$. For instance in the case of liquid $GeSe_2$ treated in section 6.2.2, we compare FPMD data and NDIS experiments based on three total neutron structure factors ${}^{N}_{N}F(k)$, ${}^{70}_{N}F(k)$, ${}^{73}_{76}F(k)$ obtained via identical diffraction experiments. With N the natural isotopic abundance, ${}^{N}_{N}F(k)$ is obtained by taking ${}^{N}Ge{}^{N}Se_2$, ${}^{70}_{N}F(k)$ via ${}^{70}Ge{}^{N}Se_2$ and ${}^{73}_{76}F(k)$ via ${}^{73}Ge{}^{76}Se_2$. Concerning that specific comparison (the first carried out between our own FPMD data and the NDIS experiments of the Bath's team) it is interesting to observe the quite strong and yet stimulating statements employed in [40] to criticize the performances of FPMD calculations due to the *failure of the simulations to reproduce the large FSDP in* $S_{GeGe}(k)$ at $\sim 1 Å^{-1}$ (see figure 2.11). The behavior of the Ge subnetwork in the Ge-based chalcogenides has been ever since a

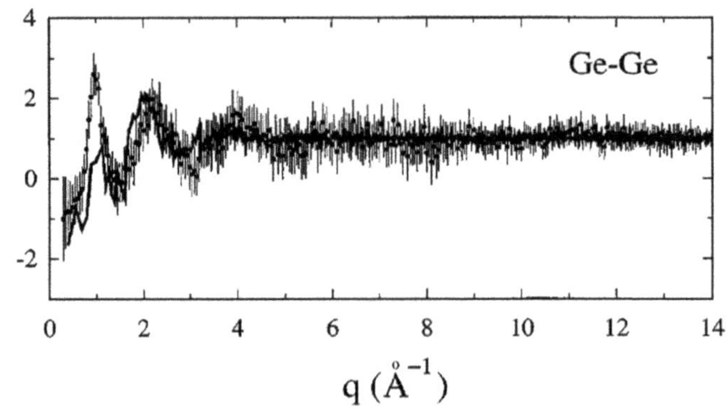

Figure 2.11. NDIS data (dots with error bars) for the Ge–Ge partial structure factor compared to FPMD calculation (black line) [41]. Experiments are from [38]. The level of comparison is the one based on the FPMD results available at that time. Reprinted figure with permission from [41]. Copyright (1998) by the American Physical Society.

matter of investigation and improvements from the FSDP standpoint, as it will be detailed later in several parts of this monograph.

- Sticking for a while to reference [40], information is provided as an alternative way of exploiting NDIS measurements by using only two total structure factors, giving the so-called first-order difference functions. Se–Se correlations are eliminated if one takes $\Delta F(k)_{\mathrm{Ge}} = {}^{70}_{N} F(k) - {}^{N}_{N} F(k)$. The same holds for Ge–Ge correlations with the expression $\Delta F(k)_{\mathrm{Se}} = {}^{73}_{76} F(k) - \frac{b^2({}^{73}\mathrm{Ge})}{b^2({}^{N}\mathrm{Ge})}{}^{N}_{N}F(k)$, where it has to be intended that the proper scattering lengths have to be used $b({}^{N}\mathrm{Ge}) = 8.185$ fm, $b({}^{70}\mathrm{Ge}) = 10.0$ fm, $b({}^{73}\mathrm{Ge}) = 5.09$ fm, $b({}^{N}\mathrm{Se}) = 7.97$ fm and $b({}^{76}\mathrm{Se}) = 12.2$ fm. Similar expressions have also been used more recently in the framework of a NDIS study [42] of glasses $GeSe_3$ and $GeSe_4$ (note that for quite some time these concentrations within the Ge_xSe_{1-x} family were not accessible to the NDIS technique).
- In introducing the option **C** in section 2.3 for calculating the pair correlation functions from the partial structure factors we mentioned that this procedure is the closest in spirit to experiments since based on what is available directly from diffraction data. This is why equation (2.13) contains an upper integration range, related to the highest value of k, k_{max} allowed by the diffractometer. In more general terms, this argument should apply from the outset to the total neutron structure factor, for which equation (2.2) (rewritten by taking $g_{\mathrm{T}}(r) - 1 = G_{\mathrm{T}}(r)$ and $S_{\mathrm{T}}(k) - 1 = F(k)$) should also include a modification function $M(k)$

$$G_{\mathrm{T}}(r) = \frac{1}{2\pi^2 \rho r} \int_0^{\infty} dk\, k\, [F(k)M(k)] \sin(kr) \tag{2.27}$$

such that $M(k) = 1$ for $k \leqslant k_{\mathrm{max}}$ and $M(k) = 0$ for $k > k_{\mathrm{max}}$. The issue of the spurious contribution affecting the experimental data have been touched upon when commenting option **C** in section 2.3. It is worth stressing in this context that when comparing FPMD and NDIS data as will be shown in some of the applications, the partial structure factor calculated via FPMD has been obtained by applying and comparing most of the times the different options **A**, **B** and **C** discussed in section 2.3. As to the presence of Fourier transform artifacts in the experiments, we refer the interested reader to references [40, 43].
- I cannot end this section and this chapter without recommending, with no intention to produce an exhaustive list, some fine papers (mostly review ones, well adapted to be quoted in this book) issued from the team of P S Salmon [44–49][10]. The foundations of the neutron scattering experiments and, in particular, of the NDIS method are recalled in a highly pedagogical form and the right space is given to references of modelling, treated very fairly in spite

[10] Some of them have been already quoted and others will be referred to in the chapters devoted to applications.

of a slightly higher tendency to point out their limits (I would tend to say to stimulate improvements and predictive power, as it has been in my case).

References

[1] Moss S C and Price D L 1985 Random packing of structural units and the first sharp diffraction peak in glasses *Physics of Disordered Materials. Institute for Amorphous Studies Series* ed D Adler, H Fritzsche and S R Ovshinsky Physics (Boston, MA: Springer)

[2] Faber T E and Ziman J M 1965 *Philos. Mag.: J. Theor. Exp. Appl. Phys.* **11** 153–73

[3] Waseda Y 1980 *The Structure of Non-crystalline Materials: Liquids and Amorphous Solids* (New York: McGraw-Hill)

[4] Bhatia A B and Thornton D E 1970 *Phys. Rev.* B **2** 3004–12

[5] Salmon P S 1992 *Proc. R. Soc. Lond. A* **437** 591

[6] Massobrio C, Pasquarello A and Car R 2001 *Phys. Rev.* B **64** 144205

[7] Massobrio C and Pasquarello A 2008 *Phys. Rev.* B **77** 144207

[8] Ori G, Massobrio C, Bouzid A, Boero M and Coasne B 2014 *Phys. Rev.* B **90** 045423

[9] Hansen J P and McDonald I R 2006 *Theory of Simple Liquids* 3rd edn (Amsterdam: Elsevier)

[10] Allen M P and Tildesley D J 2017 *Computer Simulation of Liquids* 2nd edn (Oxford: Oxford Scholarship)

[11] Massobrio C, Pasquarello A and Car R 1999 *J. Am. Chem. Soc.* **121** 2943–4

[12] Errington J R and Debenedetti P G 2001 *Nature* **409** 318–21

[13] Chau P-L and Hardwick A J 1998 *Mol. Phys.* **93** 511–8

[14] Wezka K *et al* 2014 *Phys. Rev.* B **90** 054206

[15] Bouzid A, Le Roux S, Ori G, Boero M and Massobrio C 2015 *J. Chem. Phys.* **143** 034504

[16] Le Roux S, Zeidler A, Salmon P S, Boero M, Micoulaut M and Massobrio C 2011 *Phys. Rev.* B **84** 134203

[17] Wannier G H 1937 *Phys. Rev.* **52** 191–7

[18] Resta R 1998 *Phys. Rev. Lett.* **80** 1800–3

[19] Marzari N, Mostofi A A, Yates J R, Souza I and Vanderbilt D 2012 *Rev. Mod. Phys.* **84** 1419–75

[20] Marzari N and Vanderbilt D 1997 *Phys. Rev.* B **56** 12847–65

[21] Billas I M L, Massobrio C, Boero M, Parrinello M, Branz W, Tast F, Malinowski N, Heinebrodt M and Martin T P 1999 *J. Chem. Phys.* **111** 6787–96

[22] Silvestrelli P L, Marzari N, Vanderbilt D and Parrinello M 1998 *Solid State Commun.* **107** 7–11

[23] Car R and Parrinello M 1985 *Phys. Rev. Lett.* **55** 2471–4

[24] Ori G, Massobrio C, Bouzid A, Boero M and Coasne B 2014 *Phys. Rev.* B **90** 045423

[25] Fischer H E, Barnes A C and Salmon P S 2005 *Rep. Progress Phys.* **69** 233–99

[26] Price D L, Moss S C, Reijers R, Saboungi M L and Susman S 1988 *J. Phys. C: Solid State Phys.* **21** L1069–72

[27] Elliott S R 1991 *Phys. Rev. Lett.* **67** 711–4

[28] Nemanich R J, Solin S A and Lucovsky G 1977 *Solid State Commun.* **21** 273–6

[29] Bridenbaugh P M, Espinosa G P, Griffiths J E, Phillips J C and Remeika J P 1979 *Phys. Rev.* B **20** 4140–4

[30] Boolchand P, Grothaus J, Bresser W J and Suranyi P 1982 *Phys. Rev.* B **25** 2975–8

[31] Bresser W J, Boolchand P, Suranyi P and de Neufville J P 1981 *Phys. Rev. Lett.* **46** 1689–92

[32] Sugai S 1987 *Phys. Rev.* B **35** 1345–61
[33] Boolchand P and Bresser W J 2000 *Philos. Mag.* B **80** 1757–72
[34] Susman S and Volin K J 1990 D. G. Montague, and D. L. Price *J. Non-Cryst. Solids* **125** 168–80
[35] Vashishta P, Kalia R K, Rino J P and Ebbsjö I 1990 *Phys. Rev.* B **41** 12197–209
[36] Vashishta P, Kalia R K, Antonio G A and Ebbsjö I 1989 *Phys. Rev. Lett.* **62** 1651–4
[37] Vashishta P, Kalia R K and Ebbsjö I 1989 *Phys. Rev.* B **39** 6034–47
[38] Penfold I T and Salmon P S 1991 *Phys. Rev. Lett.* **67** 97–100
[39] Salmon P S 1994 *Proc. R. Soc. Lond. A* **445** 351
[40] Salmon P S and Petri I 2003 *J. Phys.: Condens. Matter* **15** S1509–28
[41] Massobrio C, Pasquarello A and Car R 1998 *Phys. Rev. Lett.* **80** 2342–5
[42] Rowlands R F, Zeidler A, Fischer H E and Salmon P S 2019 *Front. Mater* **6** 133
[43] Salmon P S 2006 *J. Phys.: Condens. MatterJ. Phys.: Condens. Matter* **18** 11443–69
[44] Salmon P S 2007 *J. Phys.: Condens. MatterJ. Phys.: Condens. Matter* **19** 455208
[45] Salmon P S *et al* 2019 *J. Non-Cryst. Solids:* X **3** 100024
[46] Salmon P S and Zeidler A 2019 *J. Stat. Mech.:J. Stat. Mech.: Theory Exp* **2019** 114006
[47] Salmon P S and Zeidler A 2015 The atomic-scale structure of network glass-forming materials *Molecular Dynamics Simulations of Disordered Materials (Springer Series in Materials Science* vol 215) ed C Massobrio, J Du, M Bernasconi and P Salmon (Cham: Springer)
[48] Salmon P S and Zeidler A 2015 *J. Phys.: Condens. Matter* **27** 133201
[49] Salmon P S and Zeidler A 2013 *Phys. Chem. Chem. Phys.* **15** 15286–308

IOP Publishing

Carlo Massobrio

Chapter 3

Molecular dynamics to describe (amorphous) materials

The methodological foundations of molecular dynamics are implemented by aiming at an atomic-scale description of real systems. To be valuable and prone to be employed as a trustworthy area of knowledge, the entire framework has to satisfy the following criteria: time trajectories sufficiently extended, systems sizes well adapted to the dimensions of interest, and realistic forces to describe bonding between the atoms. Focusing on this last point, the notion of interatomic potential is addressed by showing, in the case of metals and iono-covalent systems, how it can be extended to go beyond simple, two-body expressions. Some examples are given to elucidate the occurrence of possible shortcomings. As a second ingredient of any basic molecular dynamics (MD) application, I introduce the idea of temperature control via a thermostat that can be made compatible with a Newtonian description via the Nosé formalism. The rest of the chapter is entirely devoted to a presentation of the first-principles molecular dynamics (FPMD) theory as developed within the Car–Parrinello scheme. Concepts like extended Lagrangian, adiabaticity, fictitious electronic kinetic energy, together with the role of the total Kohn–Sham energy as analogous of the potential energy in classical MD are revisited and explained in a step-to-step fashion. Also, the different parts of the total energy expression are introduced by avoiding any unnecessary complexity, within the spirit of a 'guided' tour well suited to beginners and more experienced users. Finally, we address the issues of time and size limitations for molecular dynamics simulations of amorphous systems, by pointing out the obvious objections to this methodology and the way to cope with them without falling into the (opposite) traps of prejudice or unmotivated belief.

doi:10.1088/978-0-7503-2436-6ch3

3.1 Molecular dynamics: what for?

According to its current definition, commonly employed and accepted by any scientific community producing results based on the behavior of collections of atoms moving in time, molecular dynamics (MD) amounts to the implementation on a computer of the solution of the equations of motion. The idea is simple and stems from the Newton expression coupling forces to accelerations through the atomic masses

$$\mathbf{F}_i = -\frac{\partial E_{\text{pot}}}{\partial r_i}(r_1, \ldots, r_N) = m_i \frac{\mathrm{d}^2\mathbf{r}_i}{\mathrm{d}t^2}. \tag{3.1}$$

The only requirement implied in equation (3.1) is the existence of a potential energy surface visited by N atoms[1] moving in time and the availability of some expression for E_{pot} to be derived with respect to the coordinates of each atom i. Also, we are relying on the hypothesis of a potential energy surface depending on the atomic positions only and on the validity of the Newtonian, classical equations of motion for the problem under consideration.

One might wonder why the practical implementation of such an elementary concept nowadays universally accepted and widely used as a research technique was initiated so late in the past century as an academic tool (in the 1950s–60s only for ideal systems and in the late 1970s–80s as a realistic counterpart to experiments) [1, 2]. There are two underlying reasons: the first has to do with the way statistical mechanics and thermodynamics have been taught for a long time. While the concept of time averages to obtain a macroscopic behavior appeared as a natural outcome of statistical mechanics, there was no sense in pursuing this goal in practice due to the very high number of degrees of freedom involved. This was also due to the availability of thermodynamics to extract an average behavior, the only one considered to be meaningful as a link between microscopic and macroscopic scales when we are dealing with extended systems. The fathers of statistical mechanics (Boltzmann and Gibbs in particular) were able to propose valuable concepts to address this issue, ranging from the idea of an irreversible macroscopic behavior resulting from reversible mechanics through an ergodic exploration of the phase space and the notion of statistical ensemble defined with respect to independent thermodynamic variables [3]. Both rely on the attainment of equilibrium, corresponding to an average stationary state meaning that the sampling through atomic movement of all possible configurations is reliable enough to be representative of the macroscopic properties. Needless to say, the foundations of statistical mechanics as a discipline fostering thermodynamics were established when computers were essentially non-existent as scientific tools, thereby excluding any possible use of

[1] Hereafter the number of atoms in a given system will be denoted via N or N_{at}. Warning: the example of section 4.5 devoted to SiN refers to N as the nitrogen atom.

them as problem solvers when trying to bridge the gap between macroscopic observable quantity and atomic-scale objects.

In the second half of the past century [4, 5] scientists[2] began asking themselves whether atomic trajectories could be indeed produced via the iterative solution of the equation of motion by using computers accomplishing two kinds of calculations: (1) forces interacting on each individual atom and due to its neighbors (to be clearly identified) and (2) time derivatives of the atomic position under the action of the above forces. This awareness led to the first pioneering implementations of MD, that have been celebrated in great detail in several papers and monographs [6, 7] to which we refer the interested reader. The revolution of MD rests on its intrinsic ability to establish a correspondence between the time spent by a computer to calculate the necessary quantities and the real physical time that becomes proportional to the number of iterations. In principle, any system is prone to be treated by this modern and unconventional theoretical recipe, provided three criteria are met:

- time trajectories long enough to describe the phenomenon of interest and/or bring the system to equilibrium,
- system size compatible with the characteristic dimensions of the system/process under consideration and, last but not least,
- forces calculated realistically and truly reflecting the chemical bonding keeping the atoms together.

Computational material scientists active in MD have been facing this challenge on a daily basis over the past half-century, by achieving a spectacular predictive power when comparing MD results and experimental data.

Of the three criteria mentioned above, one might consider at first sight that the first two are merely related to computer power and it can be circumvented by exploiting the increasing performances of the modern computers. However, they are both intimately related to the complexity of the interatomic forces (third criterion) needed to move the atoms and trigger the iterative scheme. This point is one somewhat limiting the applications of MD and it is often pointed out as the biggest, unavoidable pitfall of the whole approach. This can be understood by observing that, by definition *classical molecular dynamics (CMD)* corresponds to interatomic forces expressed as a function of the atomic positions only. CMD is based on the assumption that, for any system, an analytical form avoiding any explicit contribution of the electronic structure is sufficiently accurate to achieve a realistic description of a given system, both at zero and at finite temperature. However, this postulate is far from being true for any system. As a matter of fact, interatomic

---

[2] Two names have to be invoked when talking about the early days and the foundations of molecular dynamics: Berni Alder and Aneesur Rahman. While many others contributed, these two scientists can be considered as the pioneers and the 'inventors' of a new scientific discipline.

potentials employed in CMD are of qualitative nature, being derived from parameterization based on experimental quantities and hardly reproducing correctly properties not included in the original constructions[3]. This is particularly true for any bonding situation differing from a two-body scheme, adequate for rare gases at most. This is exactly the model employed to address generic issues related to the stability of an ideal monotomic glass in chapter 5, section 5.1 of this book.

3.2 Beyond two-body potentials

It is important to realize that the notion of two-body potential limits the range of applications in terms of chemical bonding quite drastically. *Two body* means that the force acting between atom i and atom j depends on their interatomic distance only. These conditions are clearly not met in most compounds but they have not prevented widespread use in atomic-scale modeling at the crossover between statistical mechanics and material science. When the focus is on testing behaviors and physical laws on ideal systems, a two-body potential stands out as easy to control and of minimal analytical complexity. It would be unfair to consider two-body potentials as mere objects devoid of any realistic character. It turns out that they are customarily applied in CMD to mimic the repulsive part (due to the unpenetrable character of electronic clouds) regardless of the n-body nature of the attractive part of the potential, employed when the emphasis is on the prediction or reproduction of real physical properties (for systems differing from rare gases). The transition from two-body to n-body potential within the framework of CMD is considered and exemplified in this book by relying (again, in line with a global strategy) on my direct experience. One of the most striking examples of failure of two-body potentials is their intrinsic inability to model metallic bonding in terms of elastic properties and vibrational properties. More intuitively, the delocalized character of electrons makes the atoms interacting more collectively than what is expressed via the sum of pair interactions.

Regardless of any conceptual origin underlying a specific choice for a given n-body expression of potential and forces in metals, one can think of a simple, two-body repulsive part added to an attractive part taking the form of a functional $F_i(\rho_i)$ of an atomic dependent 'pseudo' electronic density ρ_i. This latter can also be composed of two-body pair interactions as follows (with all sums running from 1 to N atoms in the system)

$$\rho_i = \sum_{j(\neq i)} \phi_{ij}(r_{ij}) \qquad \rho_j = \sum_{k(\neq j)} \phi_{jk}(r_{jk}) \tag{3.2}$$

$$E_{\text{pot}} = \frac{1}{2}\sum_i \sum_{j(\neq i)} \gamma_{ij}(r_{ij}) + \sum_i F_i(\rho_i) \tag{3.3}$$

[3] This statement has to be softened if one considers interatomic potentials derived within one of the so-called 'Machine Learning' schemes. In that case the richness of the input data base describing the potential energy surface within density functional theory is much higher, greatly improving the realistic character of the calculated properties. For an example see [8]. More on this issue on section 4.1.

$$E_{\text{pot}} = \frac{1}{2}\sum_{i}\sum_{j(\neq i)} \gamma(r_{ij}) + \sum_{i} F_i\left(\sum_{j(\neq i)} \phi_{ij}(r_{ij})\right) \tag{3.4}$$

where we have assumed that all two-body interactions γ_{ij} and ϕ_{ij} (i.e. depending on the distance between i and j only) are functions of r_{ij}, that is the absolute value of $\mathbf{r}_{ij}$. We found it convenient to show two equivalent expressions for ρ_i and ρ_j to stress the character of functional variable pertaining to a given atom i, a feature that is crucial to capturing the very nature of n-body potentials when compared with respect to two-body ones. Potentials of the kind made explicit in equation (3.3) were first proposed in the 1980s as a result of distinct developments differing by the analytical form of the functional F_i (ρ_i), the final goal being the optimal reproduction of a set of experimental properties, essentially the cohesive energy and its first (zero pressure condition at equilibrium) and second order (elastic constants) derivatives. Available parameters are adjusted accordingly, following prescriptions depending on the exact analytical form adopted but having as the same purpose the best agreement with experiments at room temperature (or, better to say, zero temperature, fits being performed on static configurations). Such n-body potentials were found to perform very well in terms of properties not included in the fitting procedure. A wealth of examples are available in the literature [9–20]. A short, non exhaustive list of those most popular and fostering continuous refinements over the years features the 'tight-binding' potentials (TB) based on the second moment approximation of the tight-binding scheme to describe the electronic density of states [9–12], the Finnis–Sinclair potentials [13, 14], the so-called 'glue model' firstly devised for studying gold properties [15], the embedded atoms potentials (EAM) [16] also available in its most recent versions [17], and the effective medium theory potentials [18]. Some recipes were also intended to account more explicitly for the electronic character of bonding [19, 20]. Here we shall refer mostly to results obtained for amorphous metallic alloys by employing the second moment approximation of a tight-binding scheme, resulting in a choice for the functional form $F_i(\rho_i)$ taking the form of a square root, namely

$$E_{\text{pot}} = \frac{1}{2}\sum_{i}\sum_{j(\neq i)} \gamma(r_{ij}) + \sum_{i} \sqrt{\left(\sum_{j(\neq i)} \phi_{ij}(r_{ij})\right)}. \tag{3.5}$$

It is instructive at this point to follow the analytical derivation of the forces acting on each atom i to capture the origin of the n-body nature of the atomic interactions. The derivation is quite general and holds for any functional form F_i, marking a first step of increased complexity in a conventional classical molecular dynamics code. As a personal note, I would like to add that all the results presented in chapter 5 were obtained after introduction of n-body forces in an otherwise quite standard Fortran program, no packages being easily available at that time for a generic expression of the functional form. While a few additional loops became necessary, I have no memory of an unaffordable increased computational cost when switching from two-body to n-body interatomic potentials and forces. By separating E_{pot} in a repulsive part E_{potR} and in an attractive part E_{potA} one obtains for the force acting on atom i (with all sums running from 1 to N atoms in the system):

$$\frac{\partial E_{\text{potR}}}{\partial r_i} = \sum_{j(\neq i)} \frac{\partial \gamma_{ij}(r_{ij})}{\partial r_{ij}} \tag{3.6}$$

$$\frac{\partial E_{\text{potA}}}{\partial r_i} = \frac{\partial F_i}{\partial \rho_i}\frac{\partial \rho_i}{\partial r_i} + \sum_{j(\neq i)}\left(\frac{\partial F_j}{\partial \rho_j}\frac{\partial \rho_j}{\partial r_i}\right) \tag{3.7}$$

$$\frac{\partial \rho_i}{\partial r_i} = \sum_{j(\neq i)} \frac{\partial \phi_{ij}(r_{ij})}{\partial r_{ij}} \qquad \frac{\partial \rho_j}{\partial r_i} = \sum_{k(\neq j)} \frac{\partial \phi_{jk}(r_{jk})}{\partial r_{jk}} \tag{3.8}$$

$$\frac{\partial E_{\text{potA}}}{\partial r_i} = \frac{\partial F_i}{\partial \rho_i}\sum_{j(\neq i)} \frac{\partial \phi_{ij}(r_{ij})}{\partial r_{ij}} + \sum_{j(\neq i)} \frac{\partial F_j}{\partial \rho_j}\sum_{k(\neq j)} \frac{\partial \phi_{jk}(r_{jk(k=i)})}{\partial r_{jk}(r_{jk(k=i)})} \tag{3.9}$$

$$\frac{\partial E_{\text{potA}}}{\partial r_i} = \frac{\partial F_i}{\partial \rho_i}\sum_{j(\neq i)} \frac{\partial \phi_{ij}(r_{ij})}{\partial r_{ij}} + \sum_{j(\neq i)} \frac{\partial F_j}{\partial \rho_j}\frac{\partial \phi_{ji}(r_{ji})}{\partial r_{ji}} \tag{3.10}$$

$$\frac{\partial E_{\text{potA}}}{\partial r_i} = \sum_{j(\neq i)} \frac{\partial F_i}{\partial \rho_i}\frac{\partial \phi_{ij}(r_{ij})}{\partial r_{ij}} + \sum_{j(\neq i)} \frac{\partial F_j}{\partial \rho_j}\frac{\partial \phi_{ji}(r_{ji})}{\partial r_{ji}} \tag{3.11}$$

$$\frac{\partial E_{\text{potA}}}{\partial r_i} = \sum_{j(\neq i)} \frac{\partial F_i}{\partial \rho_i}\frac{\partial \phi_{ij}(r_{ij})}{\partial r_{ij}} - \sum_{j(\neq i)} \frac{\partial F_j}{\partial \rho_j}\frac{\partial \phi_{ij}(r_{ji})}{\partial r_{ij}}. \tag{3.12}$$

The above equations show clearly that any n-body potential results in a dependence of the force acting on atom i that cannot be expressed as a superposition of pair contributions involving i and all the neighbors j. In the period 1988–91, the above scheme was implemented and employed to study amorphization by chemical disorder in the $NiZr_2$, as will be detailed in chapter 5. Interestingly, the schema worked very well also for a monatomic system (zirconium) for which we were able to follow at the atomic scale the mechanism of the transition from the *bcc* to the *hpc* structure (see figure 3.1) [10, 21].

3.3 Potentials for iono-covalent systems

Another example of chemical bonding first modeled via two-body interactions and then improved via the inclusion of higher order contributions accounting, in some empirical form, for electronic structure effects is provided by systems predominantly ionic, that is to say made of species having distinctly different levels of electronegativity. To this family belongs purely ionic compounds (as NaCl, the prototype of system that can be thought to a large extent as made by anions and cations) and a

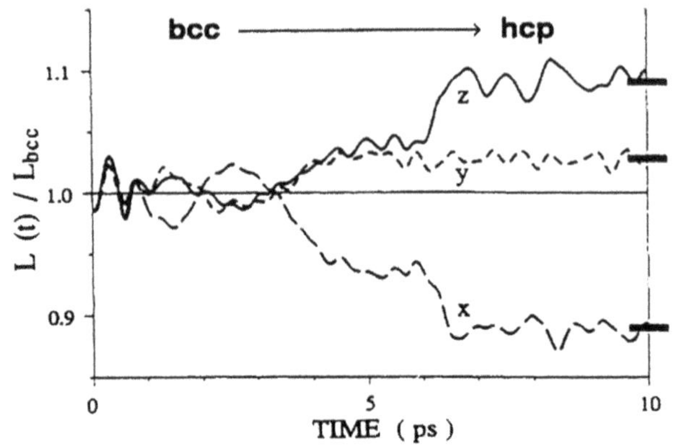

Figure 3.1. Temporal evolution of the cell edges normalized to the average *bcc* values, showing the phase transition at $T = 1500$ K. On the right-hand vertical axis, values for a perfect *hcp* structure are indicated. Reprinted figure with permission from [21]. Copyright (1989) by the American Physical Society.

large share of oxides and chalcogenide compounds for which a covalent character (bonds sharing electrons) also exists but it can, in a very first crude approximation, be neglected[4]. Typically, the construction of such two-body potentials relies on intuitive concepts. In a ionic system atoms become point charges, the structural stability being ensured by the net effect of repulsion between charges of the same sign and attraction between charges of opposite signs. The interaction between point charges is correctly accounted for in the Coulombic form implemented in a periodic system through the Ewald summation technique [24, 25]. In addition, it is sensible to include some form of repulsion as it is customary in any interatomic potential model to avoid collapse at very high densities and obtain an effect reminiscent of the non superposition of electronic clouds. Having sketched a plausible potential construction in these generic terms, it is instructive, for future purposes and in line with the presentation strategy followed in this monograph, to focus on the the selected case of Ge_xSe_{1-x} systems and, in particular, on the stoichiometric composition $GeSe_2$. Within the ionic approximation, Ge^{4+} cations are bound to Se^{2-} anions. This choice is based on a moderate and yet sizeable difference of electronegativity $\Delta_{ElNeg,Ge-Se} = 0.55$ (2.0 for Ge and 2.55 for Se) and it was at the origin of early achievements in modeling liquid and amorphous $GeSe_2$ by the team of P Vashishta [26–28]. In that context, $GeSe_2$ was treated much in the same way as SiO_2 [29] for which the ionic approximation is more legitimate in view of the much higher value of $\Delta_{ElNeg,Si-O} = 1.5$. In practical terms, P Vashishta and his team described molten $GeSe_2$ as a system made of Ge^{4+} and Se^{2-} bonded via a Coulombic interaction E_{coul}, supplemented by a charge–dipole interaction E_{CD} and a steric, repulsive term E_{ST}. Charge–dipole interactions are invoked to account for the electronic polarizability, a

[4] Reviewing the state-of-the-art of interatomic potentials of oxides is out of the scope of the present book. However, it is worth quoting at least two masterpiece references issued at different times and representative of general implementation strategies, see [22, 23].

physical property that plays an important role in the construction of suitable potentials going beyond the mere rigid ion charge point approximation.

$$\phi(r_{ij}) = E_{\text{coul}} + E_{\text{CD}} + E_{\text{ST}} \tag{3.13}$$

$$\phi(r_{ij}) = \frac{Z_i Z_j}{r_{ij}} + \frac{1}{2}\frac{\left(Z_j^2 + Z_i^2\right)}{r_{ij}^4} e^{-r/r_s} + A_{ij}\left(\frac{\sigma_i + \sigma_j}{r_{ij}}\right)^{\eta_{ij}}. \tag{3.14}$$

Parameters were fitted to available experimental properties and readjusted to obtain an acceptable melting temperature for the model, thereby leading to a potential capable of performing well also at finite temperatures. To improve upon the agreement with experimental data (in particular, on the total neutron structure factor) and provide a description of the resulting network in line with the experimental evidence available at that time, the interaction potential expressed in (3.14) was further completed by the inclusion of a three-body term depending on the angles between triads of atoms [26]. This terms was bound to reinforce the tetrahedral character of the network by containing in its analytical expression appropriate target values for the angles of Se–Ge–Se and Ge–Se–Ge triads within a tetrahedral network. When the structural analysis on disordered $GeSe_2$ came out, there were no reasons to believe that the main differences could be found between the description obtained from this model and actual structure. However, in 1991, Penfold and Salmon produced striking evidence of a deficiency present in the $GeSe_2$ model by relying on a set of measurements based on the technique of isotopic substitution within neutron diffraction, giving access to the partial structure factors and, by Fourier integration, to the partial pair correlation functions [30], see also chapter 2, section 2.6. Experiments highlighted the presence of homopolar Ge–Ge and Se–Se bonds via the appearance of unmistakable peaks at distances very close to Ge–Se bonding ones in both the Ge–Ge and Se–Se pair correlation functions. This difference is the sign of a profound shortcoming inherent in interatomic potential models of references [26, 27] since no sign of such feature on both Ge and Se subnetworks had ever appeared in the corresponding structural study. This fact exemplifies the impact of modeling studies on chalcogenides to obtain information of general validity. To dwell on this issue let's mention that about 20 years after the work by Vashishta quoted above, the limits of a two-body interatomic potential for liquid $GeSe_2$ improved by the account of n-body polarization effects (in addition to a largely predominant Coulombic term) were still detectable in a work I had to opportunity to perform with the team of M Wilson [31]. In that paper, the addition of polarization terms was found crucial to achieve better structural properties (when compared to a purely ionic attractive interaction) [32]. This is due to their effect on the Ge–Se–Ge angle via the dipole induced on the Se atoms that screens the repulsion between the positively charged Ge atoms.

Despite the clear pedagogical value of such model potentials, able to control the network topology via a single model parameter (the polarizability of the more

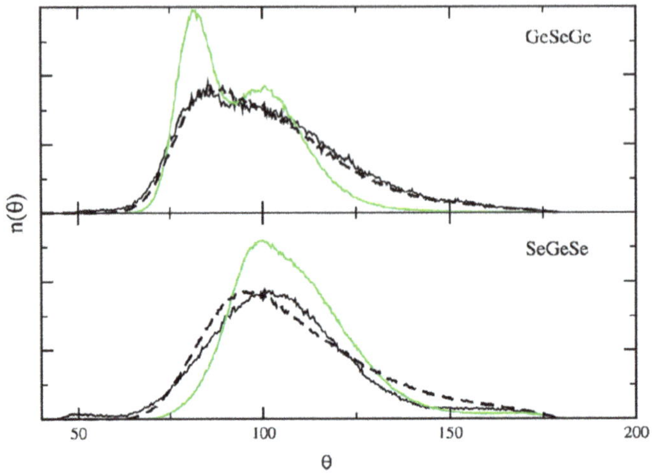

Figure 3.2. Bond-angle distributions θ_{GeSeGe} (upper panel) and θ_{SeGeSe} (lower panel) of liquid $GeSe_2$ as obtained by FPMD ([33], black lines) at T = 1050 K and by the polarizable ion model employed in [31] at T = 3000 K (green line) and at T = 7000 K (dashed line). Reprinted figure with permission from [31]. Copyright (2008) by AIP Publishing.

anionic species), it remained true that absence of any departure from chemical order (no homopolar bonds and no deviations from the fourfold tetrahedral Ge coordination) resulted in a large overestimate of the melting temperature, T_m = 2900 K against the experimental value T_m = 1025 K [31]. This is exemplified in figure 3.2 showing the bond-angle distributions. A good agreement between the FPMD and the potential model is obtained at the expenses of a huge difference in the reference temperature. These flaws further highlight the inadequacy of atomic-scale approaches not considering explicitly the electronic structure in the description of bonding, at least for cases based on analytical expressions largely empirical and fitted on a necessarily limited number of experimental parameters.

3.4 Thermostats for molecular dynamics

Controlling the temperature of a system within MD has been one of the first objectives pursued ever since the advent of this methodology as a viable tool to extract macroscopic properties from statistical mechanics. Since the most natural description of a collection of atoms evolving in time is the one provided by the microcanonical ensemble at constant energy and volume, the problem of imposing values for a quantity (the temperature) that is well defined only as an average is non-trivial and, as such, it has driven valuable efforts since the 1970s. In what follows, the focus will be on temperature control techniques for which I have direct experience as a user, nurtured first by close contacts with one of the main contributors to this field, W G Hoover, my supervisor at University of California Davis in the period 1984–85 and later by the needs of mastering temperature control in the framework of first-principles molecular dynamics (FPMD) (introduced in section 3.5 and in further sections of this chapter, and also treated in chapter 4)[5].

The simplest way of circumventing the problem of imposing the value of an averaged properties by acting on a trajectory built stepwise at each instant of (discretized) time is to impose a constraint on the corresponding instantaneous quantity. By defining a set of vectorial coordinates r_i, velocities $\dot{r}_i$ and momenta $p_i = m\dot{\mathbf{r}}_i$, (masses m assumed to be equal for all atoms and scalar product indicated by the $\langle\rangle$ notation) this means the time derivative of the kinetic energy $\frac{1}{2}\sum m\langle\dot{\mathbf{r}}_i\dot{\mathbf{r}}_i\rangle$ should be set to zero

$$\sum\langle\mathbf{p}_i\dot{\mathbf{p}}_i\rangle = 0 \tag{3.15}$$

where each sum has to be intended over all atoms i. Equation (3.15) can be satisfied by introducing into the Newton equations a friction constant α taking the form

$$\dot{\mathbf{p}}_i = \mathbf{F}_i - \alpha\mathbf{p}_i \qquad \alpha = \frac{\sum\langle\mathbf{F}_i\mathbf{p}_i\rangle}{\sum\langle\mathbf{p}_i\mathbf{p}_i\rangle} \tag{3.16}$$

resulting in a constant kinetic energy since

$$\sum\langle\mathbf{p}_i\dot{\mathbf{p}}_i\rangle = \sum\langle\mathbf{p}_i\mathbf{F}_i\rangle - \frac{\sum\langle\mathbf{p}_i\mathbf{F}_i\rangle}{\sum\langle\mathbf{p}_i\mathbf{p}_i\rangle}\sum\langle\mathbf{p}_i\mathbf{p}_i\rangle = 0. \tag{3.17}$$

3.4.1 The breakthrough of S Nosé

While it remains true that a constraint on the instantaneous kinetic energy produces an average kinetic energy that is trivially constant, this cannot be taken as a proper realization of a canonical ensemble in which the average kinetic energy (that is to say, the temperature) is the target constant value in the thermodynamic limit. While other somewhat empirical modified equations of motion were available to control the temperature in a less abrupt way, a breakthrough in this direction was achieved in 1984 by Suichi Nosé [36, 37]. Nosé demonstrated rigorously how a canonical ensemble can be obtained within a molecular dynamics trajectory by coupling a system of N_{ind} independent degrees of freedom to a new dynamical variable s. This is compatible with the Hamiltonian character of the extended global dynamical quantity, provided the time t_{v} inherent in (3.18) can be considered as a virtual physical parameter, linked to the real time t_{r} by the equality $t_{\text{v}} = st_{\text{r}}$:

$$\mathcal{H}_{\text{Nosé}} = \sum_{I=1}^{N}\frac{p_i^2}{2ms^2} + E_{\text{pot}}[r_i] + (N_{\text{ind}} + 1)K_bT^{\text{TG}}\ln s + \frac{p_s^2}{2\eta}. \tag{3.18}$$

[5] With respect to the early stages of thermostats implementation within MD, I will not mention (at least) a couple of other approaches that played a stimulating role in that context [34, 35]. Again, the underlying reason being the overall strategy of highlighting techniques and calculations for which I had a direct experience, being persuaded that the spectrum covered is large enough to ensure some generality.

In equation (3.18) (where $\langle \mathbf{p}_i\mathbf{p}_i\rangle = p_i^2$) the s variable appears with its conjugated momentum p_s^2 and mass η, $(N_{\text{ind}} + 1)K_bT^{\text{TG}} \ln s$ playing the role of a potential energy acting in combination with the kinetic energy $\frac{p_s^2}{2\eta}$ to impose on the average the target temperature T^{TG}. In view of the impact of the Nosé technique on the entire field of molecular dynamics and, in particular, on applications devoted to disordered materials for which there is a continuous need of temperature control due to the presence of repeated cooling and heating cycle, it is important to review the essence of this methodology step by step. By avoiding the mere reproduction of what can be found in the available literature, I will underline some specific details stemming from my experience of early user acting on the classical and FPMD contexts. In 1985, W G Hoover, rewrote first, in terms of t_{r}, the equation of motions that can be derived from (3.18),

$$\frac{\partial \mathbf{r}_i}{\partial t_{\text{r}}} = s\frac{\partial \mathbf{r}_i}{\partial t_{\text{v}}} \tag{3.19}$$

to obtain (with $\mathbf{F}_i$ the force acting on atom i as derived from the potential energy $E_{\text{pot}}[r_i]$)

$$\dot{\mathbf{r}}_i = \frac{\mathbf{p}_i}{ms} \qquad \dot{\mathbf{p}}_i = s\mathbf{F}_i \tag{3.20}$$

$$\dot{s} = \frac{sp_s}{\eta} \qquad \dot{p}_s = \sum_{i=1}^{N}\frac{p_i^2}{ms^2} - (N_{\text{ind}} + 1)K_bT^{\text{TG}}. \tag{3.21}$$

It is important to realize that by expressing the equations of motions in terms of t_{r} one looses their Hamiltonian character, this meaning that equations (3.20) and (3.21) cannot be directly obtained from the expression in equation (3.18). However, one has to keep in mind that the notion of a quantity conserved in time holds true even in the absence of a direct link between it and the equations of motion. This feature can be fully appreciated by following the further changes made by Hoover to equations (3.20) and (3.21) [38]. The key equalities, leading to the introduction of a friction force fully compatible with Nosé dynamics are

$$\frac{\dot{s}}{s} = \frac{p_s}{\eta} = \beta. \tag{3.22}$$

With the help of some easy algebra one ends up with a set of equations of motion in which the variable s has disappeared and β plays the role of a friction parameter

$$\dot{\mathbf{r}}_i = \frac{\mathbf{p}_i}{m} \qquad \dot{\mathbf{p}}_i = \mathbf{F}_i - \beta\mathbf{p}_i \tag{3.23}$$

$$\dot{\beta} = \left(\sum_{i=1}^{N}\frac{p_i^2}{m} - N_{\text{ind}}K_bT^{\text{TG}}\right)/\eta. \tag{3.24}$$

To make the connection between this set of equations and a conserved quantity consistent with the original proposal by Nosé (equation (3.18)) let's recall that $t_v = st_r$ and insert the new variable β so that

$$\mathcal{H}_{\text{Nosé–Hoover}} = \sum_{i=1}^{N} \frac{p_i^2}{2m^2} + E_{\text{pot}}[r_i] + (N_{\text{ind}})K_b T^{\text{TG}} \int \frac{\dot{s}}{s} \mathrm{d}t + \beta^2 \eta/2 \tag{3.25}$$

in which N_{ind} has replaced $N_{\text{ind}} + 1$ to ensure consistency between equations (3.18) and (3.23). It turns out that the presence of a time integral in equation (3.25) is somewhat cumbersome for practical purposes, knowing that the stability of the conserved quantity along a time trajectory is the strongest test to monitor the implementation of the equations of motion in any homemade or largely distributed code. This slight inconvenience of the Nosé–Hoover approach was not pointed out in the original paper but it can be removed by following the approach detailed in [39] where the idea of thermostat control of temperature is employed within first-principles MD (FPMD). For this reason, I do prefer to come back extensively on this point later. However, one can anticipate that by replacing β by $\dot{\alpha}$ equation (3.25) remains perfectly consistent provided $(N_{\text{ind}})K_b T^{\text{TG}} \int \frac{\dot{s}}{s} \mathrm{d}t$ and $\beta^2 \eta/2$ take the place of $(N_{\text{ind}})K_b T^{\text{TG}} \alpha$ and $\dot{\alpha}^2 \eta/2$, respectively.

The formalism by Nosé and Hoover has fostered not only thousands of applications but also on the methodological side, many refinements and a wealth of attempts to improve it and/or to cure some of its possible (slight) inadequacies for certain situations[6]. An important extension of the original version (introducing the notion of 'chains of thermostats') allows a better handling of ergodicity and it is routinely used in first-principles MD codes [41], as detailed, for instance, in [42]. Again, both for the sake of simplicity and to stick to a guideline of presentation based on direct involvement, I will limit any further reference to the temperature control to the original Nosé and Hoover formulation [36–38].

3.5 First-principles molecular dynamics via the Car–Parrinello method

3.5.1 Basic ideas

The conceptual problem of accounting for the role played by electrons to keep the atoms together is a task of huge complexity, as proved by the wealth of underlying theoretical treatments developed in the area of electronic structure and chemical bonding calculations. Within the context of the present chapter, this consideration amounts to rewriting the Newton classical equations of motion by including a dependence of the interatomic forces on electronic variables via a suitable theoretical prescription. In what follows, we shall refer almost exclusively (unless explicitly

[6] Quite recently, Hoover has made available a review on thermostats within MD [40] that, despite an exceedingly high preference for model systems at the expenses of real applications, has the advantage of collecting a lot of conceptual thoughts in this area.

mentioned for specific cases) to Density Functional Theory (DFT) in the Kohn–Sham formalism [43, 44] able to reduce the n-body problem of the electron–electron interaction to a one-body approximation in which orbitals can be taken as independent, provided exchange and correlation effects are added *a posterori* in the global total energy Hamiltonian (for more details, see section 3.6). What is the main effect of this extension and why does this mark the transition from CMD to FPMD? The higher complexity of the analytical expression contained in the derivatives at the right-hand side of equation (3.1) does not affect, in principle, the classical character of the equations of motion, since the trajectory remains fully deterministic and does not contain any ingredient that could be reminiscent of the quantum motion invoked for instance, in a process like quantum tunneling. However, quantum mechanics is fully present within DFT through the representation of electrons in orbitals, their being cast in single or double occupied states (when spin is not considered) and their being treated as quantum objects, with presence in space having a probabilistic character. Such considerations have legitimated the label of first-principles molecular dynamics assigned to any molecular dynamics accounting for forces extracted from quantum mechanical Hamiltonians, DFT being the most popular among them. The extension of equation (3.1) to the consideration of potential energies and forces based on the electronic structure is indicative of a huge upscale in computational time. This can be discouraging when considering that, in principle, the electronic ground state has to be calculated for any atomic (ionic) set of positions, and this for systems and time trajectories sufficiently extended to make possible a connection with macroscopic reality via statistical mechanics.

In 1985, Roberto Car and Michele Parrinello [45] proposed a recipe for including forces having a quantum mechanical accuracy within a classical molecular dynamics scheme in a manageable way that avoids costly calculations of the electronic structure for each ionic configuration. This has marked the advent of FPMD as a potentially exploitable tool in computational material science. The starting point is the Lagrangian of collection of ions of coordinates $\mathbf{R}_I$ with masses M_I and electrons described via orbitals $\psi_i(r)$. μ is a mass (not having the physical dimensions of an atomic mass) that will be described in detail later. Here I refers to the atoms and i to the occupied orbitals (or eigenfunctions). We can also assume, without loss of generality, that the total energy is derived within the density functional Kohn–Sham framework[7]

$$\mathcal{L}_{\mathrm{CP}} = E_{\mathrm{fic}}(\psi_i(\mathbf{r})) + E_{\mathrm{kin}}(\{\mathbf{R}_I\}) - E_{\mathrm{tot}}[\{\psi_i\};\ \{\mathbf{R}_I\}] + E_{\mathrm{const}}(\psi_i(\mathbf{r})) \tag{3.26}$$

$$\begin{aligned}\mathcal{L}_{\mathrm{CP}} = \mu \sum_i \int |\dot{\psi}_i(\mathbf{r})|^2 \mathrm{d}\mathbf{r} + \frac{1}{2}\sum_{I=1}^{N} M_I \dot{\mathbf{R}}_I^2 - E_{\mathrm{tot}}[\{\psi_i\};\ \{\mathbf{R}_I\}] \\ - \sum_{ij} \lambda_{ij}\left(\int \psi_i^*(\mathbf{r})\psi_j(\mathbf{r})\mathrm{d}\mathbf{r} - \delta_{ij}\right).\end{aligned} \tag{3.27}$$

[7] Hartree atomic units are employed throughout this book.

Quite often in the literature equations (3.26) and (3.27) are given with the expression for $E_{\text{tot}}[\{\psi_i\}; \{\mathbf{R}_I\}]$ taking the form $E_{tot}[\{\psi_i\}; \{\mathbf{R}_I\}, \alpha_l]$. This accounts for the presence of additional l dynamical variables, as the thermostats we have described in section 3.4.1. This convention will be followed hereafter, by making explicit in the case of temperature control the actual form of $E_{\text{tot}}[\{\psi_i\}; \{\mathbf{R}_I\}, \alpha_l]$. To complete the description of equation (3.27), it has to be noted that a dynamical constraint implying orthogonality between the wavefunctions (fourth term on the right-hand side) has to be added to comply with one of the 'golden rules' of quantum mechanics.

3.5.2 The Car–Parrinello method step by step

In what follows we shall present the Car–Parrinello (CP) approach in a step-by-step fashion by avoiding the unnecessary complexity of a wealth of technical details but insisting on those aspects necessary to use the method effectively. Our target reader has some experience with CMD and is not uncomfortable with the notion of chemical bonding and electronic structure. The intention here is to mark the transition between CMD and FPMD without resorting to an overwhelming number of conceptual notions and subtleties, largely available in specialized textbooks and reviews [42, 46]. When considering the Lagrangian given in equation (3.27) it is convenient to make the connection with the classical MD expression by looking for analogies. The quantity $\frac{1}{2}\sum_{I=1}^{N} M_I \dot{\mathbf{R}}_I^2$ is easily recognizable as the kinetic energy of the ions, depending on the coordinate $\mathbf{R}_I$ only. In equation (3.27) the role of the potential energy is played by the density functional total energy $E_{\text{tot}}[\{\psi_i\}; \{\mathbf{R}_I\}, \alpha_l]$ that contains all interactions involving ions, electronic degrees of freedom (one can also refer to them as electrons, even though in what follows it will be more convenient and correct to talk about orbitals or wavefunctions) and additional dynamical variables. It should be stressed that the challenge of finding a valuable expression for the FPMD potential energy amounts to solving the electronic structure problem for a collection of atoms in interaction. This covers an entire field of research somewhat independent on the existence of FPMD and in continuous evolution. It has to be kept in mind that many scientists work intensively on the electronic structure of condensed matter systems in a way not necessarily related to the idea of making atoms *moving at finite temperatures* within a sound model as realistically as possible.

When it comes to facing the potential energy in equations (3.26) and (3.27) (Kohn–Sham total energy) some beginners have a tendency either to get discouraged by the difficulties inherent in the formulation of a DFT total energy or to give in to the choice of taking $E_{\text{tot}}[\{\psi_i\}; \{\mathbf{R}_I\}, \alpha_l]$ as a black box likely to be provided by some expert in electronic structure calculations. My advice goes in the direction of mastering most of the technical features underlying $E_{\text{tot}}[\{\psi_i\}; \{\mathbf{R}_I\}, \alpha_l]$ without necessarily going through all the literature about them, especially if the focus is the production of reliable FPMD trajectories and the study of the structural/dynamical properties to be extracted from them.

By taking for granted at this stage $E_{\text{tot}}[\{\psi_i\}; \{\mathbf{R}_I\}, \alpha_I]$ (to which we shall come back later in section 3.6), it is necessary to note that a dynamical trajectory of quantum mechanical accuracy (since obtained via the DFT potential energy of equation (3.27)) can be produced, in principle, by exploiting the two terms described above, namely *the kinetic energy of the ions and potential energy accounting for the full electronic structure problem.* This corresponds to solving the static problem of finding the ground state electronic structure (at fixed ionic positions) and then propagating the trajectory according to Newtonian dynamics with forces on each ion derived as gradients of the total DFT (that is, potential) energy. *This strategy is termed Born–Oppenheimer (BO) MD since based on the electronic structure strictly converged on the BO surface* [47] as a prerequisite to calculate the forces acting on each atom. From the standpoint of a conserved dynamics, the BO approach poses the problem of the optimal conservation of the total (kinetic plus potential) energy since the minimization of the DFT based potential energy entails an external perturbation not included (and not compatible) with the Newtonian evolution. These considerations drive our attention to a first, substantial difference between a scheme reminiscent of CMD (though based on a potential energy fully rooted on the electronic structure of the system) and the one made explicit in equation (3.27).

Let's consider now $E_{\text{fic}}(\psi_i(\mathbf{r}))$ in equations (3.26) and (3.27).

$$E_{\text{fic}}(\psi_i(\mathbf{r})) = \mu \sum_i \int |\dot{\psi}_i(\mathbf{r})|^2 \mathrm{d}\mathbf{r}. \tag{3.28}$$

This term has the form of a kinetic energy in which one recognizes a mass μ and the square of time derivatives $\dot{\psi}_i$, where ψ_i are the orbitals pertaining to the electronic states i. By attributing a dynamical nature to ψ_i, one allows the orbitals to depart from the BO surface so as to adjust to the movement of the ions. Provided μ is small enough, this engenders no essential changes with respect to a full electronic ground state minimization of the energy for that particular set of ionic coordinates. The attribution of a dynamical character to the wavefunctions has a fictitious nature since $E_{\text{fic}}(\psi_i(\mathbf{r}))$ has strictly nothing to do with a real physical quantity, neither through its mass μ nor via the scalar product of the orbitals time derivatives. Its role, though crucial in the implementation of equation (3.27), is limited to ensure a proper control of the departure from the BO surface while making possible a combined dynamical evolution of the extended set of degrees of freedom made by ions and the newly created *fictitious electronic degrees of freedom* ψ_i.

3.5.3 Two families of degrees of freedom in non-equilibrium

Having established that $E_{\text{fic}}(\psi_i(\mathbf{r}))$ is expected to play a major role, one has to understand which are the practical implications of its use and under which conditions the methodology inherent in equation (3.27) is bound to work. As a first, historical consideration, we mention that the idea of Car and Parrinello has its roots into the family of extended Lagrangians containing ad hoc dynamical variables imposing some external conditions in a way compatible with the existence of a conserved quantity. To the case of temperature thermostats treated in section 3.4.1 one can add the control of pressure, pioneered in [35, 48] and referred to briefly in chapter 5, section 5.1. These schemes have marked the evolution of MD from the mere application of the microcanonical, constant energy ensemble to other, more elaborate forms leading to the conservation, in average, of quantities as temperature or pressure. *However, in additional to a totally new methodology including the account of the electronic structure, there is a profound difference between the recipe contained in equation* (3.27) *and previous realizations of the extended dynamical variable concept.*

In typical applications of CMD combining the atomic degrees of freedom to a thermostat, for instance, the mass attached to this variable, say μ_{ther} (we called it η in section 3.4.1) is, by construction, such that the dynamical frequencies of the thermostat are comparable to those of the system. This warrants some effectiveness to the coupling between the atomic motion and the action of μ_{ther}, to be fine-tuned by optimizing its values. The case of equation (3.27) stands for the opposite situation. Here one targets a separation between two families of degrees of freedom, since the electronic structure has to adapt to changes of the ionic configuration without a sizeable transfer of energy promoting a departure from the BO surface. This condition is representative of an adiabatic behavior for the motion of the variables ψ_i and has its equivalent, in classical dynamics terms, to a separation between two families of degrees of freedom evolving on distinct ranges of dynamical frequencies. As a personal note, I would like to share a convenient line of thought I employed to understand this issue of dynamical families of degrees of freedom evolving together and yet, not thermalizing on accessible time scale. The analogy is purely classical but rich of significance and it can be of some help especially for beginners[8].

- Let's suppose we have a system of M diatomic molecules evolving in time. Each one of these molecules is made of pair of atoms connected by a very stiff spring, so that $2M = N$, with N number of atoms. When counting the number of degrees of freedom one can also consider focusing on the molecules, for which we have $3M$ (translations) + $2M$ (rotational) + M (vibrational, the spring) = $6M$ = $3N$ degrees of freedom as it should. If we start to run MD with the diatomic molecules at their equilibrium distance, by imposing an initial temperature to all degrees of freedom (for instance by acting on the kinetic energy of each atom)

$$\frac{1}{2}\sum_i m_i \langle v_i^2 \rangle = (3N - 3)\frac{k_B T}{2} \tag{3.29}$$

[8] Incidentally, a model based on a separation between translational/rotation and vibrational degrees of freedom was proposed and employed in [49] to study molecular mixtures of monomers, rigid dimers and flexible dimers.

(k_B being the Boltzmann constant and v_i the velocity of atom i) and by recalculating the temperature in terms of translational, rotational and vibrational contributions T_{tr}, T_{rot} and T_{vib} we do observe after relaxation and equilibration that T_{tr} and T_{rot} can take close values fairly quickly. On the other hand, T_{vib} remains quite small as if its own phase space were disconnected from the one visited by T_{tr} and T_{rot}. This kind of behavior stems from the drastic difference of characteristic frequencies hampering the transfer of kinetic energy from the trans/rotational to the vibrational motion. For the system under consideration, the choice of springs to mimic a chemical bond between two atoms was early recognized as a choice to be improved, giving rise to smarter methods to simulate bonds based on the theory of constraints as applied to molecular dynamics [50]. However, the example of the diatomic molecules remains instructive since it stands for an unwanted effect that becomes, in contrast, very much sought and to be realized, in practice, within the CP methodology. For all these reasons, a choice of μ in equation (3.27) being much smaller that the ionic masses is a prerequisite to fulfill two crucial conditions.

- First, ions and fictitious electronic degrees of freedom can evolve in time without achieving, at least on time scales accessible to FPMD, full thermal equilibrium. If this were the case, it would mean that the electronic structure has moved away from its ground state value, at odds with the main requirement of the adiabatic condition behind the solution of the Kohn–Sham equations for each set of $\mathbf{R}_I$. While it is certain that, in the long run, all degrees of freedom will come to equilibrium at the same temperature, it remains true that a non-equilibrium situation featuring distinct thermalization at different temperature can hold for trajectories long enough to allow for calculations of statistically significant time averages.
- Second, the simultaneous solution of the equations of motion derived from equation (3.27)

$$\mu\ddot{\psi}_i(\mathbf{r}, t) = -\frac{\delta E_{\text{tot}}[\{\psi_i\}, \{\mathbf{R}_I\}, \alpha_l]}{\delta\psi_i^*(\mathbf{r}, t)} + \sum_j \lambda_{ij}\psi_j(\mathbf{r}, t) \tag{3.30}$$

$$M_I\ddot{\mathbf{R}}_I = -\nabla_I E_{\text{tot}}[\{\psi_i\}, \{\mathbf{R}_I\}, \alpha_l] \tag{3.31}$$

makes available a viable scheme where the costly search of the electronic ground state for each set of ionic coordinates is replaced by an adiabatic evolution of the electronic structure fully compatible with a well conserved Lagrangian.

3.5.4 A first summary and some practical considerations

So far, we have introduced the concepts of

1. Extended Lagrangian in which the density functional total energy plays the role of potential energy and depends explicitly on the atomic position and on the electronic structure of the system via the wavefunctions (orbitals).
2. Fictitious electronic kinetic energy based on the dynamical character attributed to the orbitals. This quantity is a measure of the departure from the exact BO ground state of the wavefunctions pertaining to a set of ionic coordinates.
3. Adiabatic conditions satisfied provided the dynamics of the two subsystems (ions and fictitious electronic degrees of freedom) correspond to well separated ranges of frequencies. This prevents full thermal equilibrium to be reached on time scales extended enough to allow taking meaningful equilibrium averages.
4. Electronic structure readily available at each ionic configuration since evolving in time and adjusting to changes in ionic configurations.

3.5.4.1 Small but unavoidable: the Car–Parrinello timestep

In order to achieve an objective and manageable view of the whole FPMD/CP methodology, one has also to account for a set of disadvantages that can be summarized by invoking the role of the fictitious mass μ. Having established that this quantity has to be much smaller than the atomic masses to ensure adiabatic behavior, the challenge is to integrate a dynamics evolving on frequencies way too high when compared to the ionic (atomic) ones. Since experience suggests for μ values at least 100 times smaller than M_I, the resulting drawback is a timestep Δt_{CPMD} for the integration of the equations of motion much smaller (at least 10 times) than those employed in CMD for an equivalent system, Δt_{CMD}. Such short timesteps have been pointed out as a discouraging feature of the method since its early years, together with the fact (obvious but not welcome by many practitioners) of accounting explicitly for the electronic structure, this meaning less affordable calculations and longer waiting times. Focusing on the 'too small' timestep issue, it is somewhat amazing to realize that any criticism of this kind (still very much persisting over the years as an irrefutable demonstration of intrinsic limits bound not to be overcome) has been overshadowed by the continuous increasing of computational power and speed, undermining statements expressed vigorously 20 or 30 years ago and become totally outdated nowadays. In other words, while it remains true that a relatively small timestep (say 10^{-16} s) penalizes some applications of equation (3.27) especially for large systems, this drawback has been largely offset by progresses in high performance computing advancing at a pace much faster than the supposedly high $\frac{\Delta t_{\mathrm{CMD}}}{\Delta t_{\mathrm{CPMD}}}$ ratio.

3.5.4.2 Some first indications to make the good choices for Δt_{CPMD}

Here we would like to provide a very first 'rule of thumb' for the choice of Δt_{CPMD}. This can be obtained straightforwardly since users concerned with the timestep issue to achieve the best optimized dynamical evolution when using equation (3.27) have gained a substantial experience since the early applications of the FPMD scheme. With the exception of some situations involving very light atoms or trajectories drastically out of equilibrium (see chapter 4 for more details on such cases) we can legitimately state that timesteps in between 3 and 5 a.u. (1 a.u. $= 2.418\,9 \times 10^{-17}$ s) are very much likely to be appropriate for any use of the CPMD methodology described here. In certain cases (low temperature simulations and heavy atomic masses), the conservation of the constant of motion intrinsic in the above Lagrangian can even tolerate timesteps as large as 20 a.u. In any event, Δt_{CPMD} should be the best compromise between a steady, not diverging dynamics and the need to optimize the computational resources. As to μ, we mentioned above that values should be ~100 times smaller than M_I. For instance, if we refer to germanium (atomic mass 72.64, close to 133 000 in a.u.) a choice for μ close to 1000 a.u. is appropriate, even though a somewhat lower value would strengthen adiabaticity with little or none reduction of Δt_{CPMD}.

The notion of adiabaticity deserves some more attention. It might be difficult to grasp its importance in terms of 'trajectory departing (or not) significantly from the BO surface'. Some insight can be gained by observing that $E_{\mathrm{fic}}(\psi_i(\mathbf{r}))$ depends linearly on μ and, to avoid any unwanted equipartition of energy (meaning thermalization), its value has to be much smaller that the kinetic energy associated to the ions. However, there is no point in arbitrarily reducing μ. First, this would make Δt_{CPMD} unaffordably small. Second, one needs to employ the machinery of equation (3.27) in a truly dynamical sense, by allowing the electronic degrees of freedom to follow the ions, this implying a sizeable kinetic energy $E_{\mathrm{fic}}(\psi_i(\mathbf{r}))$. Having mentioned such intuitive considerations, one can also rely on the simple example of a dimer to understand what adiabaticity means in practice.

In figure 3.3 we show the behavior of the ionic kinetic energy and of the fictitious kinetic energy (in red) for the Si dimer. The oscillations in the movement of the

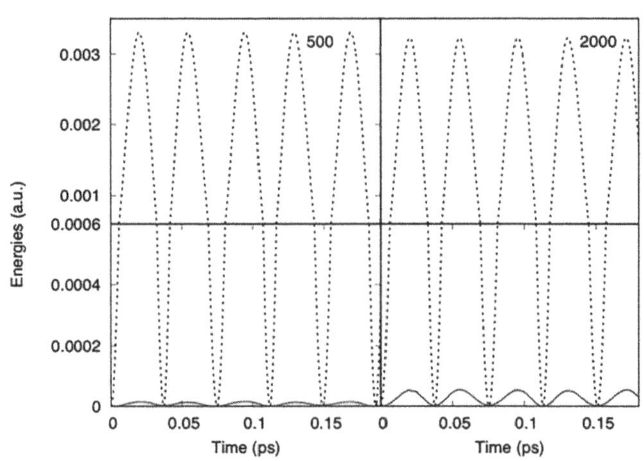

Figure 3.3. Ionic kinetic energy (black line) and fictitious kinetic energy (red line) for a Si dimer dynamics. Note that the vertical scale has been adapted in two sections so as to visualize the entire amplitude range. Courtesy of E Martin (ICube, Strasbourg).

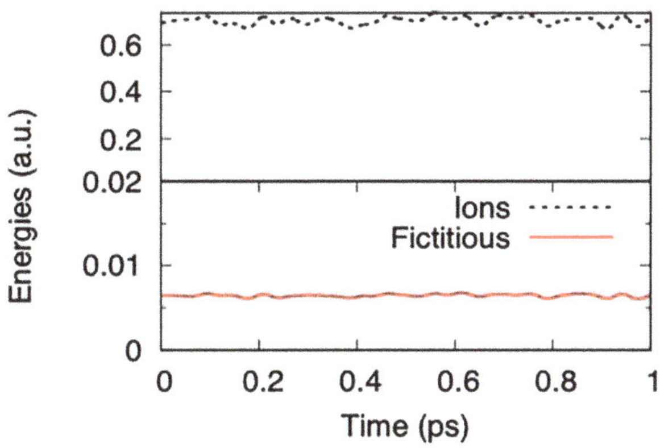

Figure 3.4. Ionic kinetic energy and fictitious kinetic energy (in red) for a disordered $GeSe_3$ system at $T = 300$ K. Note that the vertical scale has been adapted in two section so as to visualize the entire amplitude range. Courtesy of E Martin (ICube, Strasbourg).

dimer are perfectly consistent with the idea of adiabacity as expressed above, as shown by the fact that for each complete oscillation, $E_{\mathrm{fic}}(\psi_i(\mathbf{r}))$ finds back its initial (zero) value. The amplitudes of the fictitious kinetic energies $E_{\mathrm{fic}}(\psi_i(\mathbf{r}))$ are those corresponding to the selected values of μ, namely 500 a.u. and 2000 a.u. Both $E_{\mathrm{fic}}(\psi_i(\mathbf{r}))$ are much smaller than the largest ionic kinetic energy, the ratios of the amplitudes being $\sim$100 ($\mu = 500$ a.u.) and $\sim$20 ($\mu = 2000$ a.u.). In this case, either one of them can be employed, by knowing that $\mu = 2000$ a.u. will allow using a larger value for the timestep.

What is the pattern taken by the same quantities under adiabatic conditions for a real system modeling a disordered network? Figure 3.4 shows a clear separation on the same units of energies between the ionic and fictitious kinetic energies at room temperature. By referring to chapter 4 for a more systematic presentation of typical FPMD setups aimed at simulating disordered materials, figure 3.4 is quite illustrative of a troubleless dynamical situation fully in line with the 'golden rule' statement *trajectory not departing from the BO surface.*

3.5.5 The role of thermostats within FPMD

FPMD as presented in equations (3.27), (3.30) and (3.31) is by far *not* an universal recipe applicable to any system under any physical condition, and this regardless of the complexity of the total energy expression or to the important investment required in terms of computational power. The main limit is intrinsic to the method, for which adiabaticity is the essential ingredient.

- Since we are dealing with quantum mechanics, the eigenstates of the system correspond to distinct energy levels, piled up so as to reach the highest occupied molecular orbital (HOMO). If the electronic structure of the system (its electronic density of states) features a gap between the highest occupied energy level and the lowest unoccupied energy level (HOMO and LUMO in the language of quantum chemistry, top of the valence band and bottom of the conduction band for solid-state physicists), there are in principle no difficulties in keeping at very low values $E_{\text{fic}}(\psi_i(\mathbf{r}))$. This is because populating a higher energy level unfilled at the ground state is essentially forbidden on very long time scales. Also, the number of occupied states will stay constant in time, ensuring a proper dynamical evolution for a well defined number of degrees of freedom.

- The dynamical evolution inherent in equation (3.27) couples the orbitals to the ionic coordinates leading to an (unwanted) full thermal equilibrium among all degrees of freedom involved within a finite time. If we stick to zero or room temperature behavior, such finite time is by no means accessible to the simulation time scale (even the largest, say ns) provided there is an energy gap preventing the dynamical orbitals from leaving at will their lowest energy states, those closer to the BO surface. However, since we are targeting substantial ionic dynamics covering drastic structural changes such as melting or phase transformations, *gaps can also be affected by temperature*, their reduction increasing the probability that energy states can find their way into the conduction band.
- The consequences for the CP dynamics can be sketched by invoking some prototypical cases, easily traced back to the nature of bonding as taught in textbooks. The safest situation is the one corresponding to gaps as large as some eV as in insulators or in finite systems (molecules or clusters). In that case, $E_{\text{fic}}(\psi_i(\mathbf{r}))$ shows no tendency to any increase even on extremely extended trajectories with fairly standard choices of μ and Δt_{CPMD}. At the opposite site we find metallic bonding, for which the CPMD strategy as expressed in equation (3.27) is not applicable and, as of today, not appropriate in view of what one expects in terms of gaps and adiabaticity. Somewhere in between, but I would say closer to the first case, are systems (that one can term semiconductors for sake of convenience) with small gaps (less than 0.5 eV). *To this class of systems belong most of the binary and ternary compounds we shall be considering when describing disordered chalcogenides.* For them, the condition of adiabaticity can be less straightforward to obtain especially at high temperature, even though we have been able to apply equation (3.27) successfully well above the melting point. I refer to chapter 4 for some examples related to the issues invoked here.

3.5.5.1 The need of temperature control: extending its meaning

The idea of controlling the temperature associated to specific sets of degrees of freedom can be used to prevent $E_{\text{fic}}(\psi_i(\mathbf{r}))$ from taking arbitrarily high values much in the same way as it can be done for the ions when one fixes a target temperature as explained in section 3.4.1. This means the inclusion of a specific variable α_l among those acting in $E_{\text{tot}}[\{\psi_i\}; \{\mathbf{R}_I\}, \alpha_l]$ and playing the role of a thermostat for the dynamical orbitals in $E_{\text{fic}}(\psi_i(\mathbf{r}))$. Originally, the search of a temperature control to ensure adiabaticity in gapless (or small gaps) situation was presented in [39] as a possible methodology to cope with the divergence of $E_{\text{fic}}(\psi_i(\mathbf{r}))$ even in the case of metals, the application presented focusing on the case of liquid aluminum at the melting point and featuring a well conserved behavior of $E_{\text{fic}}(\psi_i(\mathbf{r}))$ over 1 ps. It is of interest at this point to follow in detail the derivation of the ideas presented in [39] since, when combining the concept of thermostat with the method by Car and Parrinello, it is appropriate to modify slightly the original expressions of both schemes.

1. Our starting point is to make explicit in the expression of $E_{\text{tot}}[\{\psi_i\}; \{\mathbf{R}_I\}, \alpha_l]$ the additional dynamical variables α_l, in analogy with the idea of an extended Lagrangian allowing the introduction of thermostats (or barostats, or other convenient dynamical variables) in a way compatible with the existence of a constant of motion. So one can first rewrite equation (3.27) in this form

$$\begin{aligned}\mathcal{L}_{\text{CP}} = \mu\sum_i \int |\dot{\psi}_i(\mathbf{r})|^2 \mathrm{d}\mathbf{r} + \frac{1}{2}\sum_{I=1}^{N} M_I \dot{\mathbf{R}}_I^2 - E_{\text{tot}}[\{\psi_i\}; \{\mathbf{R}_I\}, \alpha_l] \\ - \sum_{ij}\lambda_{ij}\Big(\int \psi_i^*(\mathbf{r})\psi_j(\mathbf{r})\mathrm{d}\mathbf{r} - \delta_{ij}\Big) + \frac{1}{2}\sum_l \eta_l \dot{\alpha}_l^2.\end{aligned} \tag{3.32}$$

   Having in mind the control of the ionic temperature via a thermostat as a first external condition to be satisfied by the equations of motion, one can express the dependence of $E_{\text{tot}}[\{\psi_i\}; \{\mathbf{R}_I\}, \alpha_l]$ on α_l by using an appropriate function $F(\alpha)$ much in the same spirit of the Nosé–Hoover conserved quantity of equation (3.25), α_l playing the role of the friction constant appearing in the equations of motion given by equations (3.23) and (3.24).

$$\begin{aligned}\mathcal{L}_{\text{CP}} = \mu\sum_i \int |\dot{\psi}_i(\mathbf{r})|^2 \mathrm{d}\mathbf{r} + \frac{1}{2}\sum_{I=1}^{N} M_I \dot{\mathbf{R}}_I^2 - E_{\text{tot}}[\{\psi_i\}; \{\mathbf{R}_I\}] \\ - \sum_{ij}\lambda_{ij}\Big(\int \psi_i^*(\mathbf{r})\psi_j(\mathbf{r})\mathrm{d}\mathbf{r} - \delta_{ij}\Big) + \frac{1}{2}\eta_I \dot{\alpha}_I^2 + F(\alpha_I).\end{aligned} \tag{3.33}$$

   The use of variable α_I featuring a kinetic term of the form $\frac{1}{2}\eta_I\dot{\alpha}_I^2$ implies that

$$F(\alpha_I) = 2E_{\text{kin}}^{I-\text{TG}}\alpha_I \tag{3.34}$$

   the variable $\dot{\alpha}_I$ playing the same role as β_I in the original Nosé–Hoover formalism, with $2E_{\text{kin}}^{I-\text{TG}}$ being the targeted average ionic kinetic energy, typically expressed via the target temperature so as $2E_{\text{kint}}^{I-\text{TG}} = N_{\text{ind}}k_B T^{\text{TG}}$, N_{ind} being the number of degrees of freedom[9].
2. So far, we have only extended the Nosé–Hoover ideas to the CP ion dynamics, with no impact on the control of the fictitious electronic kinetic

[9] It should be recalled that only the **average** value of the kinetic energy can be compared to a temperature, taken to be a thermodynamic quantity. However, quite often, in the context of MD, both kinetic energy and temperature are used as instantaneous and average quantities for the sake of convenience.

energy. This can be achieved as a natural extension within the following expression

$$\mathcal{L}_{CP} = E_{fic}(\psi_i(\mathbf{r})) + E_{kin}(\{\mathbf{R}_I\}) - E_{tot}[\{\psi_i\}; \{\mathbf{R}_I\}, \alpha_I, \alpha_e] + E_{const}(\psi_i(\mathbf{r})) + E_{ext}(\alpha_I) + E_{ext}(\alpha_e) \tag{3.35}$$

$$\mathcal{L}_{CP} = \mu\sum_i \int |\dot{\psi}_i(\mathbf{r})|^2 d\mathbf{r} + \frac{1}{2}\sum_{I=1}^{N} M_I \dot{\mathbf{R}}_I^2 - E_{tot}[\{\psi_i\}; \{\mathbf{R}_I\}] - \sum_{ij}\lambda_{ij}\left(\int \psi_i^*(\mathbf{r})\psi_j(\mathbf{r})d\mathbf{r} - \delta_{ij}\right) + \frac{1}{2}\eta_I\dot{\alpha}_I^2 + 2E_{kin}^{I-TG}\alpha_I + \frac{1}{2}\eta_e\dot{\alpha}_e^2 + 2E_{kin}^{e-TG}\alpha_e \tag{3.36}$$

in which E_{kin}^{e-TG} is bound to be the target value of $\mu\sum_i \int |\dot{\psi}_i(\mathbf{r})|^2 d\mathbf{r}$. The equations of motion ensuring the time conservation of equation (3.36) together with a dynamical evolution in the 'double' canonical ensemble encompassing the ions and the fictitious electronic degrees of freedom are

$$\mu\ddot{\psi}_i(\mathbf{r}, t) = -\frac{\delta E_{tot}[\{\psi_i\}, \{\mathbf{R}_I\}]}{\delta\psi_i^*(\mathbf{r}, t)} + \sum_j \lambda_{ij}\psi_j(\mathbf{r}, t) - \mu\dot{\psi}_i\dot{\alpha}_e \tag{3.37}$$

$$M_I\ddot{\mathbf{R}}_I = -\nabla_I E_{tot}[\{\psi_i\}, \{\mathbf{R}_I\}] - M_I\dot{\mathbf{R}}_I\dot{\alpha}_I \tag{3.38}$$

$$\eta_I\ddot{\alpha}_I = 2\left(\frac{1}{2}\sum_{I=1}^{N} M_I\dot{\mathbf{R}}_I^2 - \frac{1}{2}N_{ind}k_B T^{TG}\right) \tag{3.39}$$

$$\eta_e\ddot{\alpha}_e = 2\left(\mu\sum_i \int |\dot{\psi}_i(\mathbf{r})|^2 d\mathbf{r} - E_{kin}^{e-TG}\right). \tag{3.40}$$

3. When thinking of the control of $E_{fic}(\psi_i(\mathbf{r}))$ via the combination of equations (3.37) and (3.40) it should be said that this strategy, although formally unmistakable, has been superseded by the observation, over the years, that its mere application cannot cure the lack of adiabaticity on extended intervals. This occurs when the wavefunctions can depart from their ground state for finite temperatures due the nature of the bonding, with the striking example of gapless systems as metals. Our experience with chalcogenide systems goes in the same direction, although the shortcomings occurring when not using a thermostat on the fictitious electronic degrees of freedom are much less severe, for the simple reason that valence and conduction bands are separated at least at not too high temperatures. In practice, divergences of $E_{fic}(\psi_i(\mathbf{r}))$ have been observed for all binary and ternary systems referred to in the next sections as FPMD applications. Let's suppose we take as the starting point of a thermal cycle composed of heating and cooling (the so-called melt and quench technique to produce amorphous

structures) a well relaxed structure at room temperature for which both ionic and electronic degrees of freedom bear no residual kinetic energy. In this case, $E_{\text{fic}}(\psi_i(\mathbf{r}))$ will not diverge significantly at low temperature and its value will be the one compatible with the notion of orbitals following adiabatically the ions. For higher temperatures corresponding to significant atomic diffusion, we found convenient to stabilize $E_{\text{fic}}(\psi_i(\mathbf{r}))$ with an appropriate thermostat, characterized by a target $E_{\text{kin}}^{e-\text{TG}}$ and a mass η_e. As to the choice of these values, the recommendation is to set η_e so as to obtain frequencies $\omega_{\text{fic}} = \sqrt{E_{\text{kin}}^{e-\text{TG}}\eta_e}$ well above the ionic ones (in line with the fact that $\mu \ll M_I$). For $E_{\text{kin}}^{e-\text{TG}}$ the idea could be to monitor the behavior of $E_{\text{fic}}(\psi_i(\mathbf{r}))$ at the first temperature where some departure from adiabaticity manifests itself and choose as target the last plateau value before the onset of a divergence. A more rigorous strategy to select an appropriate value for $E_{\text{kin}}^{e-\text{TG}}$ is described in [39] and in section 4.4.3. We shall come back to these issues in more practical terms when discussing the implementation of FPMD simulations in chapter 4.

4. After all these explanations, recommendations and warnings, I felt necessary to show a clear and unambiguous first example of what the thermostats really perform when equations (3.37)–(3.40) are implemented. This is shown in figure 3.5 for the case of a disordered $GeSe_4$ system for which an initial 'ad hoc' structure was selected (an extensive rationale on the choice of the initial configurations as a prerequisite to any MD simulation of amorphous systems will be presented in section 4.3). Convergence to the target values of $T = 300$ K and $E_{\text{fic}}(\psi_i(\mathbf{r})) = 0.04$ a.u. is readily obtained, showing clearly that the control of these quantities makes a real difference in terms of efficiency and handling of the whole FPMD scheme.

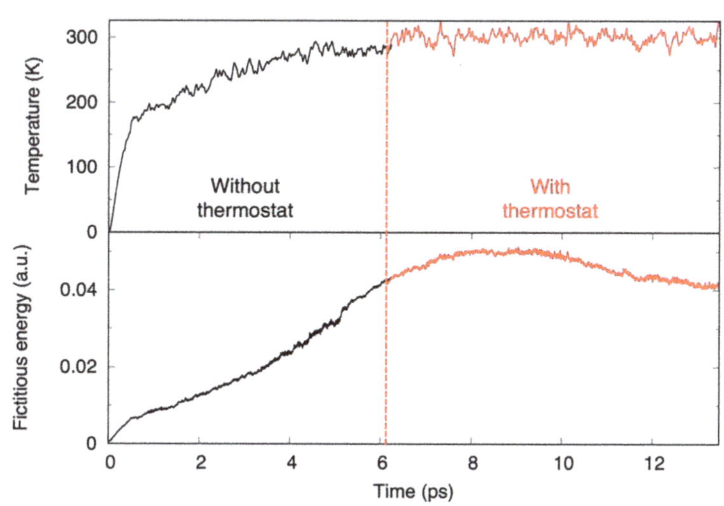

Figure 3.5. Behavior of the ionic temperature and of $E_{\text{fic}}(\psi_i(\mathbf{r}))$ at the beginning of a FPMD simulation for a disordered $GeSe_4$ system. Courtesy of E Martin (ICube, Strasbourg).

3.6 Getting acquainted with the total energy

The main adjustment to face when making the transition from classical to FPMD rests on the handling of the potential energy, that is to say the quantity customarily named 'total energy' in the field of electronic structure calculations. This convention is followed within the formalism underlying the expression in equation (3.27) and all developments thereof. In what follows, I will take the reader through the complexity of $E_{tot}[\{\psi_i\}; \{\mathbf{R}_I\}]$.[10] The ambition is to reach a compromise between a sufficiently clear explanation of all different parts composing this quantity and the intention of avoiding aspects that can be skipped if the main purpose is to run FPMD effectively and knowledgeably. Needless to say, this book has no intention to rewrite formulas largely available in specialized monographs (such as [42, 51–56]). The aim is to provide a quite simplified track to achieve successful use of total energy concepts and precise understanding of them.

It is important to realize that the expression for $E_{tot}[\{\psi_i\}; \{\mathbf{R}_I\}]$ has to be tractable and somewhat affordable since bound to be calculated iteratively within a temporal trajectory, calling for some small but unavoidable sacrifices in terms of absolute accuracy. Some choices have to be made by keeping in mind that we want to go quantitatively but we need to bridge the gap between atomic-scale events and macroscopic properties through the use of statistical mechanics. DFT is the most viable and popular framework to meet the goals of predictive power combined to adaptability in terms of best use of available resources, these latter continuously increasing with time and redefining the boundaries of what is possible in terms of time and space scale. One can start by realizing that, within DFT, $E_{tot}[\{\psi_i\}; \{\mathbf{R}_I\}]$ is known as the Kohn–Sham energy [43, 44], the minimum of which is the ground state energy of a system of interacting electrons for a given position $\mathbf{R}_I$ of the classical ions (nuclei). By dropping the explicit dependence on $\mathbf{R}_I$ for simplicity one has

$$E_{tot}\,[\{\psi_i\}] = K_s\,[\psi_i] + \int U_{ext}(\mathbf{r})\rho(\mathbf{r})d(\mathbf{r}) + \int U_H(\mathbf{r})\rho(\mathbf{r})d(\mathbf{r}) + E_{xc}(\rho(\mathbf{r})) \qquad (3.41)$$

where $\rho(\mathbf{r})$ is the electronic density of the system of which $E_{tot}[\{\psi_i\}]$ is a functional via the set of ψ_i. To allow practical implementation of equation (3.41) within the FPMD scheme and circumvent some stumbling blocks related to the quantum nature of electrons and to their mutual interaction, a certain number of issues have to be addressed. One can describe them through their significance in the different parts composing $E_{tot}[\{\psi_i\}]$ in equation (3.41).

3.6.1 Electronic kinetic energy: better avoiding confusions!

The wavefunctions (orbitals) ψ_i are orthonormal wavefunctions having a one-particle nature, this corresponding to a set of orbitals piled up in energy so as to be decoupled as if they were uncorrelated and non-interacting. This leads to an expression for

[10] In the rest of this section we come back to a definition of $E_{tot}[\{\psi_i\}; \{\mathbf{R}_I\}]$ not including external dynamical variables, of no use in this context.

the electronic density and the electronic kinetic energy $K_s[\psi_i]$ taking the form (sums over the occupied states)

$$\rho(\mathbf{r}) = \sum_i f_i\,|\psi_i(\mathbf{r})|^2 \qquad K_s[\psi_i] = \sum_i f_i\,\langle\psi_i| -\frac{1}{2}\nabla^2|\psi_i\rangle. \tag{3.42}$$

The simplification inherent in the choice of decoupling the wavefunctions is enormous and marks the boundary between DFT and quantum chemistry methods that are devised to treat the electronic correlation from the outset. Having in mind the dynamical nature that the CP approach confers to the ψ_i, it is now easier to understand why DFT fits in so well with the extended dynamical strategy built in equation (3.27). Each ψ_i becomes a dynamical degree of freedom moving adiabatically out of its BO surface to (slightly) depart from its ground state without affecting the self-consistent character of the solution of the electronic structure problem. However, care must be exercised not to mistake $K_s[\psi_i]$ (the electronic kinetic energy that contributes to $E_{\text{tot}}[\{\psi_i\}]$, the potential energy of the system in the language of classical mechanics) for $E_{\text{fic}}(\psi_i(r))$ (the fictitious electronic kinetic energy) that is bound not to affect in any manner the calculation of a given property resulting from a time average on a FPMD trajectory. This is why it is important (especially in the course of formal or even informal oral exchanges) to take some time (a few seconds more ⋯) to invoke correctly the fictitious electronic degrees of freedom and its related kinetic energy, without giving in to the temptation of making it short with an inaccurate (and wrong) 'electronic kinetic energy'.

3.6.2 The most convenient basis set: plane waves

Having assumed that the wavefunctions characterizing the electronic states within DFT can be taken as individual functions, decoupled and uncorrelated from each other, one has to represent them via a suitable basis set. This will take the form of an expansion on analytical functions having some tractable form and summing up in a weighted manner to achieve the desired convergence. By convergence with respect to a basis set one means that a given property (or a set of properties) issued from a calculation does not experience any more changes for a required level of precision when the basis set is increased.

Plane waves are very much appealing in this context since one can easily control the convergence without any use of physical or chemical intuition but simply increasing the number of them in the expansion up to a cutoff value. This marks the attainment of convergence, that is the absence of any relevant changes in the calculated properties (at least those to be considered as representative) when adding extra members in the expansion. This basis set goes along with the existence of a periodic system, in a way fully compatible with the periodicity inherent in bulk systems. In the following definition,

$$\psi_i(\mathbf{r}) = \frac{1}{\Omega}\sum_{\mathbf{G}} c_i(\mathbf{G})e^{i\mathbf{G}\cdot\mathbf{r}} \tag{3.43}$$

one has dropped any dependence on the k points of the Brillouin zone invoked in solid-state physics to describe the band structure of crystals since, for disordered systems sufficiently large and (even better) of non metallic nature the inclusion of k points in the Kohn–Sham expression has never proved to be necessary; i.e. the simulation cell has no point group symmetries. In equation (3.43), $\mathbf{G}$ are reciprocal space vectors compatible with the volume Ω and $c_i(\mathbf{G})$ are coefficients to be determined through the search of the ground state for $E_{\text{tot}}[\{\psi_i\}]$ and the corresponding $\rho(\mathbf{r})$ via equation (3.42). Having in mind the framework of FPMD and the notion of dynamical variables attached to $\psi_i(\mathbf{r})$, equation (3.43) allows for attributing the same meaning to the set of $c_i(\mathbf{G})$ but in reciprocal space. This is very convenient for planning all operations related to the calculation of $E_{\text{tot}}[\{\psi_i\}]$ and its derivatives either in real or in reciprocal space depending on the manageability of the analytical expressions.

3.6.3 Introducing the notion of pseudopotentials

By continuing our analysis of equation (3.41), we can now turn to $U_H(\mathbf{r})$

$$U_{\text{H}}(\mathbf{r}) = \int \frac{\rho(\mathbf{r}')}{|\mathbf{r} - \mathbf{r}'|} \, \mathrm{d}\mathbf{r}' \tag{3.44}$$

that goes along with the equation of Poisson, namely $\nabla^2 U_{\text{H}}(\mathbf{r}) = 4\pi\rho(r)$ and $U_{\text{ext}}(\mathbf{r})$

$$U_{\text{ext}}(\mathbf{r}) = -\sum_{I} \frac{Q_I}{|\mathbf{R}_I - \mathbf{r}|} + \sum_{I\langle J} \frac{Q_I Q_J}{|\mathbf{R}_I - \mathbf{R}_J|} \tag{3.45}$$

where Q_I are the nuclear charges associated with the ions, the first member on the right-hand side accounting for the electrons–nuclei interaction and the second one for the ion–ion interactions. While this second one poses no problems besides those related to the account of a Coulombic interaction between point charges in the periodic structure, the first part entails a great deal of methodological and conceptual developments that are crucial to ensure the success of DFT altogether and of the FPMD schemes in particular.

- If one is willing to consider explicitly all the electrons and the corresponding orbitals in the expression for $\rho(\mathbf{r})$, the high localization of electrons in the inner region close the nuclei (core electrons) results in prohibitively high values of the electron kinetic energy $K_s[\psi_i]$. As a mere consequence of the Heisenberg uncertainty principle, and by the very expression of $K_s[\psi_i]$, one can deduce that very sharp oscillations of the wavefunctions in reduced regions of space are beyond the reach of a plane wave expansion since implying an unrealistic number of components. Since it is essential to stick to plane waves to ensure the tractability of a scheme based on extensive calculation of forces on discretized temporal trajectories, one needs to address the issue without modifying the essence of the description of chemical bonding while making the calculations feasible. This is exactly where pseudopotentials come into play.

- Due to the negligible involvement of core electrons into chemical bonding, the idea consists in 'making life easier' to the plane waves, by freezing all core electrons since they are active in regions where there is no overlap between wavefunctions localized on different atoms. Then, one replaces the rapidly diverging potential with an analytical 'ad hoc' object named pseudopotential, $U_{\mathrm{pp}}(\mathbf{r})$. The pseudopotential thereby obtained has finite value at zero distance, while recovering the original form for distance beyond a cutoff value r_{ps}. This cutoff value marks the boundary between the real potential and pseudopotential, as well as, by construction, between the real wavefunctions and the pseudo ones.
- What I have summarized in a few lines reflects intensive theoretical and computational work aimed at establishing the best recipes to include viable pseudopotentials into the Kohn–Sham formulation, so as to reach the best compromise between an affordable number of plane waves (easily achievable by extending r_{ps}) without affecting the electronic structure (as one succeed in doing by reducing r_{ps}). In practice, this is done by selecting one specific recipe for the pseudopotential construction, together with one choice for the exchange–correlation functional (see below for this part of $E_{\mathrm{tot}}[\{\psi_i\}]$) and acting with respect to reference results obtained for a single atom by including all electrons, i.e. both core and valence ones.
- The introduction of $U_{\mathrm{pp}}(\mathbf{r})$ induces a substantial rewriting of equation (3.41) that can be further extended by taking into account the explicit form of the plane waves basis set, as is highly convenient for FPMD applications. In practice, to complete the minimal information to be retained on pseudopotentials, it is essential to know that they appear as the sum of a local (depending only on the distance from the ionic core) and of a non-local part. This separation can be pretty tricky since involving some judicious technical choices when calculating the corresponding parts appearing in equation (3.41). More specifically, when referring to the plane wave basis set, it is customary to express $U_{\mathrm{pp}}(\mathbf{r})$ in the form

$$U_{\mathrm{pp}}(\mathbf{r}) = U_{\mathrm{loc}}(\mathbf{r}) + \sum_{l,m}^{l_{\mathrm{max}}} U_l(\mathbf{r})\mathrm{Pr}_{l,m}(\Omega) \tag{3.46}$$

where l, m are angular indexes and $\mathrm{Pr}_{l,m}(\Omega)$ suitable angular operators of variables Ω. This separation in local and non-local parts allows easy handling of the local contribution $\int U_{\mathrm{loc}}(\mathbf{r})\rho(\mathbf{r})\,d\mathbf{r}$ that appears now inside equation (3.41) as part of $U_{\mathrm{ext}}(\mathbf{r})$. The local part has to be specified upfront when using a pseudopotential, for instance by following the convention that the sum in equation (3.46) includes only values of l compatible with the reference electronic configuration of the atom and for the largest of them $U_{l_{\mathrm{max}}}(\mathbf{r}) = U_{\mathrm{loc}}(\mathbf{r})$. More complex is the calculation of the total energy relative to the non-local part of $U_{\mathrm{pp}}(\mathbf{r})$, that is the object of some specific treatments well documented in the literature and to which we shall refer in the context of chapter 11, section 11.2.

As personal note, I would like to add the following. When it comes to pseudopotentials, users tend to have different attitudes, ranging from the strict recovery of tools made available in some data libraries to (less frequent, though) earnest attempts to create their owns. As of today, available pseudopotentials cover essentially all periodic table and come to the public in connection with popular plane waves electronic structure codes (CPMD being the one employed for all calculations referred to in this book). One assumes that all of them have been tested and substantiated by relevant publications. This is certainly true and encourages the use of them almost carefree. However, it would be worthwhile, at least a few times, to reproduce a pseudopotential already available by following the guidance of a researcher specialized in this very special kind of 'craftwork'. In principle, the idea of following one of the available recipes and achieve a solution meeting the required goals (norm conservation as compared to the all-electron solution, suitable cutoff radius allowing for substantial reduction of plane waves without sacrificing chemical accuracy) is far from being an insurmountable stake. Finally, one very important point has to be stressed, in close relationship with the account of $E_{xc}(\rho(r))$ in equation (3.41). As a matter of fact, $U_{pp}(r)$ has to be constructed by making the same choice for $E_{xc}(\rho(r))$ in the all-electron total energy of reference. To fully appreciate this issue we have to introduce $E_{xc}(\rho(r))$ first.

3.6.4 Exchange and correlation to increase predictive power

When considering the expression for $E_{tot}[\{\psi_i\}]$, one is led to the observation that, within the approximations inherent in the Kohn–Sham formalism implemented with plane waves[11] everything is known and prone to be calculated but one term, $E_{xc}(\rho(\mathbf{r}))$, the exchange and correlation (XC) functional. The XC functional is expected to help bridge the gap between the exact solution of the problem involving all electronic correlations (with the inclusion of the exchange effects which are due to the fermionic nature of the electrons) and what DFT is able to provide. By referring to a quantity that is unknown and it is issued by a specific theoretical treatment, the most general form reads

$$E_{xc}(\rho(\mathbf{r})) = \int \rho(\mathbf{r})\epsilon_{xc}(\rho(\mathbf{r}), \nabla\rho(\mathbf{r}))d\mathbf{r}. \tag{3.47}$$

Theoreticians have been proposing schemes for $E_{xc}(\rho(\mathbf{r}))$ since the early times of electronic structure calculations, due to the obvious observation that neglecting this part brings meaningless results in terms of predictive power. In the first place it appears convenient to consider two separate terms for exchange and correlations, ϵ_x and ϵ_c respectively. For many years the most popular recipe has been to consider exchange and correlation of a very special situation, the homogeneous electron gas with no account of spatial variations for $\rho(\mathbf{r})$ ($\nabla\rho(\mathbf{r}) = 0$) in equation (3.47). This is so-called local density approximation (LDA) (or local spin density in the spin

[11] These approximations are: total electronic density expressed in terms of independent orbitals and introduction of pseudopotentials to avoid treating explicitly regions chemically inactive in between the atoms.

dependent case), for which early reference results, later on parameterized in a manageable form [57], were obtained in 1980 by Ceperley and Alder [58]. LDA solves the problem of interacting electrons as if they were at constant density in space, an idea that is reminiscent of a distribution largely unaffected by directional bonding, as is the case for metals that in a very first approximation can be modeled via nuclei embedded in an electron sea that keeps them together.

Concerning expressions including spatial variations of $\rho(\mathbf{r})$ inside equation (3.47), they are commonly known as generalized gradient approximations (GGA) [59]. These are expected to improve upon LDA for a variety of properties and bonding situations and a wealth of them have been made available, becoming essentially unavoidable in current applications. In terms of bonding, the question arises on whether GGA is able to localize more effectively valence electrons, thereby curing some of the main drawbacks of LDA, exemplified, for instance, by an unavoidable underestimate of the gap in the electronic density of states. Within the context of the present book and in line with a presentation strategy largely based on personal involvement with MD calculations on disordered structures, I prefer not to present any historical review on the genesis of GGA methods available in the literature [60–71] since, in addition, the relationship with disordered matter can be hard to grasp. For this reason, in what follows, I will provide an example of the progresses made when using GGA for a liquid chalcogenide, the selected case being a breakthrough in the area of first-principles modeling of iono-covalent systems [72]. Should the reader feel uncomfortable with such anticipation of an application of FPMD, this can be skipped or revisited after considering the methodology. In a further section (section 4.2.1), we shall come back to the exchange–correlation issue by invoking different GGA recipes that will be later compared and commented on by referring to results for given systems (section 6.7).

3.6.5 On the impact of the XC functional: the revealing case of liquid $GeSe_2$

Liquid $GeSe_2$ is a disordered network-forming system characterized by intermediate range order, as exemplified by the presence of a first sharp diffraction peak (FSDP) in the total neutron structure factor. We have already invoked this system in past sections in the need of a prototype network exhibiting intermediate range order and being sensitive to the kind of DFT atomic-scale description. For instance, it was employed in section 2.4.1 to describe the notion of coordination units and later in section 2.6 when referring to neutron scattering experiments and the comparison with MD data. It is now time to come to the origins of the role played by liquid $GeSe_2$ by focusing on its behavior when different choices of the exchange–correlation functional are adopted. From the standpoint of molecular dynamics, the task of simulating a liquid is less problematic compared to the amorphous counterpart in terms of sampling of the phase space since diffusion on significant distances (a few bonds) is observable even on limited time scales (a few ps). In 1999, a periodic system of 120 atoms (at that time compatible with the available resources) was equilibrated in the FPMD plane waves framework with the intent of comparing theoretical and experimental total and partial structure factors as well as pair

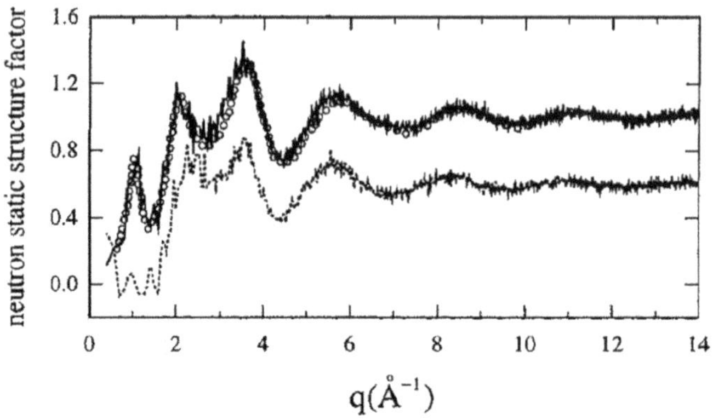

Figure 3.6. Total neutron structure factor $S_T(k)$ (see equation (2.3)) as a function of magnitude of momentum transfer (reciprocal space vector) k (termed q) for liquid $GeSe_2$, obtained within GGA (solid line [66]) and LDA (dots [57]), compared to experiment (circles) [73]. For clarity, the LDA curve is displaced downward by 0.4. Reprinted figure with permission from [72]. Copyright (1999) by the American Chemical Society.

correlation functions [72, 74]. By deferring the full comparison to a later chapter (section 6.2), let us focus on the total neutron structure factor as obtained, first, by employing for the exchange–correlation functional $E_{xc}(\rho(\mathbf{r}))$ the LDA scheme (figure 3.6).

DFT–LDA is unable to reproduce the correct behavior in reciprocal space for values of k typical of intermediate range distances, indicating that this level of theory is not sufficient to describe the structure realistically when it comes to network features beyond nearest neighbors. A striking improvement occurs when the calculations are performed on totally independent FPMD trajectories produced within the GGA approximation proposed by Perdew and Wang [66], leading to a substantial agreement between theory and experiments. This is shown by the recovery of the FSDP feature in the region around $k = 1$ Å^{-1}. These results have to be rationalized in terms of effects (on the structural properties in real space, more palatable than peaks in reciprocal space) and causes/origins (what GGA does in terms of electronic structures). Both aspects were addressed in [72]. The effect on the structural properties is the establishment of a predominant fourfold coordination, in line with the presence of $GeSe_4$ tetrahedra as the main structural motif. By predominant, one means a large majority (say, 60%–70%) of Ge atoms (Se atoms) fourfold (twofold) coordinated to Se (Ge) atoms. To capture the origins of this behavior as expressed through the bonding properties of the system, the electronic structure can be calculated by comparing its features when calculated via LDA or GGA (see figure 3.7).

By focusing on the valence charge density pertaining to a Se–Ge–Se trimer, one can highlight accumulation (depletion) of charge around the more (less) electronegative Se (Ge) species, more important in the GGA case. This effect is striking when observing the difference quantity, pointing out a clear similarity in the covalent character, the effect of GGA being of ionic nature. This result is able to correlate the better performances of GGA for a iono-covalent system to the

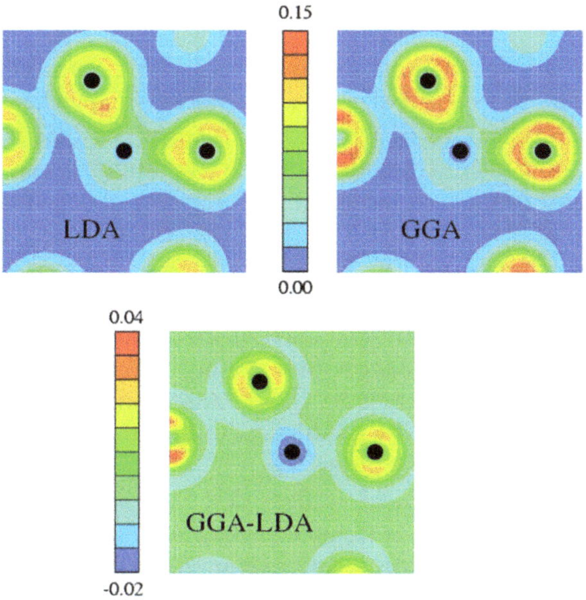

Figure 3.7. Contour plots of valence electronic charge density for a selected Se–Ge–Se trimer within liquid $GeSe_2$ in LDA [57] (upper left) and GGA [66] (upper right). A difference plot (GGA–LDA) is also shown (lower). The scales are in atomic units. Reprinted figure with permission from [72]. Copyright (1999) by the American Chemical Society.

appearance of intermediate range order in the underlying network, thereby showing that going beyond LDA can have profound consequence in the achievement of predictive power for FPMD simulations.

3.7 Glassy materials and FPMD: criteria and challenges

Having established the general framework of MD and the challenges to be faced when applying this technique to realistic systems (with particular concern for condensed matter ones), it is now time to turn to the specific cases targeted by this monograph, amorphous materials. Coming back to the 'three criteria' rule (extent of size, length of trajectories, realistic forces allowing for quantitative predictions) its fulfillment cannot be taken for granted for this class of systems. Focusing first on the *third one (realistic forces)* there is no reason to believe that interatomic potentials will be more effective when the structure of a system is disordered. In fact, the construction of a potential energy surface describing correctly the atomic environment is as crucial as in any other extended systems. This issue has been addressed in the above section, by setting the scene for a quantitative approach (FPMD) able to go beyond the qualitative description of materials (mostly inspired by statistical mechanics 'ad hoc' models) and a realistic characterization accounting for bonding and its evolution with temperature. *What can we say about the size of the systems and the length of the temporal trajectories when it comes to glassy structures?*

3.7.1 The issue of size limitations

- Concerning the size limitations, these could be circumvented to bypass the objection that the system sizes are inevitably small when compared to the macroscopic collections of 10^{23} atoms. It turns out that periodic replicas of the same arrangement (periodic boundary conditions) can offer a very good reproduction of much larger systems if one focuses on distances compatible with the size of the main unit. The crucial questions underlying this issue read: (1) are these dimensions too short with respect to our scale of interest? Is it possible to take averages representative of macroscopic properties by studying a much smaller portion replicated in space? For disordered systems, what matters is to consider distances that are not shorter than typical correlation lengths defining the topological order. Also, one has to accept that these quantities will be affected by a statistical uncertainty with averages taken over sufficiently long time trajectories.
- The notion of topological order deserves some more consideration. Disordered systems such as liquids, glasses and amorphous solids are known for exhibiting short range order, exemplified by a first peak in the pair correlation function (at the level of a specific interaction among atoms of kind A and B, not necessarily unequal). The fact that this peak lies at nearest neighbor distances does not mean that periodic boxes can have such short dimensions, since this would be reflected by an artificial periodicity imposed over the three dimensional space with a very high spatial frequency. In addition, some disordered systems (largely referred to in the following) are characterized by a network arrangement featuring connections among structural units, implying correlations extending beyond nearest neighbors in contact within the structural units itself. Accordingly, it is safer to adopt cell sizes as large as possible in a way compatible with the request of an affordable calculations and (most importantly) the account of distances several times larger (at least two!) than the largest spatial correlations expected to occur in the system.
- For amorphous networks, other thoughts have to guide the choice of the optimal size besides those invoked above. In particular, my general advice would be not to put the highest priority on 'affordable' but on 'reliable' instead, since the quality of the science produced has to overcome any attempt to obtain results regardless of their intrinsic value and impact. To develop some practical consideration, for systems having nearest neighbor distance ranging in between 2 and 3 Å, periodic boxes with size around 15 Å contain in between 100 and 150 atoms. In spite of the existence of a wealth of valuable results for amorphous systems of this size, with the inclusion of properties strongly dependent on structural order extending up to 8–10 Å, there is an overall consensus that ~500 atoms is the minimal size ensuring calculations less affected by residual artificial structural repetitions imposed by the periodicity. However, there is no strict demonstration that the essence of the basic structural properties of an amorphous system differ substantially

from say, 150 to 500 atoms. In particular, I would tend to prefer results based on several (say, 10) independent trajectories averaged for 100 atoms to a single result obtained for 500 atoms.

- The situation might change if one is interested in properties involving by construction large distances and spatial correlations. This is the case of structural units as rings, quite often encountered in network-forming materials. Also, one has to pay attention to the relative concentration of the species in the case of a multicomponent system. Producing average results based on 10–20 atoms from a single species (as in the case of Ge in $GeSe_9$ with $N_{at} = 120$, see [75] and section 9.3 for a critical analysis) cannot be taken as a reliable procedure, since the statistical error of any related properties is hard to quantify and in any case crucially affects any estimate of macroscopic properties.
- Quantities like total neutron structure factor $S_T(k)$ can provide some information on a sensitivity of basic properties, depending on distances beyond nearest neighbors, to the size of the simulation box. In direct space, insight into the size issue can also be obtained by looking at the partial pair correlation functions $g_{\alpha\beta}(r)$. Of course, one needs calculations performed on different sizes as done in [76] where amorphous $Ge_2Sb_2Te_5$ was studied for four different sizes (with some distinct shapes, having in mind thermal studies to which we refer in chapter 12), namely $N_{at} = 144$, 252, 504, 1008. The observation of figures 3.8 and 3.9 reveals that the size dependence is quite moderate and limited, at most, to some differences affecting mostly $g_{GeSb}(r)$ when moving from $N_{at} = 144$ to larger systems. While this does not rule out more important effects for other properties, I will tend to conclude that serious artifacts can be easily avoided by taking at least two system sizes in the indicated range for a given system.

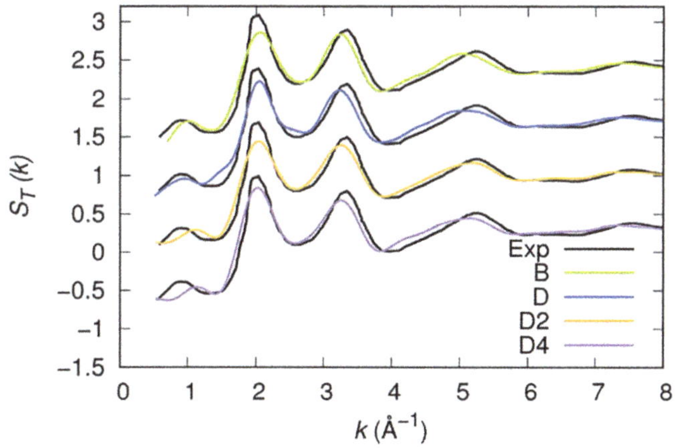

Figure 3.8. Experimental and calculated total neutrons structure factor of amorphous $Ge_2Sb_2Te_5$ (see chapter 11 for more details relative to this system). The calculated quantity has been obtained via integration in real space of the total pair correlation function (case **A**, section 2.3). System **B**, $N_{at} = 144$, system D, $N_{at} = 252$, system D2, $N_{at} = 504$, system D4, $N_{at} = 1008$. Reprinted figure with permission from [76].

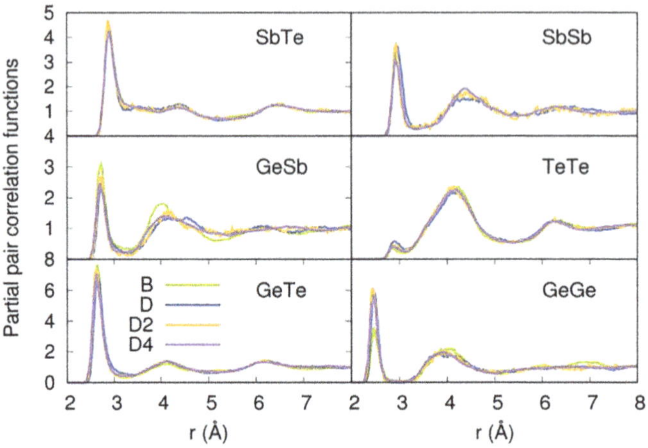

Figure 3.9. Partial pair correlation functions of amorphous $Ge_2Sb_2Te_5$ (see chapter 11 for more details relative to this system). System B, N_{at} = 144, system D, N_{at} = 252, system D2, N_{at} = 504, system D4, N_{at} = 1008. Reprinted figure with permission from [76].

3.7.2 The issue of the length of the time trajectories

Doing MD on time trajectories that cannot be longer than some ns (this restriction being particularly severe in the case of FPMD) raises the question of the adequacy of such a method when simulating systems for which any dynamical evolution occurs on much longer time intervals. This point has been briefly touched upon in the introduction and deserves some more comments here.

- For a number of scientists I met over the years, there is no point in taking seriously what comes out of MD in terms of amorphous and glassy materials, since the production intervals from the liquid state (quenching) are too short when compared to the real ones. While this statement has some legitimacy, one has to keep in mind the amount of MD data in very good agreement with experiments, especially for structural properties and for essentially any kind of chemical bonding. Also, most of the skepticism stems from the rate of cooling, namely the production steps of most amorphous phases, linking a highly mobile phase of matter (the liquid) to a much less mobile one (the amorphous state). In this context, the extent of relaxation is what matters and what is expected to characterize the amorphous state. One relevant point has to do with the memory of the initial configuration, that in principle should be lost if the final state has to be an independent state of matter fully decorrelated from the initial one.
- Many pieces of evidence have been collected to reconcile the negationists of MD and those, like myself and my closest coworkers, fully confident of the reliability of the MD tool for disordered solids. These are based, again, on

favorable comparison with experiments and the observation that deficiencies in the modeling, still present at various levels even within the more accurate FPMD, are affecting the results more than the (too fast) rate of quench. One instructive way to judge of the legitimacy of MD amorphous structure is to compare it with the corresponding liquid structure, this one being most of the time at the origin of the calculated trajectories. A first simple criteria consists in making sure that some diversities among the two cases are present besides the obvious sharpening of peaks in the pair correlation functions. By diversity one means some changes in the coordination shells and in the counting of defects. In binary systems these are deviations from a perfect chemical order, chemical order being the tendency to maximize heterogeneous bonding in a way compatible with the concentration.

- The amorphous state can never be the instantaneous picture of the liquid state. Early attempts of simulation of the amorphous state[12] were characterized by very short quenching intervals, leading to fake amorphous structures devoid of any relaxation. The fact that the time trajectories separating the liquid from the amorphous phase are much shorter that the experimental counterpart does not have to discourage the user from reducing the temperature via temporal steps as long as possible, since the energy barriers for bond breaking and reforming are not always inaccessible to the activated processes occurring when cooling the system.

In summary, there are challenges to be faced by any MD application and these are common to any collection of atoms in interaction and expected to change of configuration as a function of time and temperature. These can be listed as belonging to three main classes: **size limitations, length of trajectories and reliability of the description of interatomic forces**. By and large, the last one is as severe and demanding as in the corresponding liquid or crystalline state. Size limitations deserves a great deal of attention but there are ways (periodicity, precise account of the relevant distances) to alleviate this problem and even circumvent it. *There are no doubts that the length of the temporal trajectories underlying the relaxation process of any amorphous structure is the most critical aspect to be addressed to boost the predictive power of MD simulations, especially when the computational cost increases due to the higher accuracy built in FPMD*. In what follows, whenever useful, the examples provided will be described and commented on by considering as much as possible the three above criteria and the approach employed to minimize possible deficiencies intrinsic to the methodology. However, before turning to published examples we need to understand how the simulations of glassy structure have to be implemented, from the choice of the initial configuration to the production of the temporal trajectories and their analysis. This is the scope of the next chapter (chapter 4).

[12] There is no point in giving here specific examples by quoting published papers, since this would sound an unnecessary criticism for those behind such attempts.

References

[1] Battimelli G, Ciccotti G and Greco P 2020 *Computer Meets Theoretical Physics, The New Frontier of Molecular Simulation* (Berlin: Springer)
[2] Parrinello M 2022 *Israel J. Chem.* **62** e202100105
[3] Huang K 1991 *Statistical Mechanics* 2nd edn (New York: Wiley)
[4] Alder B J and Wainwright T E 1959 *J. Chem. Phys.* **31** 459–66
[5] Rahman A 1964 *Phys. Rev.* A **136** A405–11
[6] Battimelli G and Ciccotti G 2018 *Eur. Phys. J.* H **43** 303–35
[7] Ceperley D M and Stephen 2021 *Proc. Natl. Acad. Sci.* **118** e2024252118
[8] Behler J and Parrinello M 2007 *Phys. Rev. Lett.* **98** 146401
[9] Massobrio C, Pontikis V and Martin G 1989 *Phys. Rev. Lett.* **62** 1142–5
[10] Willaime F and Massobrio C 1991 *Phys. Rev.* B **43** 11653–65
[11] Rosato V, Guillope M and Legrand B 1989 *Philos. Mag.* A **59** 321–36
[12] Cleri F and Rosato V 1993 *Phys. Rev.* B **48** 22–33
[13] Ackland G, Sutton A and Vitek V 2009 *Philos. Mag.* **89** 3111–6
[14] Ackland G J, Tichy G, Vitek V and Finnis M W 1987 *Philos. Mag.* A **56** 735–56
[15] Ercolessi F, Andreoni W and Tosatti E 1991 *Phys. Rev. Lett.* **66** 911–4
[16] Foiles S M, Baskes M I and Daw M S 1986 *Phys. Rev.* B **33** 7983–91
[17] Jelinek B, Groh S, Horstemeyer M F, Houze J, Kim S G, Wagner G J, Moitra A and Baskes M I 2012 *Phys. Rev.* B **85** 245102
[18] Jacobsen K W, Norskov J K and Puska M J 1987 *Phys. Rev.* B **35** 7423–42
[19] Moriarty J A and Phillips R 1991 *Phys. Rev. Lett.* **66** 3036–9
[20] Moriarty J A and Widom M 1997 *Phys. Rev.* B **56** 7905–17
[21] Willaime F and Massobrio C 1989 *Phys. Rev. Lett.* **63** 2244–7
[22] Catlow C R A, Freeman C M, Islam M S, Jackson R A, Leslie M and Tomlinson S M 1988 *Philos. Mag.* A **58** 123–41
[23] Pedone A, Malavasi G, Menziani M C, Cormack A N and Segre U 2006 *J. Phys. Chem.* B **110** 11780–95
[24] Ewald P P 1921 *Ann. Phys.* **369** 253–87
[25] Wells B A and Chaffee A L 2015 *J. Chem. Theory Comput.* **11** 3684–95
[26] Vashishta P, Kalia R K, Antonio G A and Ebbsjö I 1989 *Phys. Rev. Lett.* **62** 1651–4
[27] Vashishta P, Kalia R K and Ebbsjö I 1989 *Phys. Rev.* B **39** 6034–47
[28] Iyetomi H, Vashishta P and Kalia R K 1991 *Phys. Rev.* B **43** 1726–34
[29] Vashishta P, Kalia R K, Rino J P and Ebbsjö I 1990 *Phys. Rev.* B **41** 12197–209
[30] Penfold I T and Salmon P S 1991 *Phys. Rev. Lett.* **67** 97–100
[31] Wilson M, Sharma B K and Massobrio C 2008 *J. Chem. Phys.* **128** 244505
[32] Madden P A and Wilson M 1996 *Chem. Soc. Rev.* **25** 339–50
[33] Massobrio C, Pasquarello A and Car R 2001 *Phys. Rev.* B **64** 144205
[34] Berendsen H J C, Postma J P M, van Gunsteren W F, DiNola A and Haak J R 1984 *J. Chem. Phys.* **81** 3684–90
[35] Andersen H C 1980 *J. Chem. Phys.* **72** 2384–93
[36] Nosé S 1984 *J. Chem. Phys.* **81** 511–9
[37] Shūichi N 1984 *Mol. Phys.* **52** 255–68
[38] Hoover W G 1985 *Phys. Rev.* A **31** 1695–7

[39] Blöchl P E and Parrinello M 1992 *Phys. Rev.* B **45** 9413–6
[40] Hoover W G and Hoover C G 2020 *J. Chem. Phys.* **153** 070901
[41] Martyna G J, Klein M L and Tuckerman M 1992 *J. Chem. Phys.* **97** 2635–43
[42] Marx D and Hutter J 2009 *Ab Initio Molecular Dynamics: Basic Theory and Advanced Methods* (Cambridge: Cambridge University Press)
[43] Hohenberg P and Kohn W 1964 *Phys. Rev.* **136** B864–71
[44] Kohn W and Sham L J 1965 *Phys. Rev.* **140** A1133–8
[45] Car R and Parrinello M 1985 *Phys. Rev. Lett.* **55** 2471–4
[46] Ori G, Bouzid A, Martin E, Massobrio C, Le Roux S and Boero M 2019 *Solid State Sci.* **95** 105925
[47] Born M and Oppenheimer 1927 *R. Ann. Phys., IV. Folge* **389** 457–84
[48] Parrinello M and Rahman A 1980 *Phys. Rev. Lett.* **45** 1196–9
[49] Massobrio C, Haile J M and Lee L L 1988 *Fluid Ph. Equilibria* **44** 145–73
[50] Ryckaert J-P, Ciccotti G and Berendsen H J C 1977 *J. Comput. Phys.* **23** 327–41
[51] Eschrig H 1996 *The Fundamentals of Density Functional Theory* (Berlin: Springer)
[52] Parr R G and Yang W 1980 *Density Functional Theory of Atoms and Molecules* (Oxford: Oxford Science Publications)
[53] Engel E and Dreizler R M 2011 *Density Functional Theory: An Advanced Course* (Berlin: Springer)
[54] Pang T 2006 *An Introduction to Computational Physics* 2nd edn (Cambridge: Cambridge University Press)
[55] Varga K and Driscoll J A 2011 *Computational Nanoscience: Applications for Molecules, Clusters, and Solids* (Cambridge: Cambridge University Press)
[56] LeSar R 2013 *Introduction to Computational Materials Science: Fundamentals to Applications* (Cambridge: Cambridge University Press)
[57] Perdew J P and Zunger A 1981 *Phys. Rev.* B **23** 5048–79
[58] Ceperley D M and Alder B J 1980 *Phys. Rev. Lett.* **45** 566–9
[59] Dal Corso A, Pasquarello A, Baldereschi A and Car R 1996 *Phys. Rev.* B **53** 1180–5
[60] Langreth D C and Mehl M J 1983 *Phys. Rev.* B **28** 1809–34
[61] Becke A D 1988 *Phys. Rev.* A **38** 3098–100
[62] Lee C, Yang W and Parr R G 1988 *Phys. Rev.* B **37** 785–9
[63] Perdew J P 1986 *Phys. Rev.* B **33** 8822–4
[64] Perdew J P 1986 and Wang Yue *Phys. Rev.* B **33** 8800–2
[65] Perdew J P 1992 and Yue Wang *Phys. Rev.* B **45** 13244–9
[66] Perdew J P, Chevary J A, Vosko S H, Jackson K A, Pederson M R, Singh D J and Fiolhais C 1992 *Phys. Rev.* B **46** 6671–87
[67] Ortiz G 1992 *Phys. Rev.* B **45** 11328–31
[68] Ortiz G and Ballone P 1991 *Phys. Rev.* B **43** 6376–87
[69] García A, Elsässer C, Zhu J, Louie S G and Cohen M L 1992 *Phys. Rev.* B **46** 9829–32
[70] Juan Y-M and Kaxiras E 1993 *Phys. Rev.* B **48** 14944–52
[71] Hammer B, Scheffler M, Jacobsen K W and Nørskov J K 1994 *Phys. Rev. Lett.* **73** 1400–3
[72] Massobrio C, Pasquarello A and Car R 1999 *J. Am. Chem. Soc.* **121** 2943–44
[73] Susman S and Volin K J 1990 D. G. Montague, and D. L. Price *J. Non-Cryst. Solids* **125** 168–80

[74] Massobrio C, Pasquarello A and Car R 1998 *Phys. Rev. Lett.* **80** 2342–5
[75] Le Roux S, Bouzid A, Kim K Y, Han S, Zeidler A, Salmon P S and Massobrio C 2016 *J. Chem. Phys.* **145** 084502
[76] Duong T-Q, Bouzid A, Massobrio C, Ori G, Boero M, E Martin and Adv R S C 2021 *RSC Adv.* **11** 10747–52

IOP Publishing

Chapter 4

A practical roadmap for FPMD on amorphous materials

The purpose of this section is to define a practical roadmap for running molecular dynamics (MD) simulations on amorphous materials. Our special emphasis is on first-principles MD (FPMD) applications and, in particular, on covalent or iono-covalent systems as chalcogenides as a prototype of 'non trivial' chemical bonding. This strongly motivates approaches going beyond qualitative descriptions in order to obtain (at least) the correct, basic structural properties. Some of the concepts and ideas discussed hereafter have been touched upon in chapter 3. Here the focus is on practical recommendations based on direct experience. In other words, the idea is to put to good use the methodology sketched above so as to enable sensible calculations on disordered (amorphous) systems by obtaining relevant information on their structure. Some very personal thoughts on the use of FPMD (largely based on direct experience) enriched by a sketchy presentation of the Car–Parrinello MD (CPMD) code end this chapter.

4.1 Choice of the description: classical potentials vs first-principles

Working with interatomic potentials (as those described in sections 3.2 and 3.3) in the area of materials modeling can be rewarding and reassuring. Interatomic potentials (also called force fields due to some differences in terminology between neighboring scientific areas) contain implicit information on the electronic structure based on intuition and are easy to handle analytically. They allow the use of classical molecular dynamics (CMD), as we have introduced in chapter 3. For many computational material scientists or, more generally, any scientist interested in MD as a tool to obtain macroscopic properties from scratch, interatomic potentials are unavoidable since they are affordable and tractable, and very much pleasing when it comes to collect some indications, at a very reduced cost, without waiting several weeks or months. Before dwelling on this issue by referring to my personal

doi:10.1088/978-0-7503-2436-6ch4

experience (inside and outside the domain of simulation of glasses), I would like to make clear that the notion of interatomic potential has been profoundly modified in recent years by the introduction of Machine Learning methods [1–4]. These are conceived to exploit huge data sets of properties obtained via electronic structure calculations (but not exclusively) to produce interatomic forces very much similar to those carrying explicit bonding information. Based on their rapid expansion, there are chances that in a few decades or even less, the production and use of such potentials will become so generalized that simpler, more empirical forms will be abandoned. In the present section, for sake of consistency, *I will stick to a notion of interatomic potential not including the Machine Learning developments, so as to underline their limits and how I felt when using them.*

- I have no fear of being considered hyper subjective when saying that I would discourage as much as possible from employing interatomic potentials to study materials at the atomic scale. This is because I was very much disappointed and frustrated a few times by the observation that any attempts of use outside of the range of properties employed for their construction can be as worthless and tiring as the Sisyphus efforts of mythology (carrying a rock uphill forever). Having massively used interatomic potentials in the period 1985–95, I might be argued that things have improved later on and that my record has to be updated. However, later indirect experience collected with chalcogenide glasses reinforced my first conclusions, as I have already briefly mentioned in section 3.3.
- Those that are in favor of CMD interatomic potentials against FPMD approaches have been flaunting the same story on limited affordability ever since the first applications of FPMD (1985 or so). This concept is endlessly repeated regardless of the huge advances made in computational power making possible in 2022 by FPMD (for a given system, number of atoms and length of trajectory), a calculation even longer that those carried out classically in the 80s. The real issue is not about the cost, or the time one has to wait to get things done. What matters is the quality and reliability of the results when claiming that simulations are carried out to mimic, predict or simply complement experiments.
- Overall, I think that working too much (or, worse, exclusively) on CMD has the disadvantage of restraining knowledge of computational methodologies to approaches addressing minimally the impact of electronic structure. While this helps to get things done faster, it does not contribute to an optimal formation of new generations of computational materials scientists, that might tend to neglect the importance of quantum mechanics and grow up scientifically with quite restricted horizons ahead of them.
- Incidentally, most FPMD implementations are highly pedagogical, promoting a soft approach to electronic structure, since the theoretical framework is well assessed and made popular by years of publications and textbooks (focused on the Kohn–Sham density functional theory (DFT) formalism). Also, the analytical objects to be handled are simple to work with (plane waves). Details requiring more experience and more time to be learnt (as the

construction of a pseudopotential or subtleties built in the derivation of the exchange–correlation functionals) can be skipped (but not ignored!) without affecting comprehension and effectiveness. Later in this chapter some advice on how to face these aspects of FPMD will be also given.

4.1.1 Digging out some failures of classical potentials

Coming back to the dilemma of CMD vs FPMD, the first example I would like to quote is quite revealing of a simple paradigm: never use a potential for calculating properties far from those employed in the construction of the potential itself. This is easily avoidable if one sticks to a potential built by fitting to bulk properties without exploring its validity on bonding configurations with different coordinations (in surfaces or isolated nanostructures, for instance). While this is infrequent when studying glassy systems since the main interest is on bulk behavior (even though surfaces are also quite appealing [5]), it is worthwhile reporting some shortcomings I encountered when neglecting this prescription.

- In the early 1990s, I got involved in studies of cluster deposition on surfaces for transition (coinage) metals. The idea was to collect information on the stability of very small structures. Embedded atom (EAM) potentials [6] were available for both Ag, Pt and their cross interactions. The fact that EAM diffusion barrier for Ag migration on the Pt(111) surface (0.05 eV) [7, 8] was found to be lower than both experiments (0.16 eV) and first-principles data (0.2 eV) [9] did not imply, at first sight, any other notable consequence than a lack of precision somewhat expected for an effective (or *n*-body) potential. However, it was later drawn to my attention that our findings on the Ag–Ag interatomic distance approaching the Ag bulk value for large clusters adsorbed on Pt(111) differed from experimental evidence, pointing out two-dimensional growth of Ag clusters featuring strained Ag islands, with Ag–Ag distances replicating those of the underlying Pt(111) surfaces. This stems from the underestimate of the migration barrier, favoring incommensurate against commensurate growth. Therefore, in this case, it is easy to grasp how bad numbers can lead to erroneous interpretations, clearly undermining the interatomic potential description for that specific case.
- Another example of this short gallery of drawbacks encountered with interatomic potentials for CMD has been already referred to in section 3.3 and can be summarized by recalling that Ge–Se chalcogenides are characterized by homopolar bonds wholly inaccessible to classical potentials. As such, I would even discourage from their use when trying to speed up randomization or melting of an initial configuration (as mentioned below, see section 4.3) before switching to the FPMD framework. The absence of homopolar bonds can require some quite long trajectory before being corrected within DFT, at least if the temperature is not too high. For the same reasons, classical potentials are bound to work better at least as an initial, simplified tool, in the case of oxides as SiO_2, that can be considered a disordered network-forming material devoid of homopolar bonds even in the

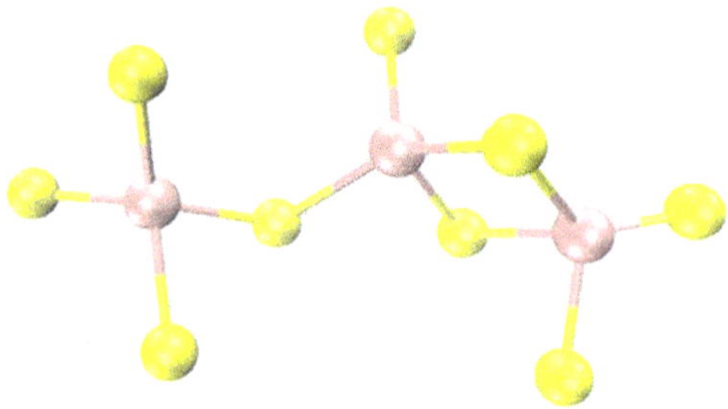

Figure 4.1. Example of corner-sharing (on the right) and edge-sharing (on the left) connections involving Ge atoms (light purple) and Se atoms (yellow). In disordered chalcogenide materials these configurations are quite general (not restricted to the specific pair of systems mentioned) and ubiquitous.

liquid state and made of only one kind of tetrahedral connections, i.e. corner-sharing (CS) (only a few edge-sharing (ES) were found at $T = 3500$ K [10, 11]) (see figure 4.1)[1]. These features have been documented in the literature in the framework of FPMD calculations [10, 11]. Within my research group, we have employed recently interatomic potentials [12, 13] to produce large initial SiO_2 structures via CMD, thereby allowing to speed up the implementation of a thermal cycle leading to an amorphous structure of SiO_2 by FPMD (see figure 4.2) and the calculation of thermal properties [14].

4.2 Methodology: the unavoidable choices to be made

Once it is decided to go for FPMD with the goal of knowing about glassy structures, one needs a code containing all the methodological ingredients sketched in the sections above. Such a code has to be user friendly to contain an input section easily manageable via keywords referring to precise options and actions. These have to correspond to the different kinds of choices for the theoretical framework and the production of MD trajectories. Over the years I found all these requirements to be nicely met by the CPMD program [15], which I have been using in very close contact with its creators, in particular with Mauro Boero who has become in recent years one of its main developers. CPMD was born in the 1990s under the impetus of M Parrinello and his team and it has been ever since (to my very subjective point the view) the best reference code for doing FPMD, at least in the CP version[2]. CPMD can be made extremely simple to use by reducing the input to a few essential instructions with the purpose of obtaining a FPMD trajectory as stable and reliable as it should be. More complex targets, related to the calculation of more specialized properties, can be achieved by adding suitable instructions without loosing its original character prone to both beginners and experienced users. Section 4.6

[1] Due to the importance of corner-sharing and edge-sharing configuration to describe the disordered network-forming materials extensively treated in this book, I have added at this point figure 4.1 that complements the visual information on this motifs provided in figure 2.5.

[2] Historically, CPMD emanated from early versions of a vectorial code conceived by R Car, M Parrinello and coworkers in Trieste containing the basic ingredients of the technique.

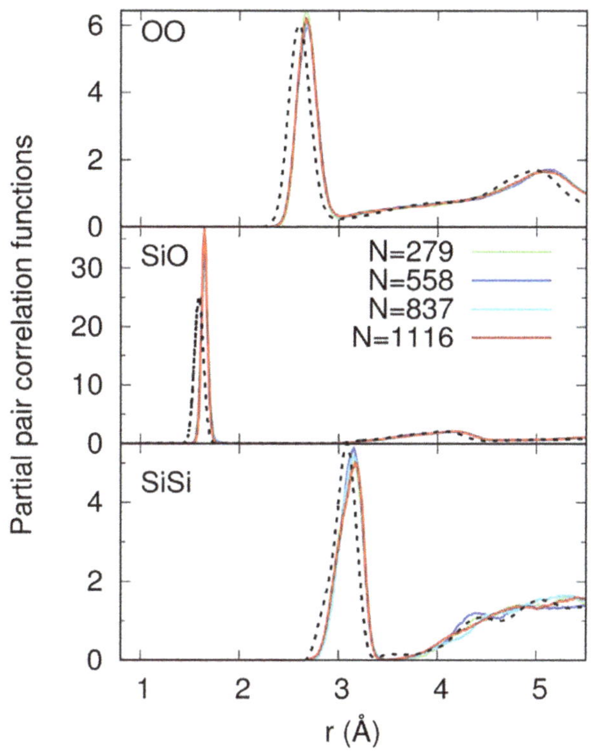

Figure 4.2. Partial structure factors for amorphous SiO_2 calculated for four different sizes of the periodic system. Dashed lines refer to the FPMD results of [11]. Note the system was prepared at room temperature by running a thermal cycle via CMD calculations with the potentials of [12, 13]. Reprinted with permission from [14]. Copyright (2022) by Elsevier.

provides some introduction and notions (deliberately not exhaustive) relative to the structure and the use of the CPMD code [15].

- In order to get started with the simulation of a given system, at a given density and for a given number of atoms so as to work with a volume ensuring statistical sampling over the required degree of structural order (short, intermediate or more), one needs to make a selection of two members built in the Kohn–Sham total energy functional (the potential energy of MD, as we have discussed in section 3.6).
- We are referring here to the exchange–correlation functional $E_{xc}(\rho(\mathbf{r}))$ and to the pseudopotential $U_{pp}(\mathbf{r})$, these two mathematical objects being strictly related, since the first determines the nature of the second, this meaning that each pseudopotential is constructed by solving an all-electron problem defined with respect to a specific $E_{xc}(\rho(\mathbf{r}))$ in the total energy. Concerning $E_{xc}(\rho(\mathbf{r}))$, the example given in section 3.6.5 is quite revealing of the impact they can have on the structure of a disordered system.
- There are a number of generalized gradient corrections available in the literature, all issued from accurate developments and bound to perform well

for iono-covalent systems intended as a broad family, this allowing to encompass most glasses as oxides and chalcogenides, leaving aside metals for simplicity. As mentioned above, I tend to recommend those I had some practice with (namely those due to Perdew and Wang (PW) [16, 17], Perdew, Burke and Ernzerhof (PBE) [18] and Becke, Lee, Yang and Parr (BLYP) [19, 20]) being convinced that at least the first two are quite general and largely applied in well-documented cases[3]. The CPMD code allows the use of almost all generalized gradient approximation (GGA) recipes made available over the past 20 years, together with their corresponding pseudopotentials. However, one important point to be retained is that they are far from being all the same. Therefore, some basic knowledge of their performances, simply gathered from the literature, should be acquired before making a choice. This diversity promotes comparative studies aimed at finding the most appropriate one for a given system (or class of systems) since improving some properties (structural, for instance) over the others.

4.2.1 More on the exchange–correlation functionals

When invoking the exchange–correlation (XC) functional it is important to show their impact on a description of a given system. This is why the experience I would like to share starts with the observation that GGA performs better than local density approximation (LDA) in the case detailed in section 3.6.5.

- We realized first that the mere introduction of GGA correction was able to improve the structural properties of liquid $GeSe_2$ due to a better account of the ionic behavior of bonding. This is coherent with the overestimate of a metallic behavior well known to affect the LDA. The notion of improved ionicity is quite important in order to link a specific improvement in a bonding feature brought about by GGA [21]. This identifies a research strategy in which GGA is a key ingredient to achieve quantitative predictions, at least for that class of chalcogenides. Having established that GGA is worth using, and knowing that our first result in this direction was achieved with the PW recipe [16, 17], the question arose on the further improvements to be obtained by searching other GGA recipes. *The idea consists in enhancing the ionic character of bonding, thereby leading to a second step of refinement characterized by higher level of agreement between theory and experiments.*
- Along these lines, our experience with chalcogenides and the quest of an improved description to be achieved by selecting the appropriate GGA illustrates the role of the XC functional as a crucial ingredient to describe materials realistically, and this regardless of any *a priori* consideration on the inadequacy of LDA and the need to go beyond it. Should there be a lesson to be thought from these considerations, it would consist in recommending not to select the first default XC functional provided by a given package for a given system without checking its performances either on well-documented results or by its own practice. The reasons driving to a specific choice might

[3] The Perdew–Wang GGA referred to in the quoted papers is also known as PW 91.

be ordinary or not having a special justification and yet, this does not prevent other GGA recipes from doing better, granting validity to some tests with them.

- In our case, we were simply both highly satisfied and unhappy with the behavior of the first GGA employed (PW) for liquid $GeSe_2$. The success of the relationship higher ionicity/improved structure was contrasted by differences found when comparing theory and experiments at the level of the Ge–Ge pair correlation functions, this meaning that the calculated Ge subnetwork was less structured in terms of shell of neighbors (more details on these results are given in sections 6.2.2 and 6.2.3). This is exactly the reason why we turned to a GGA prescription (BLYP) [19, 20] expected to be less similar, by construction, to the LDA so as to increase even more the ionic character of bonding. The choice of BLYP has at least partially met our expectations, confirming the critical role of XC functional in chalcogenides and, more generally, the usefulness of comparing the performances for a couple of them at the minimum. This is exemplified by a comparative study involving liquid and glassy $GeSe_2$ by using as GGA recipes PW and BLYP, several properties being closer to experiments in the BLYP case (see section 6.7).
- A few ideas about the conceptual foundations of the PW, PBE and BLYP XC functional are in order. In short, PW stems from LDA plus the account of second-order density-gradient expansion for the XC hole (the space surrounding the electron inside which there is zero probability of finding another electron because of electron correlation). The recipe found in [16, 17] is a fit to the numerical implementation of these ideas, further refined and improved in the PBE schemes, as detailed in [18]. BLYP follows a different strategy for the expression of the correlation energy and cannot be considered as an extension of LDA, since it does not make any reference to the uniform electron gas density from the outset.
- In summary, for the choice of the XC functional before setting up a FPMD calculation, the first recommendation is nothing but to run a literature search to make a first sensible choice. In the case of absence of information on a specific compound (this case should be rare) the PBE [18] XC functional appears the most widely used in a large number of examples, at least for the kind of systems and chemical bonding (covalent and iono-covalent) close to those treated in this monograph. However, some attention should be paid to the actual performances of the selected XC and some curiosity oriented toward extra checking and testing of alternative recipes will be a valuable bonus. In any event, for disordered network-forming materials as chalcogenides, the choice of a GGA is unavoidable at the place of the LDA. More refined treatments going beyond standard GGA (as the hybrid functionals, for one example see [22] and references therein)[4] are also very much stimulating (though much more expensive), once again confirming that

[4] This functional was employed in the framework of our study on glassy $GeSe_2$ under pressure, see section 7.2.

making a choice does not have to prevent from searching other valuable options.

4.2.2 On the selection and use of pseudopotentials

As introduced in section 3.6, building pseudopotentials for electronic structure calculation is a field of research by itself and requires skillful control of the whole DFT methodology. However, even by focusing on a specific XC functional and by restricting our rationale to norm-conserving pseudopotentials employed in a plane wave scheme, pseudopotentials are not all the same and, most importantly, they cannot be selected in a 'black box mode' since some subtleties are always haunting around the corner. This is why, at least once in a lifetime, *building a pseudopotential with the help of an expert is an experience worth going through*. However, by assuming knowledge of what they are and of how they work, there is no need to become an experienced pseudopotential manufacturer before planning FPMD calculations. This is because current packages are provided with extended pseudopotential libraries, quite often featuring at least one pseudopotential for the most common XC functionals.

- By construction, pseudopotentials determine the number of plane waves entering the calculation since, while they are acting to reduce the computational effort by smoothing out the orbitals inside the core regions, they are required to reproduce the all-electron results outside of it. This requires convergence to be attained for a set of test physical properties, reflecting the existence of a cutoff value R_c typically expressed in Rydberg and related to the number of plane waves N_{pw} by $N_{pw} \sim VR_c^{3/2}$, V being the volume. An example is given in figure 4.3 where the equilibrium distance of a dimer Si–N is plotted as a function of R_c. A value R_c of at least 60 Ry is needed to ensure convergence for this property, typically employed to extract the right value of R_c.

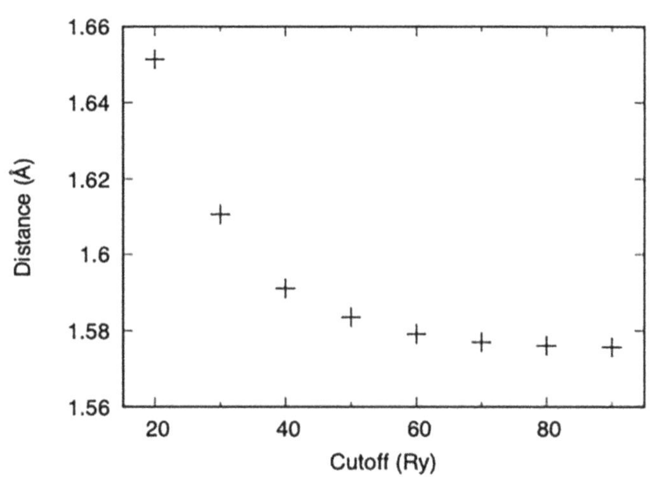

Figure 4.3. Equilibrium distance for a Si–N dimer calculated within the FPMD CP framework. Courtesy of A Lambrecht and E Martin (ICube, Strasbourg).

It appears that certain pseudopotentials are 'harder' than others, that is they require a higher number of plane waves as it happens for instance, if the core region is highly localized in space. However, after decades of first-principles calculations, the optimal R_c values for most elements and their most popular pseudopotentials are well known, even though this should be checked upfront before starting with a dimer or a small cluster calculation, for instance. Despite its obviousness, the following remarks are in order.

- One can always take R_c larger than what is customarily done, the calculations will only be more expensive and take more time. The problem arises when adopting a value significantly smaller than commonly accepted to save time and (possibly) money. If this is done for a run at equilibrium, expected to produce results on a given property and bound to be compared to experiments, such an idea is simply wrong. It might make some sense for some structural optimization bringing an initial, unstable configuration to a local minima (see section 4.3) unless the reduction is not too drastic (say, more than 30%). When referring to figure 4.3 as an example, $R_c = 40$ Ry is not adequate for production runs, while it might be for some preliminary tests.
- An additional point worth attention is the choice of the appropriate angular momentum values l_{max} and loc in equation (3.46) for the expressions for $U_{l_{max}}(r)$ and $U_{loc}(r)$. Each pseudopotential has some sensitivity to the choice made for l_{max} and loc, l_{max} marking the truncation of the sum in equation (3.46) at a given angular momentum. While these values are also tabulated with the pseudopotentials and they can be safely adopted in most cases, there are elements for which the choice is not unique and can produce sizeable differences in the results. This is the case in glassy $Ge_2Sb_2Te_5$, for which different options applied to Ge, namely $l_{max} = p$ instead of $l_{max} = d$ gave results appreciably different in terms of comparison with available structure factors. The $l_{max} = p$ choice gave results closer to experiment when calculating the deviation over the entire range of available k wavevectors for the total neutron structure factor. This issue will be extensively treated in section 11.2.

4.2.3 The quest of the best fictitious electronic mass and timestep

Having described previously the role and the significance of the fictitious mass μ, the selection of its value, in connection with the a timestep Δt_{CPMD} can be performed safely by keeping in mind the following golden rules.

- First, one needs an initial choice of μ and Δt_{CPMD} to get accustomed with the behavior of the system under the action of forces and acquire self-assurance with the methodology. We stress that (see more below in section 4.3) the initial conditions selected to produce a glass are often configurations with no special chemical stability since created with the only purpose of providing a manageable starting point and, as such, they can evolve abruptly even at low temperatures. A valuable and safe strategy consists in taking a first value for μ in the range 400–800 a.u. to be tuned depending on the ionic mass, with lower values to be considered for lighter elements.

- We recall that keeping well separated the range of frequencies of ionic and fictitious orbitals dynamics is the prerequisite for ensuring adiabatic behavior with no energy transfer driving the electronic structure away from the Born–Oppenheimer (BO) surface. These conditions are met by imposing very small values of μ when compared to the ionic masses (the mass of carbon (C), for instance being equal to 22 000 a.u.) Let us assume at this point that previous, well established knowledge of the methodology for the case at hand, enriched by extensive tests on smaller systems (as a dimer), has led us to a reliable selection of μ, hopefully more precise that the mere consideration of a range of plausible values as suggested above. What would be the reasons beyond a preference for a very small value of Δt_{CPMD}, for instance 1 a.u. (2.4×10^{-17} s), to be employed for a first set of runs devoted to testing and making sure that things are under control?
- There is strictly no waste of computational resource behind such a choice, since it allows a careful checking of total energy conservation and it will be in any case compatible with the requirement of adiabaticity for choices of μ not too small (say, $\mu > 200$ a.u.). In addition, it has the advantage of ruling out the timestep as the origin of any instability problem that might be encountered, by pointing out the existence of fully unphysical bonding configurations (i.e. atoms too close to each other) should any divergence in the conservation of the total energy occur. For most simulations up to room temperature of configurations left to evolve in the microcanonical ensemble or with a temperature control, and by focusing on disordered iono-covalent systems (chalcogenides or most oxides for instance), increasing cautiously Δt_{CPMD} up to the value of 5 a.u. is a risk free procedure that can be followed quite safely.
- In terms of energy conservation, and for particularly stable configurations (those found at the end of the thermal cycle, see below) this consideration can be extended to a $\Delta t_{CPMD} = 10$ a.u., a value that is only four times smaller than 10^{-15} s, largely adopted in many CMD simulations, especially at high (liquid) temperatures. When invoking the conservation of the total energy as expressed in equation (3.27), we are referring here to very small drifts (10^{-7} a.u. ps^{-1}) in the range of timesteps 5–10 a.u. at room temperature. In short, atomic configurations obtained by structural relaxation and with small residual forces acting on the atoms (in the interval comprised between 10^{-6} and 10^{-5} a.u.) at time $t = 0$ can evolve with high stability by ensuring excellent conservation of the corresponding Hamiltonian on long (even 100 ps) trajectories. This ensures good handling of the very first part of our targeted first-principles modeling, based on the production and further use of an initial configuration.

4.2.4 The beauty of the Verlet algorithm

Finally, we are making the assumption that there is an integration algorithm stable and fully reliable to run FPMD simulations. This is indeed the case for the

Verlet scheme, largely documented in the available literature and currently employed in the code (CPMD) we mostly refer to[5]:

$$\mathbf{r}(t + \Delta t) = 2\mathbf{r}(t) - \mathbf{r}(t - \Delta t) + \Delta t^2 \frac{\partial^2 \mathbf{r}(t)}{\partial t^2} + \mathcal{O}(\Delta t^4). \tag{4.1}$$

More than reviewing its basic features, I prefer to introduce here a personal note on the time reversible character of the Verlet scheme as I experienced in a simple case (CMD of solid argon with Lennard-Jones potentials). The exercise consists in imposing a high initial temperature (twice the melting one) to a face centered cubic argon solid via sampling of velocities pertaining to a given Maxwellian distribution. The system undergoes melting readily with redistribution of the initial energy among potential and kinetic degrees of freedom. For time reversibility to hold, one needs to change the sign of time in the equations of motion and follow the behavior of the system to make sure the initial condition is recovered: the Argon atoms are eventually back to their initial crystalline configuration! This is exactly what happens with the Verlet algorithm for a time reversal carried out after a few ps of simulation, an interval short enough to avoid roundoff errors undermining the solution of the equations of motion. The intrinsic stability of the Verlet algorithm and its capability to comply with the basic rule of Newtonian dynamics has been the driving force boosting its choice for first-principles MD applications, based on equations (3.30) and (3.31), even though its application is more tricky than what appears in equation (4.1) [23].

4.3 Creating a computer glass via MD: the initial conditions

The choice of an initial configuration is the prerequisite to any production of the desired trajectory to be averaged in time, typically at room temperature, but not only, for a glass. This means that, somehow, one has already made strategic choices about the description of the interatomic forces. In particular, for the FPMD case, one has selected the needed ingredients in the Kohn–Sham formalism (choice of the XC functional and pseudopotentials). However, what one needs to start is not only a set of coordinates but also the converged electronic structure, since this is the essential prerequisite to implement all the machinery inherent in equations (3.30) and (3.31). This goal can be obtained within the CP methodology itself by putting to good use the idea of temperature control of the electronic degrees of freedom, as made explicit in equation (3.37), that we rewrite below by replacing the variable $\dot{\alpha}_e$ by some positive friction term β_e

$$\mu \ddot{\psi}_i(\mathbf{r}, t) = -\frac{\delta E_{\text{tot}}[\{\psi_i\}, \{\mathbf{R}_I\}]}{\delta \psi_i^*(\mathbf{r}, t)} + \sum_j \lambda_{ij} \psi_j(\mathbf{r}, t) - \mu \dot{\psi}_i \beta_e. \tag{4.2}$$

[5] It should be noted that, for practical purposes, the velocity version of the Verlet algorithm is the one implemented in most codes, for the simple reason that the original version given in equation (4.1) calculates the positions of the particles at a given time t without the need of making available the velocities at the same time t. This is somewhat very smart but not convenient within FPMD.

- We learned in section 3.5.5 that the last term of equation (3.37) is the one controlling the amount of energy associated with the movement of orbitals out of the BO surface. This is exactly what can be exploited, as a particular, very unique case, to drive the electronic structure to its minimum energy configuration, by achieving in a fully dynamical sense a very complex problem of optimization. In practical terms, this can be realized by selecting any initial condition for the set of $\psi(\mathbf{r}, t)$, for instance a random set of numbers. Under the action of equation (4.2), the associated fictitious kinetic energy can be minimized at will, by leading to the electronic ground state ('zero' $E_{\text{fic}}(\psi_i(\mathbf{r}))$ with the required accuracy.
- Once this target has been reached, the next step consists in beginning an actual FPMD run in which both ions and fictitious electronic degrees of freedom evolve self-consistently, the goal being to produce temporal trajectories at the temperatures of interest (eventually, room temperature) for a disordered material bearing no memory of the initial conditions.

Two main strategies can be followed, differing by the degree of topological order of the very first configuration[6].

1. In the first case, the idea is to select an ordered system, ideally the crystalline counterpart of the targeted amorphous structure, for which the initial positions have to be known. As a prerequisite, one needs to make sure that such configuration is stable within FPMD. To this purpose, a short run of dynamics with safe values for the timestep and μ is necessary to monitor the action of those residual forces that might be present should the structural stability differ from the one assumed in the selected configuration. In what follows, we shall imply that *the system is indeed stable at very low temperatures and allows for an increase of temperature to be carried out on it*. Should this not be the case, we refer to the next strategy for the actions to be undertaken.

    - To achieve disorder, the action of heating up the system by a suitable control of the temperature in a stepwise fashion has to lead to melting and substantial diffusion. These conditions correspond to a liquid state when several interatomic distances are covered in a few ps, as it could be easily extracted from the relationship presented in equation (1.1). In principle, such a span of a phase diagram should be realized by accounting for the changes in the volume, that are accessible through the use of the constant pressure MD techniques [26, 27], devised much in the same spirit of the extended Hamiltonian devised for the implementation of thermostats[7]. This is exactly what was carried out for amorphous $NiZr_2$ (see the crystalline structure in

[6] I took the liberty to take advantage of some specific examples extracted from CMD calculations to highlight issues common to both methodologies. It appears that some of the concepts developed in what follows hold for both classical and FPMD calculations. However, differences in the strategy to be applied will appear clearly and they will be explicitly pointed out.

[7] This point will be made explicit in chapter 5 via equations (5.2), (5.3), (5.4).

figure 4.4 and the behavior of volume as a function of temperature up to melting in figure 4.5) as will be detailed in section 5.2 in the framework of a classical MD study. However, this is not strictly necessary if the final goal is the creation of a diffusive state from which to start a quenching schedule bringing the system down to room temperature by avoiding crystallization.

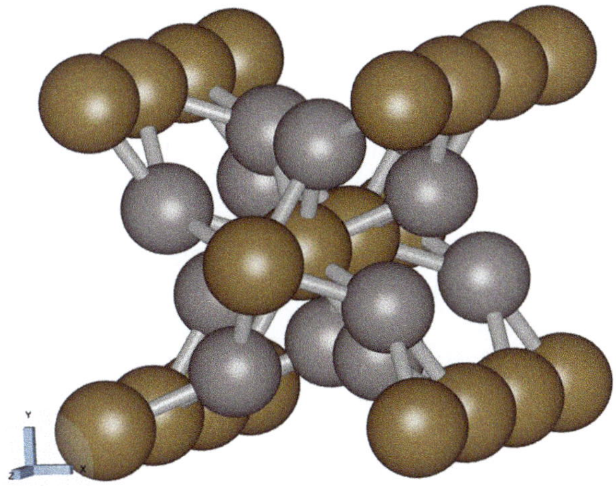

Figure 4.4. Crystalline structure of $NiZr_2$. This structure is taken as an example of system for which, within classical MD, an initial configuration fully chemically and topologically ordered was available and employed (see section 5.2 and [24]). Ni atoms are gray and Zr atoms are brown. The software iRASPA has been employed to draw the figure [25].

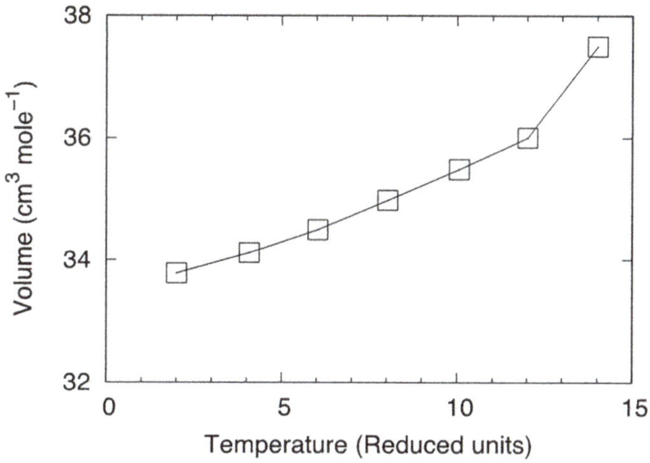

Figure 4.5. Behavior of volume for the model of crystalline $NiZr_2$ undergoing melting at high temperatures. Further details will be given in section 5.2. Reduced units are employed within a classical molecular dynamics simulation. See also [24].

- Should the interest reside entirely on the production of a liquid state, the changes in volume could be implemented, in principle, directly at the desired temperature by simple volume increase. This step is far from being trivial within FPMD, since the electronic structure of chemically identical systems having different volumes is by no means the same and has to be adapted to the ionic configuration. By deferring any further detail on this point, let us stick to a situation in which one is dealing with a disordered system, stemming from an originally ordered one, and characterized by a diffusive behavior. In this case melting followed by diffusion ensures that there is no memory left of the initial configurations, and one is ready to start quenching the structure.
- Two points are worth underlying. First, there is no need to increase the temperature to exceedingly high values (say, several times the melting temperature) on the grounds of getting quicker to a fully randomized structure with no residual topological order. This would have the unpleasant consequence of producing a structure totally different from any existing liquid, the main structural motifs, for instance, being totally disrupted.
- Second, the crystalline counterpart might not be known with the required precision (all atomic positions from x-ray) or it might be simply hosting a lot of defects. In this case, the route crystal–liquid–glass can be impractical to follow by MD. Also, I had the impression that, over the years, this particular strategy has become less popular due to its non-negligible cost when using FPMD, since it amounts to building up, at least, a kind of phase diagram at constant volume. This takes us to the second case detailed below.

2. Regardless of the existence of a crystalline counterpart, one can prefer *starting with a disordered configuration having no resemblance whatsoever to the targeted topology*. This can be a reasonable choice since we are aiming at characterizing a disordered structure and there is no obligation to go through the simulation of an ordered one. *Let us assume that we intend to work directly at the target temperature typical of a liquid.* The number of options in this context go as follows:
   (a) coordinates randomly positioned inside the periodic box,
   (b) coordinates obtained via any computer tool able to produce an equal number of atoms inside a box of required volume. This differs from the previous one since one can assume that some criteria more sensible than a mere random choice will be used.
   (c) coordinates taken from existing classical simulations for a system having close coordination features and/or close chemical composition.
   (d) coordinates taken from existing classical simulations of the same system.
   (e) coordinates taken from existing FPMD simulations for a system having close chemical nature and/or close composition. This situation includes also simulations differing by a choice of the XC functional

and/or the pseudopotential (different pseudopotentials can be constructed for the same XC functionals, as shown in section 11.2).

- Cases **(a)** and **(b)** can be highly troublesome at the beginning, due to the total mismatch between a very cumbersome initial potential energy surface and a realistic one. There is no point in attempting to follow the dynamics resulting from such initial conditions. Due to high frequencies involved with displacements out of highly non equilibrium positions, it will be quite hard to find a suitable timestep ensuring adiabaticity even for very small values of the effective mass μ. In addition, changes in the electronic structure during the rapid movement that goes along with the search of some optimal configuration will be abrupt, preventing the stability of the whole CPMD algorithm.
- These difficulties are alleviated in cases **(c)**, **(d)** and **(e)**, but not to the point of making possible a direct simulation at finite temperature stable in time. Case **(e)** is the most favorable since changes in the chemical nature and in the resulting potential energy surface are limited while still not negligible to the point of allowing a smooth transition. Is it also getting any better in case **(d)**, devised to make the connection between classical MD and FPMD for a given system? Experience shows that the temperature of the system is inevitably bound to increase due to the search of some minimal potential energy and again, unless the two potential energy surfaces are very close (as it is very rare when considering configurations issued by potentials as input of first-principles approaches) the direct, pain free production of a temporal trajectory is unfeasible: ionic temperatures too high, timestep to avoid collapse too small, fictitious electronic energy departing dramatically from the BO surface.
- *The above considerations are consistent with the conclusion that starting with a disordered configuration with the hope of bouncing directly to a finite temperature simulation makes little sense for FPMD/CP approaches since involving drastic changes in all variables depending on time.* Of course, this is perfectly feasible in a classical MD context, with some caution to be exercised to handle the first ps of structure rearrangements. To overcome any difficulty in pursuing a temporal evolution at finite temperature, the best alternative consists in *relaxing* via *minimization at zero temperature an initial structure (corresponding to one of the cases **(a)** through **(e)** above) as done for the search of the electronic structure ground state (see equation (4.2))*. This corresponds to imposing a thermostat to the ionic motion to reach very low temperatures in equation (3.38) or to replace the thermostat control variable $\dot{\alpha}_I$ by a positive friction term β_I, analogously to what we showed in equation (4.2):

$$M_I \ddot{\mathbf{R}}_I = -\nabla_I E_{tot}[\{\psi_i\}, \{\mathbf{R}_I\}] - M_I \dot{\mathbf{R}}_I \beta_I. \tag{4.3}$$

Ionic and (fictitious) electronic relaxation can be combined to produce a configuration featuring very small residual forces on the corresponding degrees of freedom, thereby leading to a viable initial configuration stable at very low (virtually zero) temperature and being fully relaxed at the electronic ground state for that ionic configuration. A first example is given in figure 4.6, where the final configuration and the initial one differ by the choice of the pseudopotential (case **(e)**, glassy $Ge_2Sb_2Te_5$). A second example is given in figure 4.7, where the initial configuration, for a disordered SiO_2 system, is obtained from a set of coordinates obtained by using the CMD

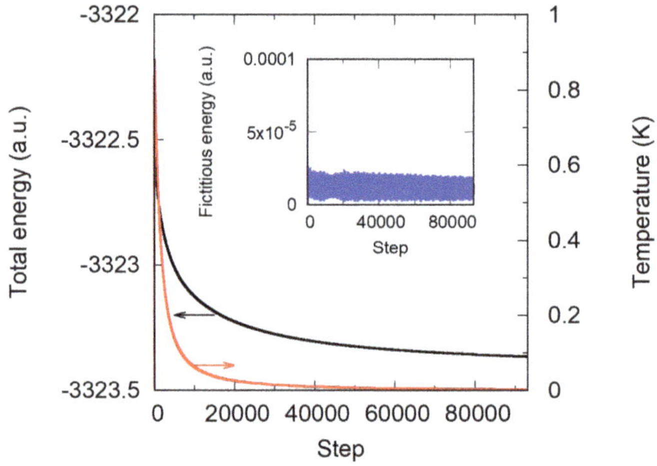

Figure 4.6. Relaxation of the total energy and of the kinetic energy in units of temperature for glassy $Ge_2Sb_2Te_5$ after switching from the BLYP pseudopotentials due to Troullier and Martins (TM) [28] to those proposed by Goedecker, Teter and Hutter (GTH) [29, 30]. Inset: behavior of $E_{fic}(\psi_i(\mathbf{r}))$. Note that the values of this quantity are very small indicating a smooth transition between the initial and the final configuration, having a limited impact on the adiabaticity. Courtesy of E Martin (ICube, Strasbourg).

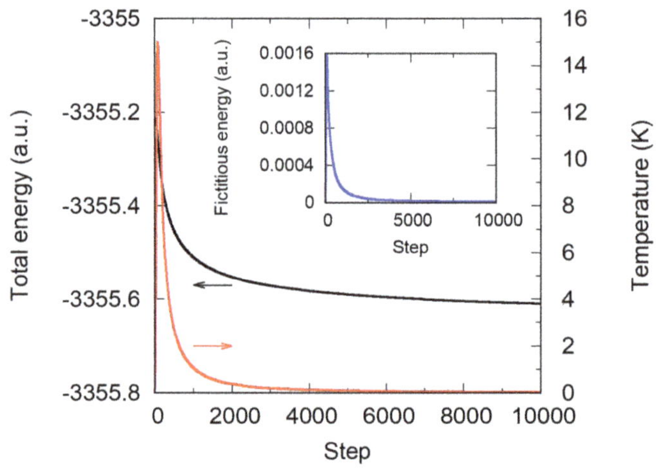

Figure 4.7. Relaxation of the total energy and of the kinetic energy in units of temperature for glassy SiO_2 after switching from the classical *n*-body potential of references [12, 13] to FPMD. Inset: behavior of $E_{fic}(\psi_i(\mathbf{r}))$. Note the rapid increase of the three quantities followed by a smooth decay to small values. Courtesy of E Martin (ICube, Strasbourg).

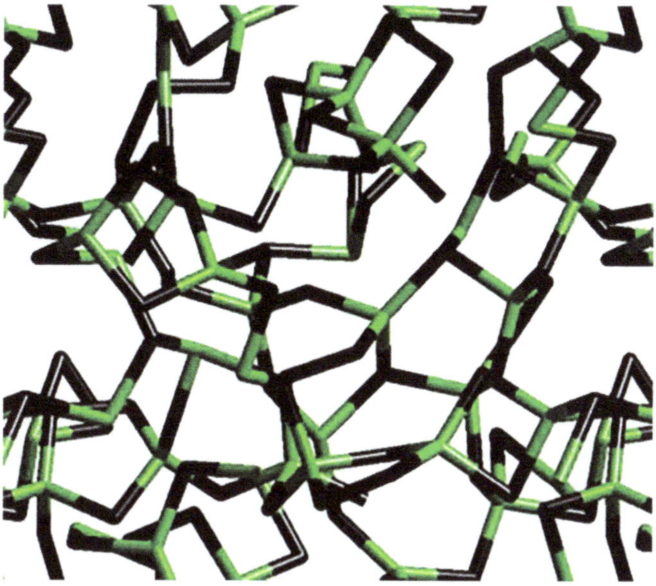

Figure 4.8. Initial structure of disordered $GeSe_2$ employed as a starting point to produce a quenching schedule aimed at obtaining the glassy state. Green sticks depart from Ge atoms and black sticks from Se atoms. Note that the predominant tetrahedral coordination of this network is altered in this specific configuration by homopolar bonds and miscoordination for both species.

potentials of references [12, 13] (case **(d)**). *The resulting configurations, characterized by vanishing values for the crucial properties to be monitored (the kinetic energy and the temperature) are now ready to be exploited to begin a thermal cycle leading to high temperatures typical of a liquid-like system. This is the prerequisite to the production of a realistic glass model.* An example of configuration suitable as initial point for the production of a glassy state is given in figure 4.8.

4.4 Production of trajectories and the setup of a thermal cycle

We are now in a position to begin our FPMD study of a disordered system by having as first target the production of a disordered phase highly diffusive, this being the prerequisite to a cooling process aimed at creating configurations structurally arrested. *We can briefly summarize the ingredients to be considered when setting up a simulation input, most of our guidelines referring to an implementation such as the one made available in the CPMD code for running CP simulations.*

4.4.1 An essential summary before hitting the road

1. Concerning the selection of the XC functionals and pseudopotentials, one has to keep in mind that these two ingredients are intimately correlated, the first determining the second since any pseudopotential has to be employed by referring to a specific XC functional. While default choices are available for

XC functionals, it is useful to be documented (at least) and test other options, since bonding properties might be dependent on subtle details differing from one XC to the other (as the extent of electronic localization reported in the example given in section 4.2.1). In principle, the same philosophy holds for pseudopotentials, despite their widespread availability for a large part of the periodic table and a variety of XC functionals. Awareness on the relationship between the nature of pseudopotentials (more or less 'hard') and the computational effort (number of plane waves) is very much required in this context.

2. On the side of timestep and fictitious mass μ, the basic recommendations are not to refrain from the use of a very small value for the timestep in the initial, preparatory parts of the simulation run, so as to escape any ambiguities on the origin of possible dynamical instabilities. Values of μ are nowadays quite well known for a large number of compounds and they can be selected in a sizeable range with no consequences, provided the timestep is taken as small enough to handle the associated frequencies. At least at room temperature, values in between 5 and 10 a.u. for the timestep should allow keeping the temporal behavior of the different quantities safely under control. We keep in mind that larger values of μ go along with smaller frequencies to be integrated, thereby allowing in principle longer timesteps.
3. To start a thermal cycle and produce a glass one needs an initial configuration. We have sketched two different strategies, the first consisting in melting a crystalline phase, the second based on the existence of some configuration more or less chemically meaningful and bound to be quickly abandoned as time evolution sets in. By preferring this second choice, we have pointed out that in this case a direct simulation at high temperature is not feasible, thereby calling for a relaxation at zero temperature in order to produce a stable configuration prone to be heated up. This can be done by using the CP scheme as a methodology to achieve optimization of both families of degrees of freedom, much in the spirit of a global optimization problem.

4.4.2 Starting to run carefully and cautiously: a mini guide

1. In what follows, by thermal cycle we mean bringing the system to a temperature high enough to allow diffusion to take place on time intervals accessible to FPMD simulations, so as to loose memory of the initial condition and be ready for a cooling process. Temperatures high enough are those typical of liquids, and in any case higher than the melting temperature. To avoid falling into the trap of an overestimate of the melting temperature for the model, combined with some overheating due to periodic boundary conditions, it is convenient to target a temperature somewhat higher than the melting temperature by 10%–20%.
2. In section 4.2.3 we have underlined the importance of taking a small timestep (even as small as 1 a.u.) at the beginning, to check the stability of the conserved quantity and get acquainted with the content of the output. As a first step of a thermal cycle, one can simply set the appropriate keywords to go for

microcanonical MD and let the system evolve from the initial configuration that we are assuming fully relaxed with minimal forces for both ions and fictitious orbitals. While unwanted, trivial mistakes in the input file can be easily detected even with a very small timestep. Therefore, this minimal integration step will allow careful checking of the changes in temperatures and potential energy, that are in any case expected for the following reasons.

3. If we are dealing with the case of a configuration lying on some local minimum, we shall observe an increase in temperature due to oscillations of the atoms around the equilibrium positions, both $E_{\text{fic}}(\psi_i(\mathbf{r}))$ and $E_{\text{kin}}(\{\mathbf{R}_I\})$ (or, the corresponding temperatures) fluctuating around some average values. However, one is perfectly aware of the fact that the initial conditions have been obtained by global optimization of some disordered structure in principle far from the targeted one, since created along the lines sketched in section 4.3. Therefore, during the first part of the temporal trajectory, the system is very much likely to escape from a very shallow minimum or a saddle point, in the search of a more stable configuration. Such behavior is predominant and it can be both observed/recorded and handled, in a parallel and instructive fashion.

- On the one hand, a system evolving freely with higher kinetic energies in the search of a lower potential energy can be the expression of an evolution toward some average initial temperature, this part of the trajectory being ideally suited for further heating to promote diffusion. This is rewarding and can be exploited provided the total energy is well conserved and $E_{\text{fic}}(\psi_i(\mathbf{r}))$ features no significant drift out of adiabaticity.
- On the other hand, one may want to drive the system to room temperature by imposing a thermostat to the ions, so as to stabilize a trajectory at that temperature and be ready to move on at higher values. Ideally, both trajectories should be followed and compared at the beginning, since a view of microcanonical and canonical evolutions carries information on the way the system is reacting to the 'ad hoc' configuration found by structural optimization. Since there are a lot of possible situations that can be encountered in this context, the best option consists in following the 'free' evolution as long as the dynamics remains stable and the temperature (averaged on a few ps) does not exceed 500–600 K. Depending on the average temperature values, the ionic thermostat can be used to obtain a trajectory at that temperature, by increasing the timestep up to 5 a.u. when fluctuations get settled at equilibrium. This situation is exemplified in figure 4.9 where the behavior of potential energy, temperature and fictitious electronic energy is followed at $T = 100$ K and $T = 300$ K as a result of the application of the thermostat at each temperature.

4.4.3 Handling adiabaticity: the gap issue

Let us focus our observations on a FPMD trajectory at a temperature in between 300 K and 600 K for which we have decided to control the temperature via a thermostat and for which we can ensure that, at least at first sight (a few hundreds of

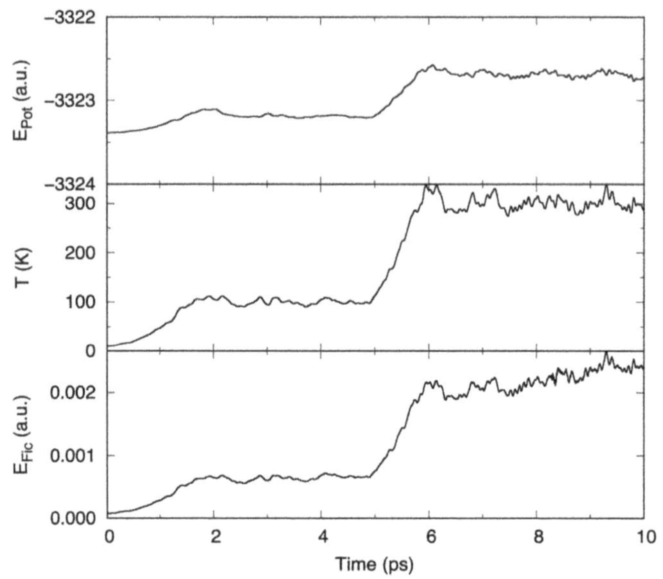

Figure 4.9. Behavior of a disordered $GeSe_3$ systems for which an initial configuration has been produced as indicated in section 4.3, case (c). The systems is brought to $T = 100$ K and $T = 300$ K by using a thermostat on the ionic degrees of freedom. Courtesy of E Martin (ICube, Strasbourg).

time steps) adiabaticity is preserved as proved by the behavior of $E_{\text{fic}}(\psi_i(\mathbf{r}))$. What can we expect for longer trajectory and/or for higher temperatures? In other words, what are the difficulties to be surmounted with respect a CMD run for which one can basically increase the temperature with no fear of instabilities (besides some possible adjustments of the timestep for high temperature)?

- In the case of a large gap system, say at least 0.5–1 eV (a few thousands of K in terms of temperature) and under the assumption that this gap does not shrink substantially at higher temperatures, one can expect a painless production of temporal trajectories up to the desired target needed for diffusive motion. This should occur on a timescale shorter than what is commonly affordable without wasting too much computational time (in between 10 and 100 ps). In this way, trajectories can be produced almost automatically, by using the ionic thermostat as a manageable parameter and with no concern for the behavior of $E_{\text{fic}}(\psi_i(\mathbf{r}))$ since bound to adjust linearly to the change of temperature by remaining small. The essence of this statement can be expressed via the relationship

$$\omega_{\min} = \sqrt{\left(\frac{2E_{\text{gap}}}{\mu}\right)} = \sqrt{\left(\frac{2(\epsilon_{\text{LUMO}} - \epsilon_{\text{HOMO}})}{\mu}\right)} \tag{4.4}$$

that correlates the smallest value of the fictitious orbitals spectrum of frequency to the HOMO–LUMO gap of the system under consideration. When $\omega_{\min}$ is large enough to be clearly separated from the ionic spectrum of frequencies, we are fulfilling adiabaticity (no thermal equilibrium between the two families of degrees of freedom), thereby ensuring full applicability and reliability of the method at all temperatures satisfying this hypothesis[8]. Unfortunately, we are all aware that gap systems (insulators) have to coexist in nature with systems less compatible with a scheme based on the strict respect of adiabaticity. By excluding for a moment metals, which clearly escape this methodology, our interest is in situations evolving with temperature so as to modify the nature of chemical bonding by reducing the band gap and, as a consequence, introducing some complexity into the FPMD implementation.

- On the practical point of view, gap reducing or gap closing behaviors with increasing temperature manifest themselves with a marked increase in $E_{\text{fic}}(\psi_i(\mathbf{r}))$, while this is not perceptible in the behavior of the ionic kinetic temperature, controlled by a thermostat and, by and large, the conservation of the total energy is not appreciably affected. The increase of $E_{\text{fic}}(\psi_i(\mathbf{r}))$ can be understood by observing (equation (4.4)) that a gap reduction can only partially be compensated by a reduction of μ, lowering ω_{min} to the point of approaching the ionic frequencies. This has the effect of favoring global thermalization of all degrees of freedom, by lowering or hampering adiabaticity, the net result being an increase of $E_{\text{fic}}(\psi_i(\mathbf{r}))$.
- As already detailed and referred to in section 3.5, the control of $E_{\text{fic}}(\psi_i(\mathbf{r}))$ via a thermostat is recommended to prevent this variable from diverging. We have already given an example of this kind in figure 3.5. A further demonstration of the effect of the thermostat on $E_{\text{fic}}(\psi_i(\mathbf{r}))$ is shown in figure 4.10. This strategy allows recovering the conditions of adiabaticity

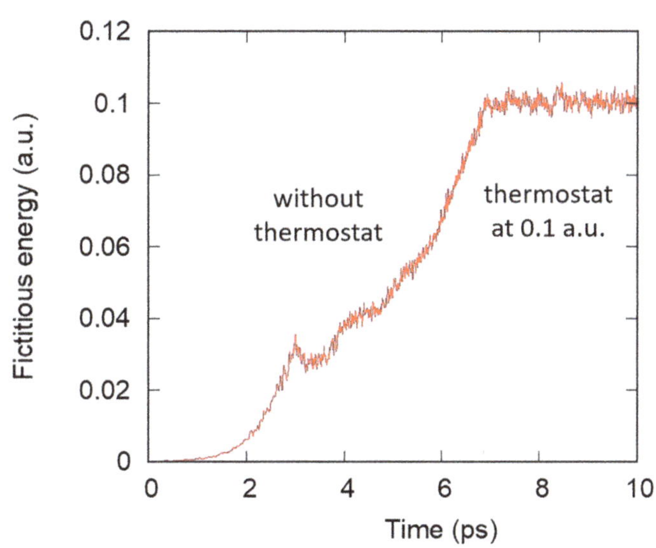

Figure 4.10. Behavior of $E_{\text{fic}}(\psi_i(\mathbf{r}))$ before and after application of a thermostat on the fictitious electronic degrees of freedom (case of a disordered SiN system). Courtesy of E Martin (ICube, Strasbourg).

[8] The first paper treating in detail the issue of adiabaticity within CP calculations via explicit examples and some analytical consideration is [31], a kind of historical reference always worth looking at.

quite effectively, provided the coupling between the variables (fictitious orbitals) and the thermostat is carried out by taking similar values for the associated masses (or slightly higher for the thermostat) and by selecting a value for $E_{\rm fic}(\psi_i(\mathbf{r}))$ in line with the analysis of [32]. A sensitive choice for the target $E_{\rm fic}(\psi_i(\mathbf{r}))$ can also consist in imposing the value taken before the occurrence of the divergence, past experience proving that it turns out to be quite close to the value hypothesized in [32].

Controlling $E_{\rm fic}(\psi_i(\mathbf{r}))$ [32]: where does it come from?

In reference [32] the idea of wavefunctions adjusting adiabatically to the movement of the atoms is exploited by establishing a relationship between the fictitious kinetic energy of the system and the velocities of the atom. This allows determining a target value $E_{\rm fic}^{\rm ad}$ for $E_{\rm fic}(\psi_i(\mathbf{r}))$ intended to ensure optimal adiabaticity. Note that $E_{\rm fic}^{\rm ad} = E_{\rm kin}^{e-{\rm TG}}$ as given in equation (3.40). The expression given in [32] reads:

$$E_{\rm fic}^{\rm ad} = 2k_{\rm B}T\frac{\mu}{M}\sum_i \langle\psi_i| -\frac{1}{2}\nabla^2|\psi_i\rangle. \tag{4.5}$$

Equation (4.5) stems from the idea of wavefunctions moving in time to follow the atoms at a velocity given by

$$\dot{\psi}_i = \frac{\partial\psi_i}{\partial\mathbf{r}}\frac{\partial\mathbf{r}}{\partial t} = \frac{\partial\psi_i}{\partial\mathbf{r}}\mathbf{v} \tag{4.6}$$

and contributing to $E_{\rm fic}^{\rm ad}$ via a kinetic energy equal to 1/2 $\mu\sum_i\dot{\psi}_i^2$. This leads to an expression for $E_{\rm fic}^{\rm ad}$ containing the square of the atomic velocities $\mathbf{v}$ (proportional to $k_{\rm B}T/M$) and the electronic kinetic energy $\sum_i\langle\psi_i|-\frac{1}{2}\nabla^2|\psi_i\rangle$, equation (4.5).

- The question arises on the capability of the double thermostat scheme to ensure effective FPMD simulations at liquid-like temperature for systems having a semiconductor nature or, in practical terms, showing a tendency to reduce drastically their band gap with increasing temperature. This statement is comprehensive of situations caused by the intrinsic limitations of DFT (underestimate of band gaps). For iono-covalent disordered network-system as those referred to hereafter (mostly chalcogenides, but also some oxides and closely related systems), we can safely affirm that the thermostat control of $E_{\rm fic}(\psi_i(\mathbf{r}))$ succeeds in providing well-behaving dynamical trajectories. In other words, all the concepts developed in section 3.5.5 can be put to good use to ensure the applicability of the exposed methodology via the thermostat control of adiabaticity. This is particularly true and needed for temperatures approaching or higher than melting point and for compositions favoring the onset of the metallic character (as, for instance, the equimolar Ge–Se system in the liquid state [33]).
- Of course, the success of CP FPMD in making possible extensive simulations (in the liquid, highly diffusive state) of chalcogenide systems rests on the

nature of their bonding character (that differs from the metallic one for the compositions and temperatures of interest) more than on the optimal performances of the method. However, this achievement is quite remarkable since for these systems adiabaticity is delicate to enforce and has to be followed very carefully. Also, it has to be kept in mind that the target value of $E_{\text{fic}}(\psi_i(\mathbf{r}))$ is linearly dependent on temperature and has to be modified accordingly. In any event, nothing has to be taken for granted when increasing the system temperature (i.e. careful monitoring always needed).

4.4.4 Some instructions to be effective when moving to high temperatures

More generally, by assuming that we are able to reach and bypass the melting temperature by preserving adiabaticity, a few issues have to be addressed on the practical side.

1. When increasing the temperature in a stepwise fashion, it is a good practice to allow for trajectories lasting at least 10 ps at each temperature[9], so as to achieve equilibrium while having enough time to check that the simulation is under control. Also, this makes possible some adjustments, as a slightly lower timestep in case of flagrant worsening in the energy conservation or changes in the masses/frequencies of the thermostats to achieve better coupling. In this context, exceedingly high/low mass values prevent any coupling as it should be for families of degrees of freedom expected to achieve equilibrium. Our primary goal in this part of the thermal cycle is to achieve substantial diffusion at a given temperature. In the first place, comparison with experiments in the liquid phase is somewhat a secondary purpose, since we are not adjusting the volume as in a phase diagram by having in mind to come back to room temperature after a randomization trajectory.
2. This requires some clarification on the properties to be calculated and monitored, knowing that, on an intuitive basis, we are simply asking the atoms to move away substantially from their initial positions. Early in the introduction (see section 1.1) we have invoked the mean square displacement and the related diffusion coefficient as key quantities to mark the difference between disordered configurations (no topological order) differing by the mobility of the atoms:

$$\langle r_\alpha^2(t)\rangle = \frac{1}{N_\alpha}\left\langle \sum_{i=1}^{N_\alpha} |\mathbf{r}_{i\alpha}(t) - \mathbf{r}_{i\alpha}(0)|^2 \right\rangle \tag{4.7}$$

and

$$D_\alpha = \lim_{t\to\infty}\frac{\langle r_\alpha^2(t)\rangle}{6t}. \tag{4.8}$$

[9] A rationale pointing out a value of 60 ps as a comfortable choice characterizing a high temperature diffusive behavior will be presented below.

3. Once again, in analogy with the rationale of section 1.1, we can estimate some average distance covered by the atoms by requiring atomic migration corresponding to several interatomic distances, definitely well above the supercooled liquid regime referred to in section 1.1, our goal being extensive randomization on accessible time scales. By taking a value of 15 Å, and with D_α equal to 10^{-5} cm^2 s^{-1}, one can extract the length of a safe trajectory to be followed to ensure diffusion and loss of the initial configuration (the one created at the beginning and relaxed, see section 4.3) via equation (4.8). The resulting value is ~60 ps. In practice, one has to follow the behavior of $\langle r_\alpha^2(t)\rangle$ at each temperature, by calculating D_α as a complement. The onset of a linear regime for $\langle r_\alpha^2(t)\rangle$ goes along with values for D_α in any case much larger than 10^{-6} cm^2 s^{-1}, signifying that the diffusive regime has been attained (see figure 4.11). At this point, it is safe to follow a trajectory lasting ~60 ps to make sure that the randomization process has acted effectively. An example of temperature behavior at equilibrium in the liquid phase for Ge_2Se_3 is given in figure 4.12 for a time interval as large as 100 ps at $T = 1000$ K.

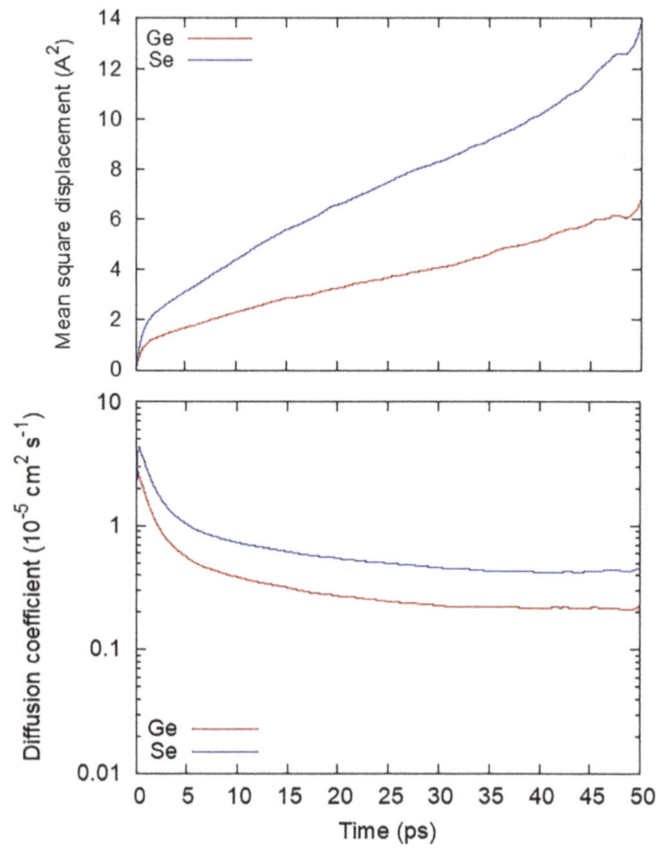

Figure 4.11. Behavior of the mean square displacement and diffusion coefficient for the liquid $GeSe_3$. Courtesy of E Martin (ICube, Strasbourg).

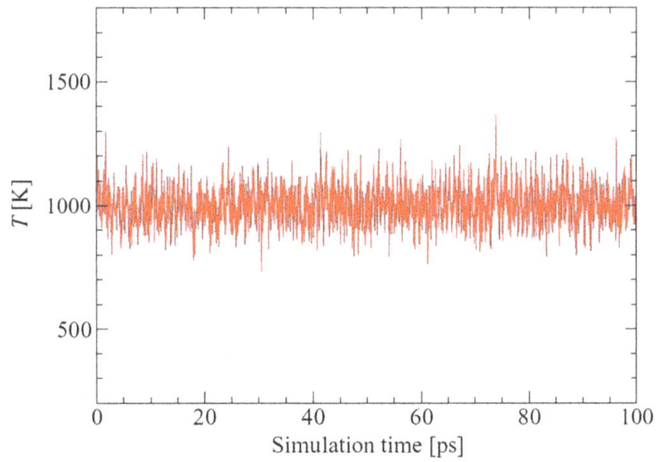

Figure 4.12. Behavior of the instantaneous temperature of liquid Ge_2Se_3 at $T = 1000$ K (see [34]). Reprinted figure with permission from [34]. Copyright (2011) by the American Physical Society.

4. An additional control of the onset of diffusion consists in comparing the behavior of sets of pair correlation functions at increasing temperature, knowing that a drastic decrease in the heights of the main peaks and a widening of their shapes is well known to characterize the liquid state when compared to the amorphous one. These pair correlation functions have to be similar (although not necessarily very close) to those pertaining to the liquid state at its equilibrium density. In this respect, we have purposely avoided considering this section of the thermal cycle as a way to produce a liquid configuration on its own, ready to be compared with the experimental counterpart whenever available. This is very much possible and feasible by adapting the density to the experimental value of the liquid, as we did for several systems [34–36].
5. In principle, MD at constant pressure can be used for producing a phase diagram at a given pressure, comprehensive of crystalline, liquid and glassy states each one at its density for a given pressure [26]. Examples are given in chapter 5 for CMD. However, since our approach to the liquid state is mainly instrumental to create a disordered state with high mobility at the targeted density of the glass, we shall not dwell on this option. It is useful to observe that the transition at high temperature from two different densities lying within a few percent can be carried out without going through the study of a phase diagram. In fact, it amounts to changing the systems size, reoptimizing the electronic structure and allowing for some structural adjustment on the liquid-like disordered structure created as detailed above.

4.4.5 Quenching down to the glassy state

The starting point of this section is the consideration of a disordered state fully compatible with the chemical composition of the system under consideration since very close to its realization in the liquid state. This is what we have obtained via the procedure described in section 4.4.4. Let's also assume that the disordered state has been thermalized for ~60 ps or so to ensure sampling of the phase space and full randomization. This means that we are ready to begin cooling the system to obtain a glassy state at room temperature.

There is a longstanding and controversial debate on the legitimacy of quenching from the liquid state with the claim of producing a reliable glassy state, at least when the purpose is to compare with samples produced on time scales inaccessible to calculations. This has to do with the quench rates that are judged to be unrealistically high, even though very fast quench rate q_{rt} are nowadays available experimentally up to the range 10^9–10^{14} K s^{-1}, leading to the production of monoatomic glasses otherwise thought unavailable to escape crystallization [37]. However, bulk glassy materials available as samples prone to structural measurements are cooled on the scale of minutes on reduced temperatures ranges. This kind of procedure is totally out of reach of atomic-scale modeling since, to be more explicit, even a q_{rt} as high as 10^9 K s^{-1} means a change of 1 K on a scale of 1000 ps, clearly not within the reach of any MD calculation. Therefore, when invoking problems with high quench rates, the main concern is the (supposed) inability to capture the correct amorphous structure since there is essentially no (or very little) thermal annealing, these drawbacks preventing the establishment of a potential energy surface significantly different from the one of the liquid. *Before planning MD studies of glasses one has to be aware that these problems are real and have to be circumvented as much as possible, in spite of the limited length of the temporal trajectories.* A sensible strategy goes as follows.

1. There is no point in simplifying the cooling process by quenching directly (or within a time span of a few ps) on grounds of using rates that will be in any case faster than in experiments. Even though some relaxation at room temperature will modify the structure of the original liquid, a drastic quench has no effect on the structure that will remain very close to the starting one. For sure, one cannot talk in this case of a glassy state but simply of an instantaneous picture of the liquid state.
2. This rationale amounts to invoking a real difference between two values such us $q_{rt_1} = 10^{14}$ K s^{-1} (100 K on 1 ps) and $q_{rt_2} = 10^{12}$ K s^{-1} (100 K on 100 ps), even though both of them could be thought pretty much equivalent at a first sight. Besides the well known fact that higher quench rates relocate at higher temperature the glass transition temperature of the system, it has to be realized that a change in temperature of 100 K on a timescale of 1 ps does not allow any thermal equilibrium to be reached. Therefore, a reduction of 1000 K on 10 ps results in the production a mere frozen liquid in all respects, with a structure that will be different from the one obtained when the temperature reduction occurs on a period, say, 100 times larger, 1000 ps.
3. *The obvious consequence of the above considerations is a strong indication to stretch the cooling process in time as much as possible, despite the fact that it will remain short on macroscopic time scales.* This will allow following all structural relaxations occurring (whenever affordable) on the ns scale, by

providing a strong argument to be substantiated at posterori via some stringent test. The argument consists in assuming that rapid quench works well to describe the structural motifs and the lengths of interest if major changes from the liquid to the glassy states are concentrated on a quite short timescale. Of course, this statement is by far more a conjecture than a certitude, its legitimacy being rooted only on the very good agreement between FPMD and experiments for structural properties for many glassy systems studied over the year. *Is this enough to adopt rapid quench as a methodology to study the glassy state? We shall leave this question unanswered for the moment, by attempting to produce further pieces of evidence to consider MD as very useful and realistic in this context.*

4. The target of reducing the temperature by cooling at a rate of 1 K each ps (10^{12} K s^{-1}) within FPMD implies the availability of resources on the scale of nanoseconds. While this might not be available to all users, it is worthwhile to work on these intervals to avoid as much as possible shortcomings due to exceedingly short relaxations. Given these conditions, cooling steps can be of various lengths, each one being long enough to achieve some sort of stationary state. By mentioning a quench rate of 10^{12} K s^{-1}, one is very much referring to a strategic perspective more to the current available examples, lying mostly around 10^{13} K s^{-1} (10 K each ps) at the best. This is because there is no consensus on planning the production of a glass via first-principles calculations on the ns scale, in the absence of a systematic study proving the importance of gaining, say, one order of magnitude. Therefore, all considerations developed therein are mostly based on experience and a double intimate conviction: (a) quench rates of MD are not so unrealistic if we can prove that some representative structural properties converge very fast to their values without requiring macroscopic relaxations times and (b) some important research investments should be planned when studying glasses by MD to make available time intervals as long as possible, and in any case not shorter than those corresponding to 5×10^{12} K s^{-1}.
5. Regardless of the distribution of the available time slots on the different steps of the cooling 'ladder', we have already underlined the importance of extending each trajectory in the search of a stationary state. This is particularly crucial at room temperature corresponding to the final relaxation step. In a paper devoted to glassy $GeSe_2$ [38] (see figure 4.13), we considered the example of the pair correlation functions $g_{GeGe}(r)$, $g_{SeSe}(r)$ and $g_{GeSe}(r)$ by comparing the results obtained for the liquid, for the first 5 ps of annealing at T = 300 K and for the last 12 ps of annealing at T = 300 K. Clearly, the results obtained for $g_{GeGe}(r)$ on the first ps at T = 300 K are very much similar to those of the liquid, proving an insufficient structural relaxation. The situations changes drastically when taking the last 12 ps at T = 300 K demonstrating that even at room temperature the potential energy surface can be effectively visited. The example refers to a situation featuring a change from T = 1100 K to T = 600 K in 22 ps (even faster than 10^{13} K s^{-1}) followed by another decrease in temperature, slightly less abrupt, from

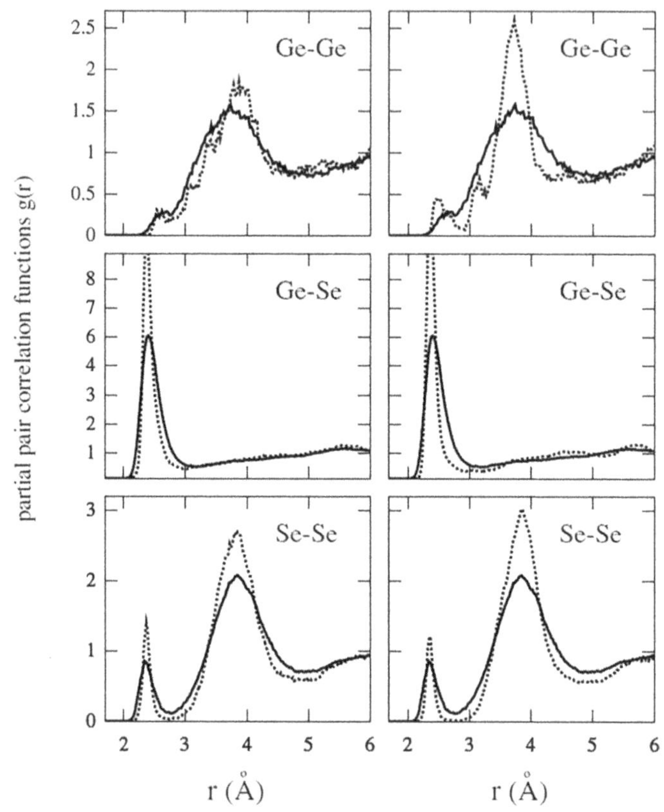

Figure 4.13. Dashed lines: pair correlation functions of glassy $GeSe_2$ at $T = 300$ K. On the left: average values $g_{GeGe}(r)$, $g_{GeSe}(r)$ and $g_{SeSe}(r)$ obtained for the first 5 ps of annealing at $T = 300$ K compared with the results for the equilibrated liquid [36]. On the right: average values $g_{GeGe}(r)$, $g_{GeSe}(r)$ and $g_{SeSe}(r)$ obtained for the last 12 ps of annealing at $T = 300$ K compared with the results for the equilibrated liquid [36]. In both cases, the averages are taken on a single temporal trajectory. Reprinted figure with permission from [38]. Copyright (2008) by the American Physical Society.

$T = 600$ K to $T = 300$ K in 22 ps. It is presumable that much longer cooling steps, as those recommended above, would have provided a structure distinct from the liquid much earlier along the quench. While being convinced this does not undermine the validity of the quoted investigation, it is a true fact that, in principle, a large share of FPMD calculations on glasses would benefit from some fresh update based on longer cooling trajectories.

To end this section, we would like to stress that there is no contradiction between a strong drive to recommend production of glassy states via cooling processes of the kind '1000 K over 1000 ps' and our own scientific production that we have developed by taking higher quench rates, typically by one order of magnitude.

First of all, the overall agreement between theory and experiments for structural properties have to be highlighted in this context, together with the awareness that the longest possible structural relaxation has to be allowed. Second, most computational material scientists active in the area have faced the same problem of reconciling the quest of knowledge with the availability of resources, without any sacrifice of quality in the description of bonding properties. Last but not least (and maybe quite subjectively) quench rates are not the only ingredients to work on when looking for an increased predictive power in the atomic-scale modeling of glasses. Nobody has ever hidden difficulties and failures in attempting to reach this goal, such intellectual honesty deserving at least some credit more than any allegation of lack of realism when using 'too high' quench rates.

4.5 Dealing with FPMD odds and ends (including non-adiabaticity): the case of SiN

In this chapter I have developed a series of considerations aimed at allowing a safe and rewarding use of FPMD in the CP scheme for disordered systems, covering methodology and practical implementation. While some examples have been provided with the help of appropriate figures I felt that the presentation of a case exemplifying most of the information conveyed was very much needed. It could also be quite useful for the reader to make sure the main messages of this chapter have been assimilated effectively and can be employed for calculations to be planned within this framework in the future. In particular, there is a point that needs to be addressed and treated explicitly since potentially undermining the entire FPMD scheme most advertised in this monograph: the notion of adiabaticity. Systems and thermodynamic conditions (temperature for instance) compatible with adiabaticity allows CP/FPMD simulations to be carried out with no fear of detecting any unwanted departure from the BO surface. We have also seen that idea of thermostat applied to the fictitious degrees of freedom can help in this context. However, we also invoked situations of gapless systems, which are very hard to be studied since not compatible with a strict separation between electronic and ionic degrees of freedom. Amorphous SiN is a representative example of system featuring different bonding natures as a function of temperature and, for these reasons, very well suited to cover most of the issues treated in this chapter [39].

For this system, we had in mind to produce a thermal cycle of the kind described in this chapter so as to obtain an amorphous state. However, when increasing the temperature to induce randomization and diffusion, we had to face non-adiabatic behavior hampering the use of the standard CP methodology. This difficulty was circumvented by employing at $T = 2500$ K the BO approach (optimization of the electronic structure for each ionic configuration as a prerequisite to the calculation of the atomic forces). Starting from this sample at high temperature, a quenching schedule was implemented for different models, leading to a meaningful description of the amorphous state, as it will be described in what follows.

4.5.1 State of the art and calculations

Interest in amorphous silicon nitride is motivated by its nature of a dielectric material employed as an insulator in several applications [40]. This system was chosen in view of our studies on thermal properties reported in chapter 12 that followed the structural investigations largely described in this monograph. As a prerequisite to calculations of thermal conductivity and related heat transport processes at the atomic scale (see, for instance the case of amorphous $Ge_2Sb_2Te_5$ [41]), amorphous SiN was selected as a representative non-stoichiometric silicon nitride Si_xN_{1-x} system [39]. By applying FPMD to SiN one can go beyond empirical potentials [42] intrinsically unable to account for the presence of homopolar bonds at non-stoichiometric compositions. Concerning the use of the BO methodology invoked at the beginning of this section to cope with the limits of the CP approach in terms of adiabaticity, the choice of BO could have been adopted from the start, instead of employing two different schemes, CP and BO, the second option being justified by the inadequacy of the first at high temperatures. However, by employing two techniques, one gains a precious insight into the advantages and drawbacks of each one of them, in a way quite instructive and fairly unique, since papers based on both FPMD (CP and BO) are rare. For these reasons the account of this work is at its right place in this chapter. Amorphous SiN had been the focus of previous FPMD results by Hintzsche *et al* [43] and Jarolimek *et al* [44] that need to be extended and improved for the following reasons. In [43] a periodic model of SiN with $N_{at} = 200$ atoms is brought at $T = 4000$ K and then to 3000 K before a further reduction of temperature ending at $T = 2000$ K. The mobility is considered as negligible at $T = 2000$ K. One could argue that describing an amorphous system on the basis of configurations at such high temperature can be different from the topology at room temperature. This point is substantiated by the notion of residual structural relaxation that can occur as a function of time even for a system that appears as configurationally arrested (see figure 4.13 and related parts in section 4.4.5). The same criticism applies to the FPMD models of [44] on hydrogenated amorphous SiN since the step at 300 K lasts only 0.5 ps, calling for more extended trajectories available at room temperature to calculated statistical averages. This is what we have achieved in [39].

4.5.2 Methodology and the appropriate FPMD schemes

We use periodic orthorombic atomic models containing N_{at} =252 and N_{at} =340 atoms with dimensions 10.0 ×20.0 ×15.0 Å^3 and 10.0 ×20.0 ×20.0 Å^3, having in mind upcoming applications of the approach-to-equilibrium molecular dynamics (AEMD) methodology [14, 41, 45–48]. In line with the CP theoretical framework described in chapter 3 we had to make some choices for the exchange–correlation part of the Kohn–Sham energy functional (we selected BLYP [19, 20]), the pseudopotentials (devised by Troullier and Martins [28]) and the cutoff for the plane waves expansion (50 Ry with expansion at the Γ point only). The mass μ is equal to to 800 a.u. and the timestep to 5 a.u. (0.12 fs). Temperature control is ensured as detailed in section 3.5.5. The initial configuration has been selected from previous sets of atomic positions made available at the FPMD level in recent years for systems of different chemical nature. In this respect, amorphous SiN falls under case **(e)** of section 4.3.

In addition to the CP methodology we have resorted to the BO one [49], briefly touched upon before in this book (see for instance a short sentence in section 4.5) but several times invoked implicitly (as the natural solution) when considering the lack of adiabaticity that might affect the CP scheme at high temperature or in the absence of a gap in the EDOS. More precisely, BO dynamics means ions evolving dynamically as a result of forces obtained, at each discretized instant of time, at the electronic ground state. It appears clearly that no fictitious electronic degrees of freedom are necessary to implement a dynamical evolution, thereby opening the way for larger time steps. What has been achieved when studying amorphous SiN demonstrates the validity of using both CP and BO, since the second allows producing FPMD trajectories at ionic temperatures too high to be treated via the CP methodology for this specific system.

The thermal cycle is performed as shown in figure 4.14 for N_{at} =340. For both Si and N the mean square displacement of the Si and N atoms above $T = 900$ K does not exceed 2 to 3 Å^2 for temperatures as high as 1600 K, with diffusion coefficient $\approx 5 \times 10^{-7}$ cm^2 s^{-1}. This means that to attain liquid-like mobilities one has to increase the temperature even more. This is exactly where the methodological transition from the CP to the BO schemes becomes necessary since at $T = 2000$ K, $E_{fic}(\psi_i(\mathbf{r}))$ (see equation (3.28)) diverges in time (bottom part of figure 4.15) despite the use of the thermostat. Therefore, the dynamical behavior of the system becomes non-adiabatic in clear contrast with the situation at T = 300 K (top part of figure 4.15). We are observing here a typical phenomena of gap closing visible in the electronic density of states as shown in figure 4.16.

Given these premises, the BO FPMD methodology is the right choice to handle high temperatures with the electronic structure evolving according to the Kohn–Sham equations. The BO part of the full FPMD thermal cycle is shown in figure 4.14. The BO dynamics start at T = 1500 K as a test temperature to calibrate the simulation parameters before moving to 2500 K. The timestep and the convergence tolerance ΔE_{tot} for the optimization of of the electronic structure at its ground state were equal to 100 a.u. and 2×10^{-6} a.u. respectively, leading to fluctuations of the total energy smaller than 5×10^{-4}. It is worth pointing out that the larger values of the timestep

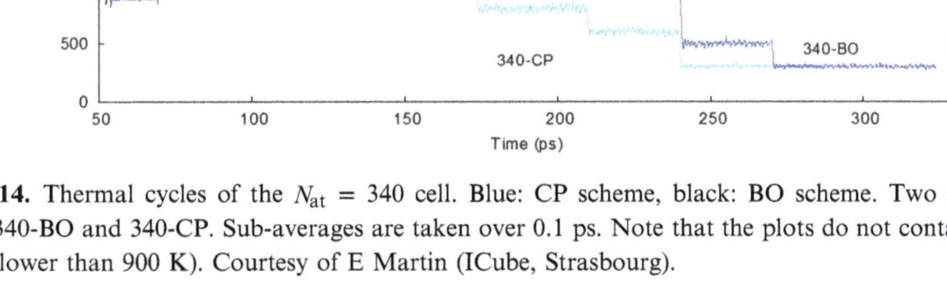

Figure 4.14. Thermal cycles of the N_{at} = 340 cell. Blue: CP scheme, black: BO scheme. Two models are created: 340-BO and 340-CP. Sub-averages are taken over 0.1 ps. Note that the plots do not contain the first 50 ps (T lower than 900 K). Courtesy of E Martin (ICube, Strasbourg).

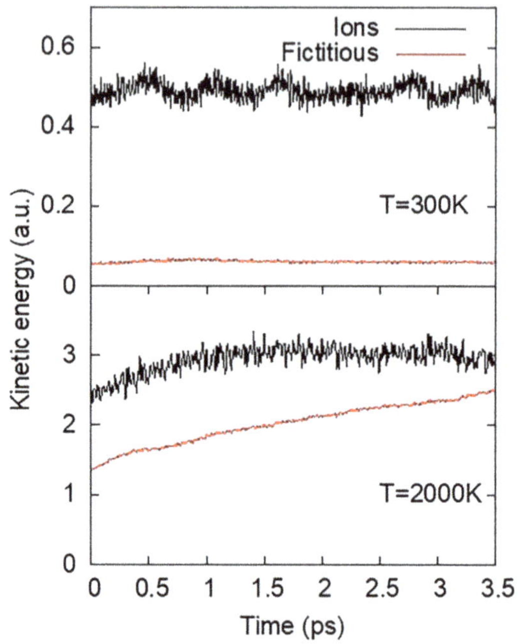

Figure 4.15. Ionic (black curves) and fictitious electronic (red curves) kinetic energies at $T = 300$ K (top) and 2000 K (bottom). $N_{at} = 340$. Courtesy of E Martin (ICube, Strasbourg).

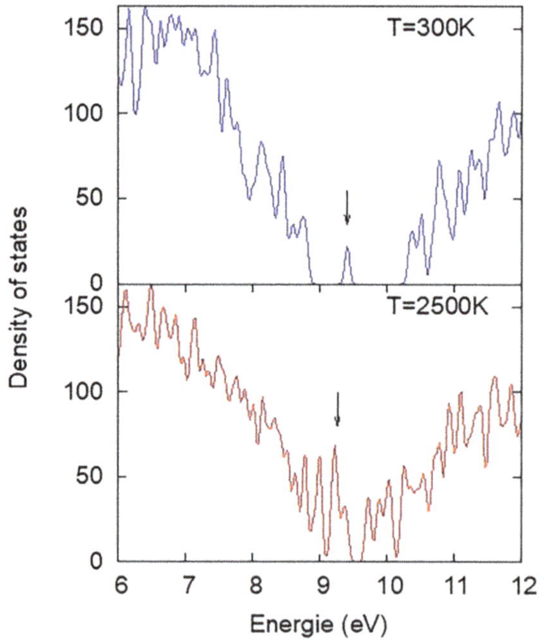

Figure 4.16. Electronic density of states at the vicinity of the gap at 300 and 2500 K. A Gaussian broadening of 50 meV has been applied. The arrows indicate the energy of the highest occupied eigenstate. The system considered contains $N_{at} = 340$ atoms. Courtesy of E Martin (ICube, Strasbourg).

when using BO framework are quite convenient with respect to the CP case. However, the minimization of the electronic structure can affect the conservation of the total energy, strongly dependent on the value of ΔE_{tot}.

At $T = 2500$ K the mean square displacement lies in between ≈3 and ≈6 Å^2 for N and ≈4 and ≈8 Å^2 for Si. The higher diffusion of the Si atoms is due to Si atoms in homopolar bonds, more mobile than in configurations featuring bonds with four N atoms. The rest of the quenching schedule takes place entirely with the CP scheme down to $T = 300$ K for a total trajectory of ≈175 ps, i.e. a quench rate of 12 K ps^{-1}. The structure obtained is labeled 340-BO. A second structure (system 340-CP) is obtained with the CP methodology by starting to quench at 2000 K. The corresponding quench rate is equal to 10 K ps^{-1}. Both 12 K ps^{-1} and 10 K ps^{-1} ensure a reliable amorphous phase at least for Si-based system, as shown in [50]. A similar quenching schedule is implemented for $N_{at} = 252$ (see [39]). In this case CPMD is performed up to $T = 2000$ K, by turning to the BO one at $T = 2500$ K during 10 ps to increase atomic diffusion. The temperature is reduced to 300 K in 110 ps for the fast (f) cooling (system 252-f) or 170 ps for the slow (s) cooling (system 252-s).

Having found the right methodology to create a thermal cycle without being affected by the intrinsic limits of the CP approach, one can safely study the atomic structure of the system by putting to good use some of the tools introduced in sections 2.3 and 2.4. The partial pair correlation functions (PCFs) $g_{SiSi}(r)$, $g_{NN}(r)$ and $g_{SiN}(r)$ are shown in figure 4.17 for the four models. The overall agreement is confirmed by the coordination numbers n_α^β reported in table 4.1.

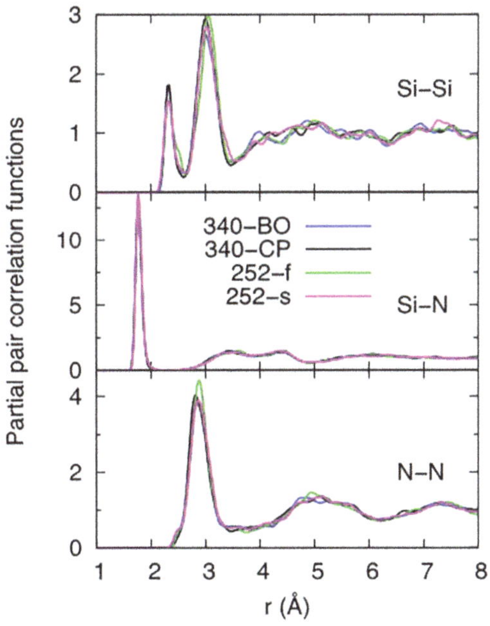

Figure 4.17. Partial pair correlation functions of the four models of amorphous SiN. Courtesy of E Martin (ICube, Strasbourg).

Table 4.1. Coordination numbers n_α^β obtained by integrating up to the first minima of the pair correlation functions $g_{\alpha\beta}(r)$. Reprinted with permission from [39]. Copyright (2022) by Elsevier.

	n_{Si}^{N}	n_{Si}^{Si}	n_{N}^{N}
Reference [39]			
252-f	3.02	1.19	0.0
252-s	3.02	1.14	0.0
340-CP	3.02	1.12	0.0
340-BO	3.02	1.12	0.0
Chemically ordered network			
	3	1	0
Previous results			
Reference [43]	2.99	1.01	0.0
Reference [44]	2.90	0.88	0.0

The coordination units n_α^β (defined in section 2.4.1) are close to those obtained within the chemically ordered network model (CON). By anticipating a concept that will be further discussed in section 6.5, the CON model gives a measure of the existence of bonds between atoms of the same or different chemical nature, and corresponds to maximizing the number of heteropolar ones. For Si_xN_{1-x} systems, stoichiometry occurs for $x = 3/7$ and, in this case (Si_3N_4), the CON model leads to each Si atom with four N neighbors and each N atom with three Si neighbors, with the absence of Si–Si or N–N homopolar bonds. Considering the $x = 0.5$ concentration, the CON model implies that only Si–N and Si–Si bonds are allowed for $x > 3/7$ while only Si–N and N–N bonds are allowed for $x < 3/7$. Note in table 4.1 that the coordination numbers found in the presence of hydrogen ([44]) are slightly smaller. The network topology can be better understood by relying on the $n_\alpha(l)$ structural units (also defined in section 2.4) where an atom of species α (say, Si or N) is l-fold coordinated to other atoms, the average being calculated over the entire trajectory. For instance we shall define 1N2Si $n_{Si}(3)$ as a unit in which a Si atom is linked to 1 N atom and 2 Si atoms and 4N $n_{Si}(4)$ means one Si atom connected to 4 N atoms.

4.5.3 Focus on the coordination units

Figure 4.18 shows $n_\alpha(l)$ for $l = 1, 2, 3, 4$ and 5 and α = Si or N for the system 340-BO. Most of N atoms are threefold coordinated in between 2 Å and 3 Å. At small r, $n_N(1)$ and $n_N(2)$ can also be found in reduced fractions, as it is the case for $n_N(4)$ and $n_N(5)$ when r is close to the first minimum of $g_{NN}(r)$. The main information of figure 4.19 concerns the structural motif 3Si $n_N(3)$ that is the most frequent for the generic structural unit for $n_N(3)$. It appears that N is coordinated as in stoichiometric Si_3N_4, by forming three bonds with silicon atoms.

The dominant coordination unit for Si is $n_{Si}(4)$ as shown in figure 4.18. What one observes for $n_{Si}(1)$ and $n_{Si}(2)$ is correlated with the shortest bonds with the closest

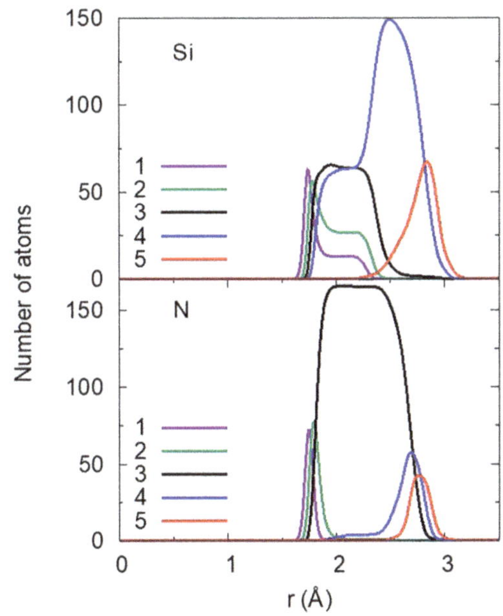

Figure 4.18. Coordination units $n_{Si}(l)$ and $n_N(l)$ atoms (bottom) for $l = 1, 2, 3, 4, 5$. System 340-BO at $T = 300$ K. Courtesy of E Martin (ICube, Strasbourg).

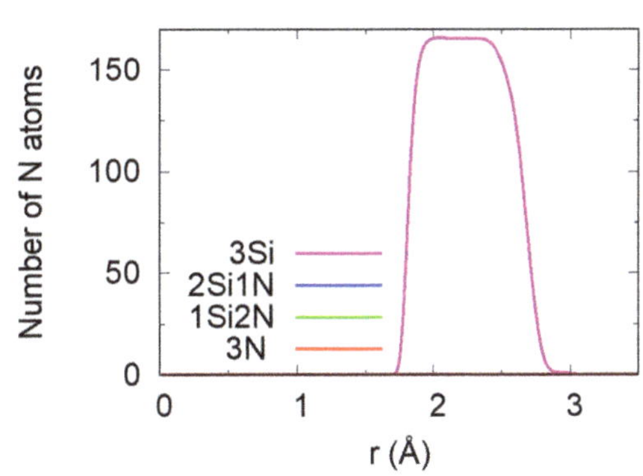

Figure 4.19. Decomposition of $n_N(3)$ by accounting for the chemical identity of the l neighboring atoms. The system considered is 340-BO at 300 K. Courtesy of E Martin (ICube, Strasbourg).

neighbors. On the other hand, the profile of $n_{Si}(5)$ is related to Si neighbors of the second shell responsible for the peak at 3 Å on $g_{SiSi}(r)$. The percentages of the different various coordination units $n_{Si}(4)$ are given in figure 4.20. About 30%–40% of Si atoms are found coordinated to 4 N atoms as for the stoichiometric composition. To summarize on the coordination units of amorphous SiN, the system is chemically ordered and made of N atoms mostly coordinated with three Si atoms (as in Si_3N_4), while Si atoms form mostly two structural units, either connecting four N atoms or three N atoms and one Si atoms, lying at larger distances.

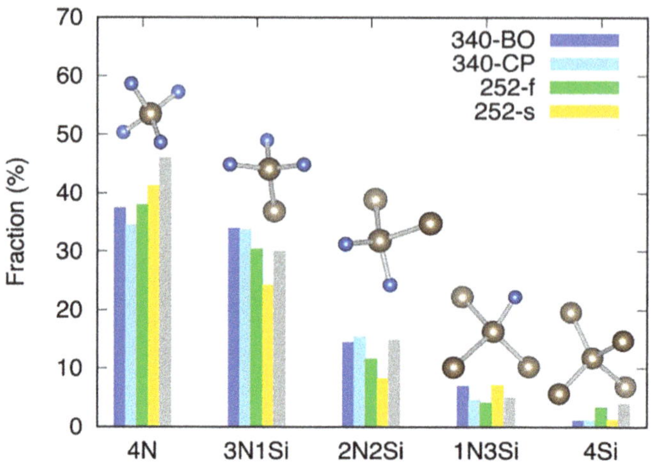

Figure 4.20. Fractions of the various environments of fourfold coordinated Si atoms in the four models and comparison with the results obtained in [43] (gray column). Si are brown and the N atoms are blue. Reprinted with permission from [39]. Copyright (2022) by Elsevier.

4.5.4 What to learn from the case of SiN?

Providing results on the structure of a disordered system like SiN at this location of the book deserves some more thoughts, even though the presence of some examples to substantiate the concepts expressed in this chapter were announced when presenting the full content in chapter 1. After dwelling on theory (chapter 3) and practical implementation (chapter 4), I realized that one complete realization would have been more compelling than the cases already presented to address specific points. Amorphous SiN is very well suited to meet this goal, since the application of FPMD to this system allows for probing the boundary of validity of the CP approach by making explicit all advantages and possible shortcomings of the methodology, at least within the theoretical framework presented in chapter 3. In short, almost all aspects treated within chapter 4 (including ways to account for lack of adiabaticity) can be traced back by following the behavior of amorphous SiN.

To summarize, we have built four models and produced a thermal cycle for each one of time. As we have seen in sections 4.3 and 4.4, a thermal cycle starts for a structure having the right chemical composition and some stable topology and consists in reaching a high temperature, achieving significant diffusion and cooling down to room temperature to obtain a disordered state. Our calculations are characterized by much longer trajectories with respect to previous FPMD results thereby highlighting the importance of constructing a thermal cycle according to the criteria exposed in this chapter. From the standpoint of methodology, the most interesting issue addressed is related to what happens when increasing the temperature. To loose track of the initial configuration and melt the system, the temperature was increased up to $T = 2500$ K. This drives the implementation of the BO approach as a methodology not requiring adiabaticity since it is based on the search of the ground state electronic structure for each ionic configurations.

4.6 The CPMD code and some thoughts on how to approach the 'code issue': an autobiographical perspective

Talking about codes can be somewhat misleading. Some people think that a code means a theory and vice versa. I tend to be aggravated when hearing about computational material scientists exchanging on the basis of the question *what kind of code are you running?* as if the concepts underlying the code could be all packaged inside informatics. The temptation of using codes by knowing little or, worse, nothing of what they contain leads to the belief that all the needed knowledge amounts to controlling the input file. This is a bad tendency to be refuted since using a MD code (especially when FPMD comes to play) cannot be assimilated to driving a car and assuming the only action to be undertaken is to fill the tank. Having launched this warning it is now the right time to say a few words about my experience with the notion and the use of codes, ending up with some introductory words to CPMD, the most invoked one in this monograph.

When I started working on CMD, almost each user, regardless of the age and status, was writing his own code as a form of unavoidable exercise. The idea was to make sure that the concepts had been well assimilated and the practical realization controlled and adapted at will. Coding was done in 'scalar' or 'vectorial mode' (this latter amounting to preferring even redundant 'do loops' so as to avoid jumping back and forth with 'go to's' unduly slowing down the sequence of operations). I remember introducing in a CMD code doing microcanonical dynamics the equations for implementing NVT and NPT dynamics together with all changes to make it compatible with the vectorial programming. I was using at that time (1986–90) French national computer centers that had adopted the VP2000 Fujitsu vectorial supercomputer, replaced a few years later by his massively parallel counterpart. This kind of 'home made' code, almost maniacally tested after any modification, was further extended and employed for the calculations detailed in section 5.2 with n-body forces. I got involved first with FPMD calculations in 1993 having as tool the early vectorial version of a CP code mostly elaborated in Trieste by the two inventors and a wealth of collaborators. I was given to learn and use on CRAY local machines the so-called CPVan code elaborated by Alfredo Pasquarello and Kari Laasonen. This was devised to account for the Vanderbilt 'ultrasoft' scheme in which the electronic density had two parts, a traditional delocalized part based on plane waves and a more localized one [51][10]. The associated input file was not particularly user friendly, containing only numbers to be specified and no keywords, but it could be easily handled once the main features of the system to study were known (such as number of particles, system size etc) and the parameters of the electronic structure (energy cutoff, number of electronic states) plus of course details on the dynamics to be implemented and a few logical links to input files determining the electronic structure (pseudopotentials). However, the notion of library containing choices for specific quantities (as the XC functionals) was lacking and each new implementation was simply added but

[10] Needless to say, CPVan could also be used as a standard norm-conserving code, as was done to obtain the results presented in chapter 6.

not systematized. I remember spending several months deciphering the CPVan code line by line in order to connect Fortran instructions and the underlying theory as much as I could. In this way, I learned a lot about both coding and theory, a step that was very much needed in my case since I was coming from statistical mechanics as a background (I had coded my own CMD tools way too simply when compared to CP/CPVan) and I had a limited knowledge of advanced programming and DFT. This training period culminated in the writing of some new routines, as required, for instance, to calculate via a self-consistent approach on frozen valence states the photoelectron spectra of Cu clusters [52].

While pursuing the use of CPVan (at the origin of all results presented in chapter 6), contacts I established with the team of M Parrinello at MPI in Stuttgart, during the period 1996–99 gave me the opportunity to get familiar with the CPMD code[11], fully parallel and much more evolved, in terms of user interface (inputs based on keywords easily adjustable to any system and simulation conditions, manual with explanations and examples available, consortium organization, libraries of pseudopotentials and exchange–correlation functionals available) than its predecessors. Mauro Boero played a key role at that time, opening up a collaboration on computational materials science issues lasting ever since and taking advantage of CPMD in its various versions as its main operational tool. I have to admit that working on CPMD with the notions acquired via the in-depth study of CPVan gave me the certitude of not being dependent on mere keywords since knowing their conceptual meaning in most cases. This is why I tend to be (unduly) hostile to the use of 'black boxes' computer codes, even though there are many knowledgeable users around that are in full control of the theory without knowing the exact content of their computational tool.

4.6.1 Inside CPMD: the essentials

CPMD is extremely friendly user, to the point that with a few clear guidelines in mind, almost any undergraduate can run at least a test calculation. In what follows I will describe the organization of the input with the help of a few examples. My only intention is to show how CPMD bridges the gap between the accessible use of a

[11] The CPMD code started as a project by IBM Zurich Research laboratory in 1993. Since version 2.0 new features were introduced including thermostats and diagonalization procedures for the explicit calculation of the Kohn–Sham eigenstates (occupied and empty). With version 2.5 the serial code became parallelized at IBM with the implementation of the MPI (Message Passing Interface) in 1994 by taking the form of a stable package making use of keywords and parameters expected to be provided in the input at the place of user-unfriendly numerical and logical (.TRUE. or .FALSE.) strings. Starting from version 3.0, the constant pressure MD within the free energy MD and the *k*-points handling was added as well as an efficient path integral version using two level parallelism. With the next release, version 3.3, Maximally Localized Wannier functions were also included. The first official version (3.5) downloadable from http://www.cpmd.org appeared in 2002. Version 3.11 featured hybrid compilations MPI/OpenMP supported on a large number of platforms worldwide and implemented both in the main code and in the QM/MM interface. Until version 3.17, the major programming language was Fortran 77. A complete restyling in terms of enhanced Fortran 90 was carried out in recent versions until the last release, version 4.3 in which Fortran 90/95 was upgraded to include all major innovations of Fortran 2003, 2008, 2018 and 2022. We recall that Fortran 2018 is regarded now as the regular standard. All information courtesy of Mauro Boero.

computer tool aimed at 'getting things done' and a tremendous amount of theoretical work combining DFT, MD and high-performance parallel programming. The tutorial available at http://www.cpmd.org can do the rest (for instance, it describes in detail the output files containing the coordinates of the atoms and other information on the electronic structure). I do advise keeping up the direct contact (questions, comments) with the developers (I do not belong to this category, while my close colleague Mauro Boero (mauro.boero@ipcms.unistra.fr) does, and his contribution in this context is invaluable)[12].

As visualized in figure 4.21, the CPMD input is composed of (at least) four mandatory sections (CPMD, DFT, SYSTEM, ATOMS). (CPMD) contains the instructions on the kind of calculation one is willing to perform. Based on what explained in the previous sections of this chapter (section 4.3) and in chapter 3, we have identified several steps to setup and achieve a structural characterization of a glassy structure (optimization of the electronic structure for a given initial configuration, optimization of an initial structure to minimize the forces, CP dynamics without and with temperature control). All these can be easily implemented in the CPMD input.

Therefore, each one of these actions corresponds to a specific input, examples being given in figures 4.22 and 4.23 for wavefunction optimization and CP dynamics with temperature controls on the ions (the other situations invoked being closely related to these two).

The (DFT) section is the one indicating how the DFT calculations will be carried out, in particular which kind of exchange–correlations (or hybrid) functionals will be

```
&CPMD
  ...
&END

&DFT
  ...
&END

&SYSTEM
  ...
&END

&ATOMS
  ...
&END
```

Figure 4.21. A schematic view of the four mandatory sections in the input file of CPMD.

[12] All figures of this section have been extracted (and modified) from a presentation kindly provided by M Boero.

```
&CPMD
  OPTIMIZE WAVEFUNCTION
  INITIALIZE WAVEFUNCTION ATOMS
  MAXSTEP
    50000
  PCG
  CONVERGENCE ORBITALS
    1.0E-05
  EMASS
    500
&END
```

Figure 4.22. Example of (CPMD) section within the input of CPMD. These instructions allow minimizing the electronic structure (search of the electronic ground state) by using the preconditioned conjugate gradient method (PCG). Minimization ends when the largest gradient on the orbitals is smaller than 10^{-5}. EMASS is the variable μ of equation (4.2). The initial choice for the wavefunctions is in this case an atomic basis set, but this can modified for instance by invoking a set of random numbers to start with.

```
&CPMD
  MOLECULAR DYNAMICS CP
  RESTART WAVEFUNCTION COORDINATES LATEST
  NOSE IONS
   300  1000
  MAXSTEP
    100000
  TIMESTEP
    4.0
  EMASS
    500
&END
```

Figure 4.23. Example of (CPMD) section within the input of CPMD. These instructions allow performing CPMD with a temperature control at $T = 300$ K and a frequency associated to the Nosé thermostat equal to 1000 cm^{-1} (see section 3.5.5). MAXSTEP is the number of iterations with a timestep TIMESTEP equal to 4 a.u. The RESTART keywords allows reading from a stored file the values of relevant variables (coordinates, velocities etc) so as to pursue the trajectory. LATEST is a keyword specifying the identity of the RESTART file.

```
&DFT
  FUNCTIONAL BLYP
  GC-CUTOFF
  0.1E-06
&END
```

Figure 4.24. Example of the (DFT) section within the input of CPMD. The indicated cutoff is used to get rid of noise beyond the accuracy of DFT when using the given XC functional.

used. It can take a very simple and descriptive form, as in figure 4.24. The (SYSTEM) section (see figure 4.25, top) contains information on the geometry of the system and the cutoff energy referred to in section 4.2.2.

```
&SYSTEM
  ANGSTROM
  SYMMETRY
    1
  CELL ABSOLUTE DEGREE
    10.00  10.00  10.00  90.00  90.00  90.00
  CUTOFF
    100.0
  &END
&ATOMS
  *GE-MT-BLYP.pps   KLEINMAN-BYLANDER
  LMAX=P  LOC=P
  1
  4.7367371096    6.6107492731    4.6811927823

  *SE-MT-BLYP.pps   KLEINMAN-BYLANDER
  LMAX=P  LOC=P
  4
  -5.9669982281    -8.7467531007    11.2938278730
   1.8904372056   -10.8283093176     4.1633837562
   5.4988512174     4.9899517884    10.6568153345
   0.0059955793    -0.1450182640     4.3887246936
&END
```

Figure 4.25. Top : Example of the SYSTEM section with a choice of the symmetry of the periodic box corresponding to a cubic shape. The size of the edges (10 Å) and the angles between the edges are provided. 100 Ry is the cutoff value for the expansion of the basis set in plane waves. Bottom: Example of the ATOMS section for a test system made of 1 Ge atom and 4 Se atoms. The pseudopotentials are defined with respect to the choice for the exchange–correlation functional, MT indicating the generation method of [28]. For the significance of KLEINMAN–BYLANDER, LMAX, LOC see sections 3.6.3 and 11.2.

Finally the (ATOMS) section (see figure 4.25, bottom) is the one describing the kind of pseudopotential used and providing the atomic coordinates for each atomic species in sequential order.

Besides the minimal description provided above, that in any case allows running FPMD calculations in the CP scheme for disordered structures with the goal of obtained coordinates evolving in time (exactly what is needed to obtain the structure), the potentiality of the CPMD code is very large. For instance, a wealth of properties can be obtained, involving correlations between structure, vibration and electronic behaviors (dipole dynamics and infrared spectra, for instance). On the side of the dynamical evolution, BO and metadynamics [53] are easily accessible among the others. In what follows, essentially all FPMD contained in chapters 7 through 11 have been obtained using the CPMD code within the CP formalism, unless specifically pointed out.

References

[1] Behler J and Parrinello M 2007 *Phys. Rev. Lett.* **98** 146401

[2] Noé F, Tkatchenko A, Müller K-R and Clementi C 2020 *Annu. Rev. Phys. Chem.* **71** 361–90

[3] Bartók A P, Kermode J, Bernstein N and Csányi G 2018 *Phys. Rev.* X **8** 041048

[4] Zhang L, Han J, Wang H, Car R and Weinan E 2018 *Phys. Rev. Lett.* **120** 143001

[5] Ori G, Massobrio C, Bouzid A, Boero M and Coasne B 2014 *Phys. Rev.* B **90** 045423
[6] Foiles S M, Baskes M I and Daw M S 1986 *Phys. Rev.* B **33** 7983–91
[7] Massobrio C and Blandin P 1993 *Phys. Rev.* B **47** 13687–94
[8] Blandin P, Massobrio C and Ballone P 1994 *Phys. Rev. Lett.* **72** 3072–5
[9] Feibelman P J 1994 *Surf. Sci.* **313** L801–5
[10] Sarnthein J, Pasquarello A and Car R 1995 *Phys. Rev.* B **52** 12690–5
[11] Giacomazzi L, Umari P and Pasquarello A 2009 *Phys. Rev.* B **79** 064202
[12] Bertani M, Menziani M C and Pedone A 2021 *Phys. Rev. Mater* **5** 045602
[13] Pedone A, Malavasi G, Menziani M C, Cormack A N and Segre U 2006 *J. Phys. Chem.* B **110** 11780–95
[14] Martin E, Ori G, Duong T-Q, Boero M and Massobrio C 2022 *J. Non-Cryst. Solids* **581** 121434
[15] Copyright MPI-FKF (1997–2001). CPMD Copyright IBM Corp. (1990–2022)
[16] Perdew J P and Wang Y 1992 *Phys. Rev.* B **45** 13244–9
[17] Perdew J P, Chevary J A, Vosko S H, Jackson K A, Pederson M R, Singh D J and Fiolhais C 1992 *Phys. Rev.* B **46** 6671–87
[18] Perdew J P, Burke K and Ernzerhof M 1996 *Phys. Rev. Lett.* **77** 3865–8
[19] Becke A D 1988 *Phys. Rev.* A **38** 3098–100
[20] Lee C, Yang W and Parr R G 1988 *Phys. Rev.* B **37** 785–9
[21] Massobrio C, Pasquarello A and Car R 1999 *J. Am. Chem. Soc.* **121** 2943–4
[22] Krukau A V, Vydrov O A, Izmaylov A F and Scuseria G E 2006 *J. Chem. Phys.* **125** 224106
[23] Marx D and Hutter J 2009 *Ab Initio Molecular Dynamics: Basic Theory and Advanced Methods* (Cambridge: Cambridge University Press)
[24] Massobrio C, Pontikis V and Martin G 1990 *Phys. Rev.* B **41** 10486–97
[25] Dubbeldam D, Calero S and Vlugt T J H 2018 *Mol. Simulat.* **44** 653–76
[26] Andersen H C 1980 *J. Chem. Phys.* **72** 2384–93
[27] Parrinello M and Rahman A 1980 *Phys. Rev. Lett.* **45** 1196–9
[28] Troullier N and Martins J L 1991 *Phys. Rev.* B **43** 1993–2006
[29] Goedecker S, Teter M and Hutter J 1996 *Phys. Rev.* B **54** 1703–10
[30] Krack M 2005 *Theor. Chem. Acc.* **114** 145–52
[31] Pastore G, Smargiassi E and Buda F 1991 *Phys. Rev.* A **44** 6334–47
[32] Blöchl P E and Parrinello M 1992 *Phys. Rev.* B **45** 9413–6
[33] Le Roux S, Bouzid A, Boero M and Massobrio C 2013 *J. Chem. Phys.* **138** 174505
[34] Le Roux S, Zeidler A, Salmon P S, Boero M, Micoulaut M and Massobrio C 2011 *Phys. Rev.* B **84** 134203
[35] van Roon F H M, Massobrio C, de Wolff E and de Leeuw S W 2000 *J. Chem. Phys.* **113** 5425–31
[36] Massobrio C, Pasquarello A and Car R 2001 *Phys. Rev.* B **64** 144205
[37] Li Zhong J, Wang H, Sheng Z, Zhang and Mao S X 2014 *Nature* **512** 177–80
[38] Massobrio C and Pasquarello A 2008 *Phys. Rev.* B **77** 144207
[39] Lambrecht A, Massobrio C, Boero M, Ori G and Martin E 2022 *Comput. Mater. Sci.* **211** 111555
[40] Burr G W *et al* 2016 *IEEE J. Emerg. Sel. Topics Circuits Syst.* **6** 146–62
[41] Duong T-Q, Bouzid A, Massobrio C, Ori G, Boero M and E Martin 2021 *RSC Adv.* **11** 10747–52
[42] Dasmahapatra A and Kroll P 2018 *Comput. Mater. Sci.* **148** 165–75

[43] Hintzsche L E, Fang C M, Watts T, Marsman M, Jordan G, Lamers M W P E, Weeber A W and Kresse G 2012 *Phys. Rev.* B **86** 235204
[44] Jarolimek K, de Groot R A, de Wijs G A and Zeman M 2010 *Phys. Rev.* B **82** 205201
[45] Lampin E, Palla P L, Francioso P-A and Cleri F 2013 *J. Appl. Phys.* **114** 033525
[46] Bouzid A, Zaoui H, Palla P L, Ori G, Boero M, Massobrio C, Cleri F and Lampin E 2017 *Phys. Chem. Chem. Phys.* **19** 9729–32
[47] Palla P L, Zampa S, Martin E and Cleri F 2019 *Int. J. Heat Mass Transf.* **131** 932–43
[48] Duong T-Q, Massobrio C, Ori G, Boero M and Martin E 2019 *Phys. Rev. Mater* **3** 105401
[49] Born M and Oppenheimer 1927 *R. Ann. Phys., IV. Folge* **389** 457–84
[50] Xue K, Niu L-S and Shi H-J 2008 *J. Appl. Phys.* **104** 053518
[51] Laasonen K, Pasquarello A, Car R, Lee C and Vanderbilt D 1993 *Phys. Rev.* B **47** 10142–53
[52] Massobrio C, Pasquarello A and Car R 1995 *Phys. Rev. Lett.* **75** 2104–7
[53] Behler J, Martoňák R, Donadio D E and Parrinello M 2008 *Phys. Rev. Lett.* **100** 185501

IOP Publishing

Carlo Massobrio

Chapter 5

Cases treated via classical molecular dynamics

By following a presentation strategy that is strongly rooted on direct experience of calculations that I conceived and carried out, this chapter is devoted to two cases treated via classical molecular dynamics (CMD), by using a two-boy potential and one of the first n-body potentials ever designed to model metallic alloys within MD. In the first case (a monoatomic Lennard-Jones argon system) the interest is mainly academic and the overall motivation of the calculation is closer to statistical mechanics than to materials science. This is because the short timescales of MD allow producing a system (glassy 'argon') configurationally arrested having (at least at the time the relevant results were produced) no corresponding experimental evidence. The effect of the quench rate on the stability of this monoatomic glass was studied in detail, as well as the effect of the difference of temperature between two adjacent trajectories. The second part of the chapter is devoted to the simulation of amorphization via introduction of chemical disorder in crystalline $NiZr_2$. In this case, an effective potential constructed along the scheme detailed in section 3.2 is employed to demonstrate that an amorphous state can be created directly from the crystal via the introduction of antisite defects. This leads to a volume expansion followed by the loss of topological order. Calculations of elastic constants are presented to substantiate these results in terms of an elastic instability.

5.1 Learning about glasses from a Lennard-Jones monoatomic system

5.1.1 Simple and instructive: a monoatomic glass model

One of the simplest (and cheapest) systems to get some generic insight into issues relative to glass formation and stability is a monoatomic collection of two-body interacting particles described via a Lennard-Jones (LJ) potential containing two parameters, ϵ and σ, conveniently fitted to argon ($\epsilon = 125\ \mathrm{k_B}$ and $\sigma = 3.45$ Å). These values for energy and distances are employed to define a set of reduced units to which we shall refer whenever useful in the following ($T^* = k_B T/\epsilon$ for the

doi:10.1088/978-0-7503-2436-6ch5

temperature, $\rho^* = N_{\mathrm{at}}\sigma^3/V$ for the density). The total potential energy of the system (purely classical, all electronic structure features being hidden into the quasi-spherical interaction between rare gas atoms) is

$$E_{\mathrm{pot}} = \frac{1}{2}\sum_{i=1}^{n}\sum_{j(\neq i)=1}^{n} 4\epsilon\left(\left(\frac{\sigma}{r_{ij}}\right)^{12} - \left(\frac{\sigma}{r_{ij}}\right)^{6}\right) \tag{5.1}$$

allowing already for extensive simulations to be carried out on timescales of thousands of ns 30 years ago, when we focused on the existence of a glassy structure and its structural stability in connection with the diffusion properties [1]. With a timestep close to 10^{-13} s ($\Delta t = 0.85 \times 10^{-14}$ s) one was able to draw conclusions based on trajectories as long as 10^{-8} s with a number of atoms $N_{\mathrm{at}} = 864$. The LJ phase diagram of argon at atmospheric pressure can be characterized in terms of a realistic melting temperature of about 85 K, a full density/temperature plot being accessible via CMD in the NPT (constant pressure constant temperature) ensemble, obtained via a generalization to constant pressure of the ideas of temperature control considered in section 3.4.1. To this purpose it is instructive to recall the set of relevant equations of motion, in which the $\mathbf{r}_i$ variable is expressed through the relationship $\mathbf{r}_i = V^{1/3}\mathbf{q}_i$ valid for a cubic cell of side $L = V^{1/3}$, adjusting in time to ensure that (on average) the pressure reaches the target value P_{ex} [2]

$$\ddot{\mathbf{q}}_i = -\frac{L}{m}\frac{\partial E(r_{ij})}{\partial \mathbf{r}_i} - \dot{\mathbf{q}}_i\frac{2V}{3\dot{V}} - \frac{\dot{s}\dot{\mathbf{q}}_i}{s} \tag{5.2}$$

$$Q\ddot{s} = Q\frac{\dot{s}^2}{s} + \sum_i \dot{q}_i^2 mL^2 - (3N_{\mathrm{at}} - 3)k_{\mathrm{B}}Ts \tag{5.3}$$

$$W\ddot{V} = W\frac{\dot{s}\dot{V}}{\dot{s}} + \frac{s^2}{3V}\sum_i \dot{q}_i^2 mL^2 + s^2\sum_i \frac{\mathbf{F}_i \cdot \mathbf{r}_i}{3V} - s^2 P_{\mathrm{ex}} \tag{5.4}$$

where Q and W are the pseudomasses associated to s and V respectively (Q being the same as η defined in section 3.4.1).

Before moving on with the account of further calculations, it is important to realize that, in principle, a monoatomic rare gas model is not expected to form a stable glass on timescales accessible to macroscopic observation (see section 4.4.5 for some comments on this issue). This makes the LJ model a good prototype of system refraining from any form of configurationally arrested disorder, due to a simpler potential energy surface than in polyatomic systems, favoring the recovery of the most stable crystalline configuration on short timescales.

5.1.2 Assessing the stability around T_{gl}

In this context, our investigation aimed at quantifying the temporal stability of the glass at a set of temperatures for two situations. First, well below the melting temperature (supercooled liquid) and then, as a final goal, around and below the glass transition temperature T_{gl}, that we located at a value close to $T = 55$ K. To this purpose, by exploiting one specific quenching schedule, we obtained a plot of density vs temperature exhibiting a pattern reconductible to two (almost) straight lines, for which one could extract an intersection temperature identified as T_{gl} (see figure 5.1).

For the production of the volume vs temperature phase diagram and the study of the phase transformation toward crystallization we adopted the following roadmap. A first choice concerned the quench rate, taken to be equal to 3.9×10^{11} K s^{-1}, a value known to ensure stable glass formation on non-negligible intervals (several ps) at temperatures well below melting and T_{gl} [2]. As a prerequisite, we investigated the stability of supercooled and glassy LJ argon by quenching first a well equilibrated liquid to $T = 65$ K and $T = 62.5$ K, for which no sign of crystallization was found over more than 700 ps. Having established that these two temperatures do not drive crystallization, we considered one further temperature above T_{gl}, $T = 60$ K, by observing the onset of a transformation toward ordering after 340 ps. Then, we moved below T_{gl}, by taking averages for the density (i.e. volume) at each temperature on short intervals (10 ps). This allows us to construct a V vs T phase diagram while extending, at each temperature, the simulations on much longer intervals with the purpose of inferring the structural stability and the possible onset of crystallization. For each quenching schedule, the quench has been carried out by reducing the temperature in a stepwise manner according to a well-defined depth of quench ΔT, that is to say the difference of temperature between two consecutive values assigned to the thermostat. The idea consists in monitoring the stability of the glass obtained at different temperatures as a function of ΔT for a fixed value of the quench rate defined above (3.9×10^{11} K s^{-1}).

Figure 5.2 gives the details of the four trajectories created at different temperatures for values of ΔT and lengths of the trajectories at constant values of T. As a first example of stability as a function of the quench mode, we consider the case (a). In figure 5.3 the pair correlation function $g(r)$ averaged over the second third of the

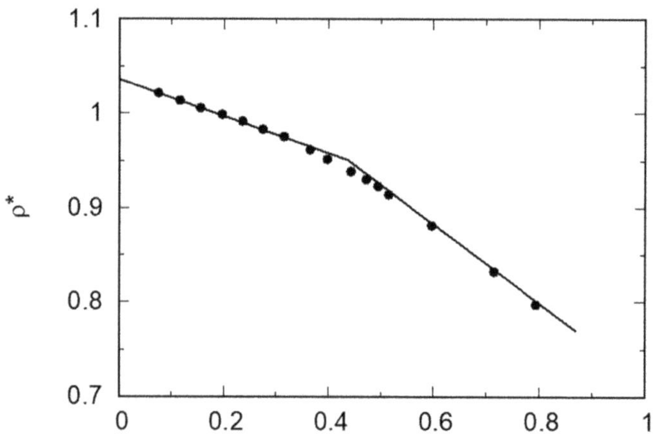

Figure 5.1. Density vs temperature phase diagram for the Lennard-Jones glass model. Straight lines are fits to high- and low-temperature results. Courtesy of E Martin (ICube, Strasbourg).

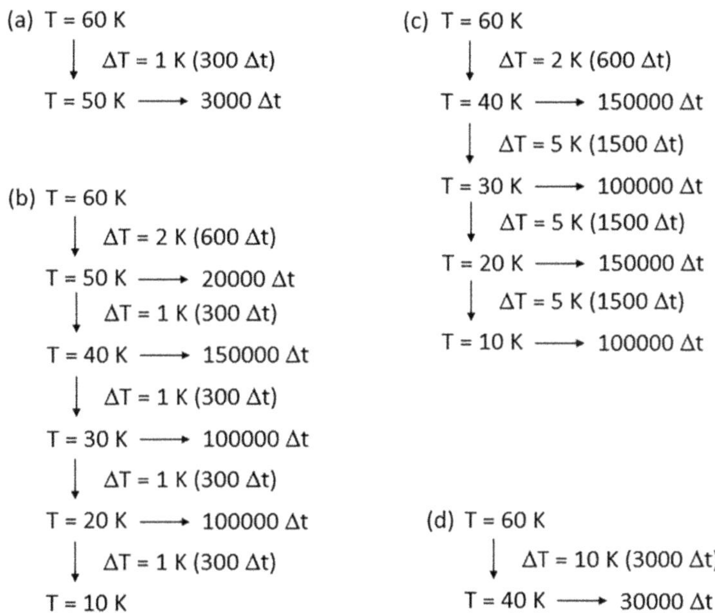

Figure 5.2. Features of the different quenching schedules employed to obtain a glass. We specify the depths of quench ΔT that bring the system to different temperatures and the length of each temporal trajectory. Courtesy of E Martin (ICube, Strasbourg).

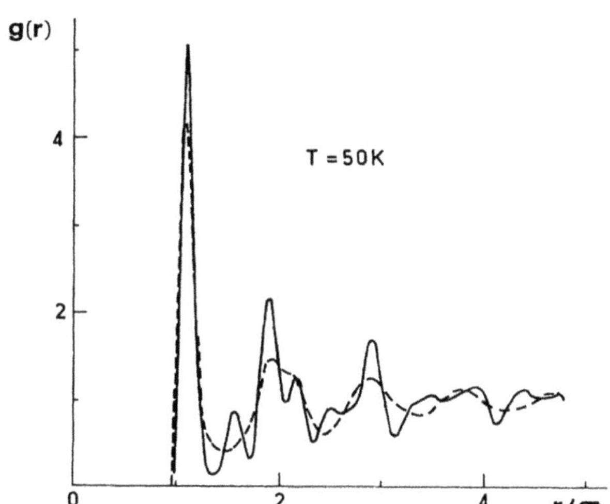

Figure 5.3. Behavior of the pair correlation function $g(r)$ on annealing at $T = 50$ K. Dashed line: averaged over the first interval of 10 000 timesteps. Solid line: averaged over the second interval of 10 000 timesteps. Reprinted figure with permission from [1]. Copyright (1989) by the American Physical Society.

trajectory (in between 10 000 and 20 000 Δt shows a clear onset of crystallization, several peaks becoming apparent as signatures of a defective *fcc* arrangement. We can conclude that the stability of the glass around T_{gl} (here we are considering $T = 50$ K, moderately lower than $T_{gl} = 55$ K) does not extend beyond 0.85×10^{-10} s.

Having in mind to investigate the same issue for temperatures well below T_{gl}, we focus on what happens at $T = 40$ K by considering processes (b), (c) and (d). Let's first consider the pair correlation functions displayed in figure 5.4 as obtained through process (b). We are interested in a glass at the early stages of relaxation and before crystallization, respectively.

The time interval separating these two averages is close to 10^{-8} s, this value providing a temporal range of stability at $T = 40$ K for the Lennard-Jones, argon-like glassy system. What is the difference in terms of stability when ΔT (the depth of quench) is made larger? Is this affecting the stability of the glass and if so, which are the underlying reasons? Valuable hints can be find in figure 5.5 featuring the behavior of the density on annealing at $T = 40$ K. One notices that the larger ΔT, the shorter is the interval of stability for the glass, as shown by process (c) ($\Delta T = 5$ K from $T = 40$ K down to $T = 30$ K and further) and (d) ($\Delta T = 10$ K from $T = 60$ K down to $T = 40$ K). In the case of (c) and (d) the ranges of stability are reduced to $\sim 0.4 \times 10^{-9}$ s and $\sim 0.1 \times 10^{-9}$ s, respectively. Globally, it appears that a sharper cooling with higher values of ΔT has a memory effect on the residual dynamics of the system, by allowing rearrangements to take place in the search of a more stable configuration.

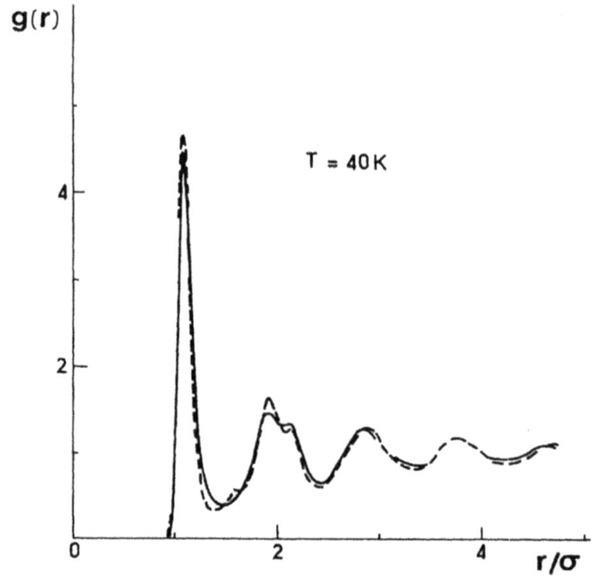

Figure 5.4. Behavior of the pair correlation function $g(r)$ for process (b) (see figure 5.2) at $T = 40$ K. Solid line: average over the first interval of 10 000 timesteps. Dashed line: average over an interval of 10 000 timesteps in between 140 000–150 000 timesteps along the full trajectory. Reprinted figure with permission from [1]. Copyright (1989) by the American Physical Society.

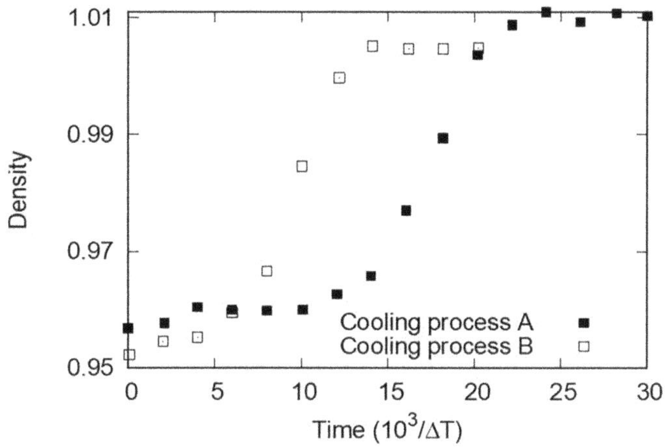

Figure 5.5. Behavior of the density on annealing at T = 50 K. The data are averages over blocks of 2000 timesteps: filled squares, cooling process (a); open squares, cooling process (b). Courtesy of E Martin (ICube, Strasbourg).

In [1] we also observed that this sensitivity to ΔT can have some dramatic and unexpected effects, as the higher stability of the glass at $T = 30$ K when produced via process (b) when compared to corresponding behavior at $T = 20$ K via process (c). *Overall, our monoatomic glass is far from being a mere qualitative model since many ingredients characterizing real systems can be recovered.* For instance, one finds high instability around T_{gl} but a rapid increase of stability at T = 40 K, extending on timescales (10^{-9} s) that remains very short when compared to macroscopic intervals but definitely larger that those necessary to evaluate equilibrium (or, better, one should say stationary) quantities via statistical mechanics.

5.1.3 Some considerations about qualitative glass models

In light of what is nowadays achievable, the results reached in 1989 could be easily improved by extending the trajectories by at least two orders of magnitude, not to mention the sizes of the systems that were considerably increased by some authors around the same period (see [3], where a strong dependence on the system size was excluded when comparing systems differing in size by two orders of magnitude, $N_{at} = 15000$ against $N_{at} = 10^6$). The paradox of producing by computer simulation a glass observable on affordable timescales (and for very large systems already in late 1980s) while knowing that an experimental counterpart cannot be realized on experimental intervals is highly intriguing. For many scientists, it is an excellent opportunity to dwell on highly valued statistical mechanical properties relevant to condensed matter theory such as relaxation and nucleation, since it is unmistakably true that a monoatomic system kept together by two-body interactions is very simple and cheap to follow and characterize when it undergoes a structural transition. Others have simply denied the existence of any stability on the laboratory timescale based on viscosity arguments [4].

I have to admit that my interest for these issues died down with my increasing involvement in the modeling of real systems and the account of chemical bonding beyond the pair interactions. However, I am convinced that the final word on the behavior of the Lennard-Jones glass model has not been said and that the establishment of some new, fresh benchmarks could be highly rewarding. In 1989 we came to the conclusion that on a timescale of 10^{-9} s we could not detect any structural changes for temperatures as low as 10 K. Regardless of this temperature being realistic for experiments, one could aim at establishing a brand new phase diagram for temperatures below T_{gl} by working with the same quench rates adopted in the past but with much larger systems and longer trajectories. To date, the precise length of the intervals of stability has not been established and this could be readily obtained with a modest computational effort. The results obtained will be far from being academic. Even by keeping the size down to some ten thousands atoms, the issues addressed (stability at very low temperature of a system known to inevitably crystallize) will be as valuable as to those having to do with the search of the nucleation time and its mechanism, concurring to enrich and complement them.

5.2 Amorphization by solid-state reaction in a metallic alloy

The idea of producing amorphous materials via a reaction driving a topologically ordered phase to a disordered one directly at the solid state [5–7] stems from the observation that rapid quench from the liquid is hardly suited to produce bulk pieces of a given compound ready to be exploited. When focusing on a phase diagram exhibiting melting, the question arises on alternative ways to reach a given point at the same temperature of the crystal (say, room temperature) but featuring a higher volume (lower density) as it occurs for a glassy phase obtained by rapid quench. We are targeting an amorphous phase obtained **without** going through heating to disrupt the structure (to achieve topological disorder) and cooling (to end up with vanishing mobility on macroscopic timescales, the amorphous state).

One is addressing here the production of an amorphous structure in a way independent on the definition of a glass transition temperature, by acting on the introduction of defects into the crystalline structure. On physical grounds, this corresponds to introducing energy into a system otherwise perfectly stable at its thermodynamic equilibrium and lying on one of its local equilibrium configurations, with vanishing chances to move away from it at least at low temperature. This process amounts to introducing chemical disorder in an ordered crystalline arrangement of a multi-component system and in following the onset of topological disorder that is typically occurring together with a volume expansion. Several experimental techniques fall within this definition (mechanical alloying and electron irradiation to quote a few) and were well established already in the late 1980s/early 90s when our first results on the amorphization by introduction of chemical disorder in crystalline $NiZr_2$ were published [8, 9].

To describe solid-state amorphization in this metallic alloy we relied on an interatomic potential taking the form described in equation (3.2), with parameters fitted to bulk properties at room temperature, the entire analytical form being validated by the very good agreement between experimental and calculated melting temperature [9]. CMD simulations were performed in the NPT scheme along the lines detailed in the previous section (see equations (5.2), (5.3) and (5.4)). In view of the expected volume change, the NPT option is the ideal scheme to be adopted to monitor the possible occurrence of a loss of topological order.

All relevant technical details on the simulations are given in [9].

1. First we concentrate on the reaction to the introduction of antisite defects by referring to the simplified scheme of figure 5.6 which helps introducing the notion of long-range order parameter S given by $(p - r)/(1 - r)$. In this definition, p is the probability of observing a given atom (of either kind, in our case Ni or Zr) at its original location in the crystalline configuration and r is the molar ratio. $S = 1$ means that the system is fully chemically ordered since the atoms are at their original sites, with the highest probability 1. The opposite situation corresponds to a probability p equal to the molar ratio r, this being the lower value compatible with any chemical composition. *An important issue to be addressed is the threshold of S leading to the loss of topological order, to be monitored by following the behavior of the pair correlation functions and the comparison with the shapes obtained when quenching from the liquid state.* The NPT methodology allows full insight into volume expansion, as visualized in figure 5.7. An equilibrated crystal of $NiZr_2$ is taken to evolve for 20 ps at room temperature and pressure. Exchanges of positions among the Ni and Zr atoms take place at $t = 20$ ps inducing various volume dilations that reflect the tendency to increase the system potential energy as a reaction to the imposed number of antisite defects. To ensure that the reaction toward the disordered state occurs and

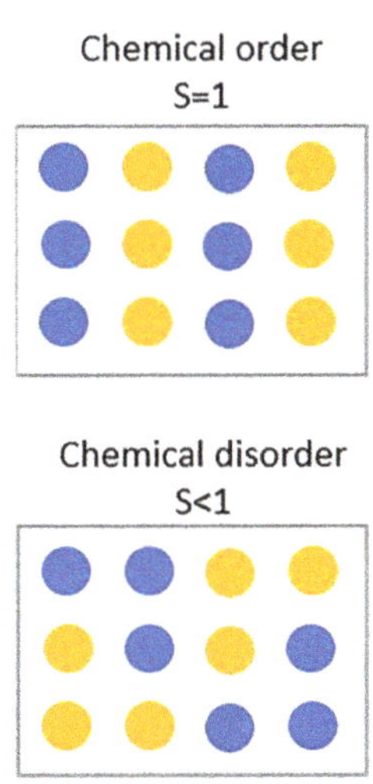

Figure 5.6. Simplified model of a system made of A and B particle chemically ordered (top part) with all atoms in their crystalline positions. Bottom part: the same system after introduction of chemical disorder. Courtesy of E Martin (ICube, Strasbourg).

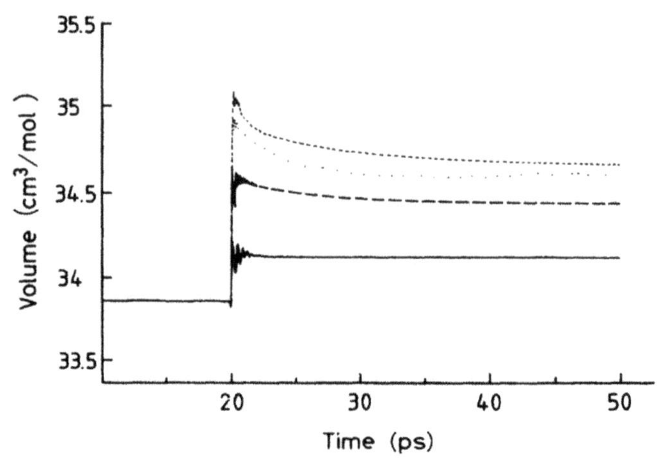

Figure 5.7. Average volume of the system at $T = 300$ K before and after introduction of chemical disorder. Solid line, $S = 0.9$; dashed line, $S = 0.6$; dotted line, $S = 0.2$; short-dashed line, $S = 0$. Reprinted figure with permission from [9]. Copyright (1990) by the American Physical Society.

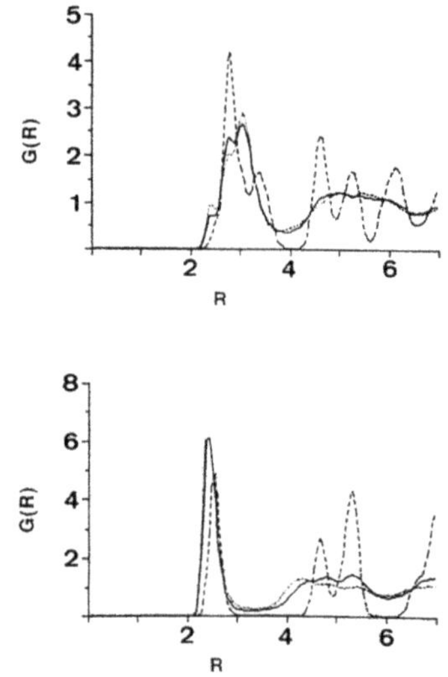

Figure 5.8. Partial pair correlation function $g_{NiNi}(r)$ (bottom) and total correlation function of amorphous $NiZr_2$ (top). Dashed line: crystalline $NiZr_2$. Solid line: after imposing chemical disorder, $S = 0$. Dotted line: amorphous obtained by rapid quench from the liquid state. R is the interatomic distance given in Å. Reprinted figure with permission from [9]. Copyright (1990) by the American Physical Society.

that we are able to obtain a system very similar to those resulting from rapid quench, it is appropriate to select the highest number of exchanges (lower value of S, $S = 0$). The comparison between the corresponding total and Ni–Ni pair correlation functions (the other partials exhibiting the same features) is given in figure 5.8, together with the results for the initial crystalline state. Having observed that we are indeed able to produce an amorphous state by following an alternative path on the V/T phase diagram, one can concentrate on the determination of the threshold value of S to be imposed in order to reach the topological disorder. This can be achieved by comparing the corresponding structural quantities of interest and looking for some plausible criterion to mark the frontier between the crystalline and the amorphous state. In [8, 9], the value $S = 0.6$ was proposed to correspond to the onset of the transformation toward the amorphous state. When critically analyzing the conclusions reached in 1990, their general validity remains compelling, even though the notion of a well-defined threshold S value could be replaced by a range of values separating the crystal from the amorphous state, moderately depending on the size of the system.

2. After proving the existence of an amorphous state obtained directly from the crystal, we focused on the mechanism of the transformation by relying on the role of the elastic constants [10]. The underlying idea views the crystal-to-amorphous transformation (c–a hereafter) in analogy to melting, since they are both driven (or, at least, characterized) by a volume expansion. Despite the fact that the c–a process is athermal unlike melting, one can hypothesize a correspondence between the two phase transformations based on the existence of a mechanical instability. This was proposed by invoking the so-called Born condition $C_{\text{shear}} = C_{11} - C_{12} = 0$, where C_{11} and C_{12} are two elastic constants, expressing second derivatives of the total energy with respect to specific deformations along given axis (the most intuitive being the bulk modulus issued via an isotropic deformation as $B = \frac{\partial P}{\partial V}$ with $P = \frac{\partial E}{\partial V}$ where E, P and V are total energy, pressure and volume respectively calculated at $P = 0$ equilibrium conditions). Calculation of the elastic constants were performed for crystals of $NiZr_2$ with an increasing number of antisite defects (decreasing values of S) and for the amorphous solid obtained by quench [10]. All systems are kept at constant temperature and pressure within the NPT framework (see section 5.1) enriched by requiring zero-stress via the application of the variable size and shape extension of MD due to Parrinello and Rahman [11].

   Figure 5.9 shows the changes of C_{shear} as a function of molar volume. By inducing an increased topological disorder via the increase of chemical disorder, C_{shear} decreases down to a volume dilation of 2%. Interestingly, the corresponding value of $S \sim 0.68$ is very close to to the threshold found by inspection of the pair correlation functions. This behavior of C_{shear} is followed by a rapid increase to the value calculated for the amorphous solid at the same molar volume, showing that the c–a transition goes along with a specific, reversible pattern of elastic softening.

   These results are suggestive of a strong analogy between melting and amorphization by solid-state reaction, even though the decrease of C_{shear} is not as dramatic as in the case of melting, its value remaining distinctly higher than zero. However, conclusions drawn on the basis of 144 atoms cannot be taken as definitive, as already suggested in [10] where a percolation approach

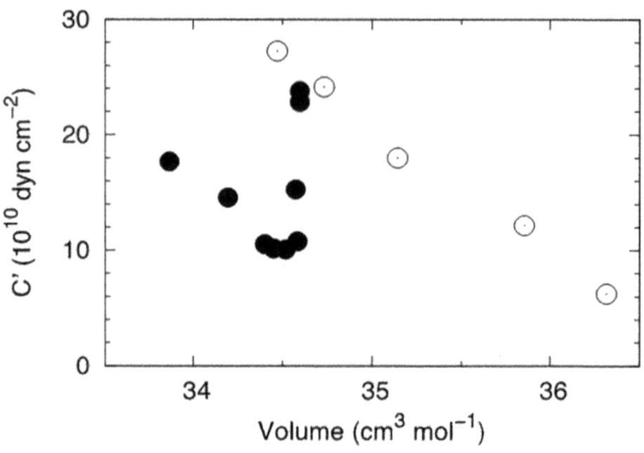

Figure 5.9. Variation of the shear modulus of crystalline $NiZr_2$ as a function of the molar volume. Solid circles: Data for values of the long-range order parameter ranging from $S = 1$ to 0.5. Open circles: Shear modulus of the amorphous solid obtained by quenching the liquid. Courtesy of E Martin (ICube, Strasbourg).

was introduced to rationalize the behavior of crystals destabilized by the introduction of defects.

3. In the context of the historical evolution of MD methods over the past 30 years, it is worth noticing that first-principles MD counterparts of models employed in this section are nowadays affordable on sizes at least comparable to those employed at that time [12]. This would make a study of amorphization by chemical reaction based on electronic-structure derived potential energies feasible and potentially capable of providing information on the effect of such transformation on chemical bonding, if any. However, it looks like most models of intermetallic alloys that have been developed and used ever since our early study are nothing but alternative recipes of the classical *n*-body potentials quoted here, each one following different derivations of the main scheme presented in section 3.2. While admitting the validity and legitimacy of all these efforts, one cannot refrain from observing that they barely improve on the overall performance of what has been already made available on the market, in terms of static and thermodynamic properties. This is especially needed since models are expected to provide the correct values for the defect formation energies, that were pointed out as the main failure of the $NiZr_2$ interatomic potentials detailed here. Significant improvements in this direction have been reported in [13]

References

[1] Massobrio C, Pontikis V and Ciccotti G 1989 *Phys. Rev.* B **39** 2640–53

[2] Nosé S and Yonezawa F 1986 *J. Chem. Phys.* **84** 1803–14

[3] Swope W C and Andersen H C 1990 *Phys. Rev.* B **41** 7042–54

[4] Shneidman V A and Uhlmann D R 1998 *J. Non-Cryst. Solids* **224** 86–8

[5] Schwarz R B 1998 *Mater. Sci. Eng.* **97** 71–8

[6] Johnson W L 1986 *Prog. Mater. Sci.* **30** 81–134
[7] Rehn L E, Okamoto P R, Pearson J, Bhadra R and Grimsditch M 1987 *Phys. Rev. Lett.* **59** 2987–90
[8] Massobrio C, Pontikis V and Martin G 1989 *Phys. Rev. Lett.* **62** 1142–5
[9] Massobrio C, Pontikis V and Martin G 1990 *Phys. Rev.* B **41** 10486–97
[10] Massobrio C and Pontikis V 1992 *Phys. Rev.* B **45** 2484–7
[11] Parrinello M and Rahman A 1981 *J. Appl. Phys.* **52** 7182–90
[12] Huang L, Wang C Z and Ho K M 2011 *Phys. Rev.* B **83** 184103
[13] Kumagai T, Nikkuni D, Hara S, Izumi S and Sakai S 2007 *Mater. Trans.* **48** 1313–21

IOP Publishing

The Structure of Amorphous Materials using Molecular Dynamics

Carlo Massobrio

Chapter 6

The atomic structure of disordered networks

Disordered $GeSe_2$ systems have been identified as prototype networks made of a predominant tetrahedral motif coexisting with miscoordinations and homopolar bonds. The pattern exhibited by the pair correlation functions and the partial structure factors is far from trivial, calling for quantitative models to be employed and carefully tested so as to ascertain their performances. In this chapter, we begin to report on first-principles molecular dynamics (FPMD) calculations on disordered chalcogenides by focusing on liquid and glassy $GeSe_2$. We describe first their structural properties, by showing that the agreement with experiments is unequal for the Ge and Se and it is particularly challenging for the Ge environment. As a second issue, we take advantage of the existence of intermediate range order in these networks to reconsider available interpretations for the origin of the first-sharp diffraction peak (FSDP) in the total and in the concentration–concentration structure factor $S_{CC}(k)$. At both levels (total and partial structure factors), we proposed two atomic-scale interpretations for the origins of this feature. A further section is devoted to the comparison, for different glassy systems, between the charge–charge structure factor $S_{zz}(k)$ and $S_{CC}(k)$. This allows extracting a criterion linking the presence of the FSDP to the topology of the network. Also, the concept of charge neutrality is rationalized. As a final item of this chapter, we show how the quest of improvements in the description of prototypical disordered chalcogenides brings us to the choice of a specific exchange–correlation (XC) functional (BLYP, introduced in section 4.2.1), able to enhance electronic localization.

6.1 General consideration: where do we start from?

There are several ways of getting interested in disordered materials from the theoretical viewpoint. One consists in focusing on a given property, feature and/or specific mark or pattern that shows up in a class of systems by attempting to understand its origin and what makes it so special for those materials. By using molecular dynamics (MD), the fact of working at the atomic-scale drives the search of

doi:10.1088/978-0-7503-2436-6ch6

all possible connections between what we are able to observe (experimental data) and the relationship with microscopic quantities accessible to the atomic-scale observation, that is to say the variable for which we can calculate a statistical average. Focusing on a specific property and on its peculiarities has the advantage of stimulating curiosity for other physical and chemical features of the targeted systems, by nurturing a rewarding approach to more extended knowledge. These considerations apply to the road of discovery that will be detailed here for chalcogenides, the class of disordered network-forming systems mostly invoked in the part of this monograph. When we started working on these systems, the key quantities attracting attention for its intriguing behavior and the implications on the atomic-scale description of disordered systems (liquids and glasses) were the partial structure factors. In particular, a feature developing at short values (typically 1 $Å^{-1}$) of the momentum transfer in reciprocal space was the object of investigations, namely the so-called first-sharp diffraction peak (FSDP) [1], indicative of intermediate range order developing on distances substantially longer than the nearest neighbors. As a feature appearing in many different systems, the FSDP was both considered a quantity to be interpreted on the basis of generic models and a signature of specific network topologies deserving accurate treatment (at the first-principles level) of chemical bonding. Our approach was mostly of the second kind, even though we had to acknowledge that there is a lot to learn on FSDP and IRO (intermediate range order) with interatomic potentials.

In what follows, we shall focus on liquid and glassy $GeSe_2$ as prototype systems to understand the interplay between atomic structure, level of the description of the potential energy surface and intermediate range order. The consideration of a liquid is part of a general analysis involving disordered network-forming systems, all ideas developed thereof applying equally well to glasses. For comparative purposes, other systems will be also touched upon, although much less extensively. When discussing the main features of the classical MD potentials in section 3.3, we mentioned that the ionic approximation inherent in equations (3.13) and (3.14) could not account for the presence of homopolar bonds in disordered $GeSe_2$. From a strictly historical perspective, it should be said that stringent evidence on the presence of homopolar bonds was not available when the results on disordered $GeSe_2$ were published, this factor contributing to their highest recognition. In addition, the total neutron structure factor was in good agreement with experiments, with the inclusion of the FSDP. While the availability of partial structure factors and the related pair correlation functions pointed out the deficiencies we referred to in section 3.3, there is a more profound reason marking the inability of point-charge potential models to compare realistically with diffraction experiments. In references [2–4] two statements were put forth, the *first* highly valuable and interesting, the *second* equally valuable but revealing of an excessive simplification in terms of chemical bonding built in the potential itself.

1. The *first* statement amounts to attributing the appearance of the FSDP to a combination of steric and charge effects, a simple charged sphere models allowing for these conditions to be met. In this way, one is able to rationalize the very good behavior of the potential when employed to reproduce the total neutron structure factor and, in real space, to promote a tetrahedral network made of connected tetrahedra, a general topological feature well accepted ever

since early experiments were made available. Note that in this context the potential based on point charges can be considered a refinement of the charged sphere models, by equally promoting the establishment of a tetrahedral network and the presence of a FSDP.

2. The *second* statements issued by references [2–4] has to do with the absence of the FSDP in the charge–charge structure factor $S_{zz}(k)$, that becomes trivially proportional to the concentration–concentration structure factor $S_{CC}(k)$ (defined in equation (2.6)) for a purely ionic system made of point charges. This was proposed to be a universal property of binary AX_2 disordered systems, by virtually ruling out any fluctuations of concentration and of charge on intermediate range distances, since for the point-charge model $S_{zz}(k)$ and $S_{CC}(k)$ are essentially equal (proportional).
3. Given this situation, one is faced with the dilemma of disproving (or confirming) that ionic models are sufficiently accurate for disordered chalcogenides by collecting any possible evidence (from experiments and/or theory) on the behavior of $S_{CC}(k)$. Therefore, our first goal was to describe liquid (and glassy) $GeSe_2$ system in the framework of FPMD and obtain properties in real and reciprocal space to be compared with experiments. This is where the partial structure factors measured by the team of P Salmon via isotopic substitution in neutron diffraction come into play [5, 6] since both investigations on the glassy and liquid $GeSe_2$ showed a clear FSDP in $S_{CC}(k)$, thereby opening an intense line of study that has marked our engagement in the area of disordered network-forming material for many years.

In what follows, we shall focus on the following issues in different sections of this chapter.

1. We describe first the structural properties of liquid and glassy $GeSe_2$, by pointing out the level of agreement with experiments and the issues left open at the time the calculations were carried out.
2. As a second point, we take advantage of the availability of FPMD calculation for several systems to discuss existing interpretations for the origin of the first-sharp diffraction peak in the total structure factor. This analysis is extended to the presence of the FSDP in the concentration–concentration structure factor, for which we achieved a substantial breakthrough by linking the FSDP to a specific structural motif.
3. As a third point, we carry out a comparative analysis of the charge–charge and concentration–concentration structure factors $S_{zz}(k)$ and $S_{CC}(k)$, leading to a general criterion for the appearance of intermediate range order based on the concept of charge neutrality at the appropriate scales.
4. We have devoted section 3.6.5 to the role played by the generalized gradient approximation (GGA) (the one by Perdew and Wang in that case) to obtain realistic structural properties for liquid $GeSe_2$. That result was at the origin of our decision to pursue the study of disordered chalcogenides by relying on GGA calculations. Here we look for further improvements by considering an alternative XC functional (BLYP, introduced in section 4.2.1), by showing that this choice brings the calculations in better agreement with experiments for certain properties, especially those pertaining to the Ge subnetwork.

6.2 The structure of liquid and glassy $GeSe_2$

6.2.1 Methodology

Here we shall focus on the FPMD model based on the PW exchange–correlation functional, by referring mostly to reference [7] for the liquid case and to reference [8] for the glassy case. These calculations share the same technical features and were published at a 7 years interval, partially because it was decided to collect much more statistics for the glassy case. In terms of technical tools, we are referring here to a plane wave code running on vectorial computers (essentially the one developed by Alfredo Pasquarello and Roberto Car at EPFL Lausanne in the 1990s) and on grants quite limited on a yearly basis, this situation imposing some limitations on the size of the system and on the temporal trajectories. Also, this explains why it took quite some time to obtain several trajectories for the glass. Parallel computers were already available at the beginning of the century but the CPMD code was not yet exploited with its parallel implementation for the cases considered here. When facing the size and time issues of the plane waves/pseudopotential scheme within FPMD, any size larger than ~150–200 atoms was unaffordable at those times, our first results based on larger sizes (say, ~500 atoms) becoming available later, after the year 2010. Given the performances of those computers, included the largest ones running on national facilities, there was no point in hoping to circumvent the strong non-linear behavior with size of the Kohn–Sham Hamiltonian, for which we recorded at that time a cost about one hundred times more important when going from $N_{at} = 100$ to $N_{at} = 500$.

- Technically speaking, it is legitimate to work with $N_{at} = 120$ to get information on intermediate range order, given the smallest wave vector compatible with the periodic cell, $k_{min} = 0.5$ Å^{-1}, smaller than the FSDP wave vector $k_{FSDP} = 1$ Å^{-1}, with as many as eight discrete k values describing the region of wave vectors around the FSDP. For these reasons, the need to switch to a larger system appeared less acute and could be postponed without sacrificing (at least as much as we could at that time) the quality of our achievements.
- As to the issue of the length of the trajectory, our strategy consisted in working with the largest time step compatible with a good energy conservation (about 0.5 fs, 20 a.u.), even though this choice appears today a bit risky in terms of adiabaticity since corresponding to fairly large values of the fictitious mass for the electronic degrees of freedom μ (5000 a.u.). Careful use and tuning of the Nosé–Hoover thermostats (see section 3.5.5) have granted success to these simulations, regardless of the fact that working with smaller time step and μ would have been much more costly but safer. For the liquid, equilibrium trajectory lasted ~20 ps, a value allowing for calculation of sensible average properties due to the effective sampling of the phase space.
- For the glass, a typical quench schedule was the one described in section 4.4.5, where we stressed the importance of following the behavior of some quantities in time before considering them as stationary property. This consideration proved valuable even in the case of quench rates faster than 10^{13} K s^{-1}. In the case of the glass, we produced $N_{tr} = 6$ trajectories by considering the average properties obtained from each trajectory as partial averages contributing to a

global one (the mean value). The error bar attributed to the mean value is taken to be $\sigma_{\text{mean}} = \sigma / \sqrt{N_{\text{tr}} - 1}$ where σ is is half the largest difference among the N_{tr} partial averages.

6.2.2 Liquid $GeSe_2$

As a first information on the structure (and the extent of intermediate range order) of liquid $GeSe_2$ we can focus on the Faber–Ziman (FZ) [9] partial structure factors (figure 6.1), introduced in section 2.3[1].

Focusing on $S_{\text{GeGe}}(k)$, the main difference between neutron scattering experiments [6] and our calculations is found at the FSDP level, with a much lower

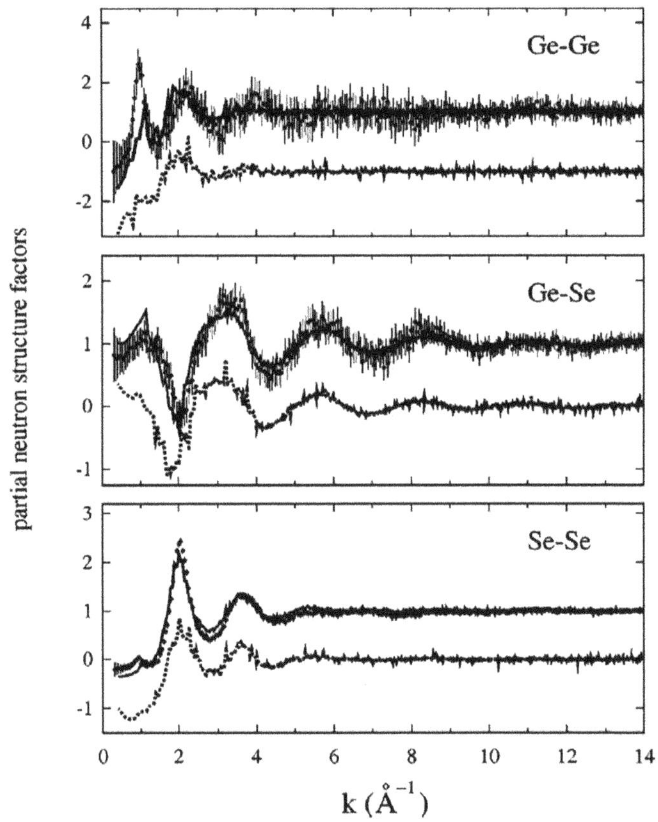

Figure 6.1. Partial structure factors for liquid $GeSe_2$: solid line: PW-GGA calculations; dotted line: LDA calculations; dots with error bars: experiments [6]. LDA results have been shifted down by 2, 1, and 1, respectively. Reprinted figure with permission from [7]. Copyright (2001) by the American Physical Society.

[1] From now on we shall drop the label FZ when indicating the partial structure factors, unless when using explicitly the Bhatia–Thornton (BT) [10, 11] partial structure factors as introduced in section 6.3. Also, we shall not comment on local density approximation (LDA) results since the limited performances of these calculations have been already highlighted previously (see section 3.6.5).

intensity in the CPMD result. This holds true even when accounting for error bars of the FSDP, estimated by taking partial averages on time periods of 2 ps and amounting to as much as 20%. This high value is mostly due to the size of our system (with as little as 40 Ge atoms) and it expresses strong fluctuations in the intermediate range properties. We shall exploit this variability when analyzing the correlation between the FSDP in $S_{CC}(k)$ and specific structural motifs (see section 6.3.2). Despite some small, discernible differences, theory is in much better agreement with experiments for the case of $S_{GeSe}(k)$ and $S_{SeSe}(k)$. This provides a first indication of a general feature of disordered Ge–Se systems modeled by FPMD (Ge–Ge correlations somewhat more problematic than Se–Se and Ge–Se ones) that will be observed throughout our studies.

It is of interest to see (figure 6.2) how this fact is reflected by the behavior of the Bhatia–Thornton [10] partial structure factors $S_{NN}(k)$ (number–number), $S_{NC}(k)$ (number–concentration) and $S_{CC}(k)$ (concentration–concentration), introduced in section 6.3, where we pointed out that for the $GeSe_2$ composition $S_{NN}(k)$ is a very

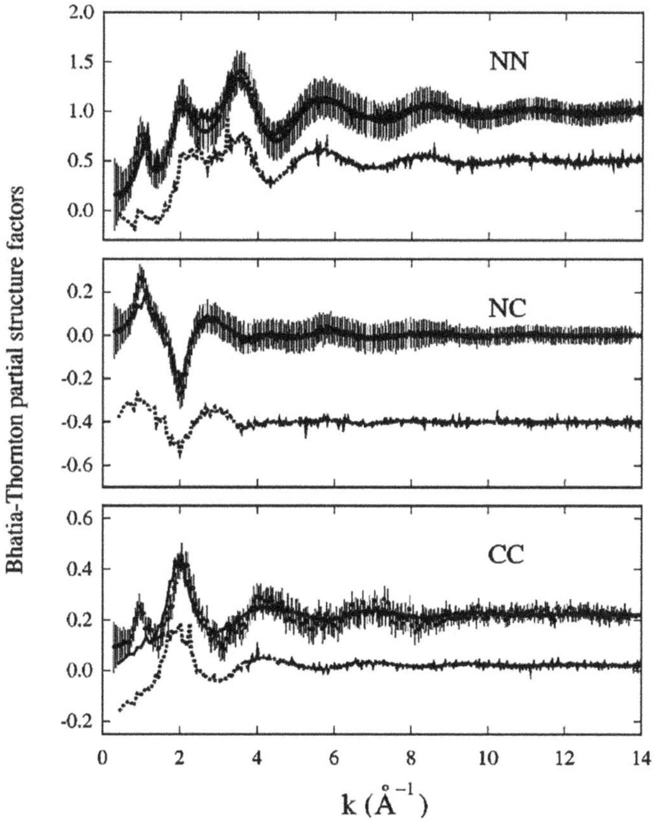

Figure 6.2. The Bhatia–Thornton partial structure factors for liquid $GeSe_2$: PW-GGA calculations: solid line; LDA calculations: dotted line; experiments: dots with error bars, [11]. LDA results have been shifted down by 0.5, 0.4, and 0.2, respectively. Reprinted figure with permission from [7]. Copyright (2001) by the American Physical Society.

good approximation of the total neutron structure factor $S_T(k)$. Both $S_{NN}(k)$ and to a smaller extent $S_{NC}(k)$ agrees very well with experiments for the entire k range, while in the case of $S_{CC}(k)$ experiments are reproduced only for $k > 1.5$ Å^{-1}. It appears that the very prominent FSDP observed experimentally is absent in the theoretical $S_{CC}^{GGA}(k)$, this difference showing that, overall, the comparison between experimental and theoretical partial structure factors is highly favorable for k values characteristic of short-range properties ($k > 2$ Å^{-1}). However, the FSDP weights in the partial structure factors are distributed differently in theory and experiment. This difference manifests itself through the behavior of $S_{GeGe}(k)$ and (very moderately) of $S_{GeSe}(k)$. Quite intriguingly these discrepancies tend to cancel with each other in the total structure factor, while they concur to produce a clear deviation at the FSDP position in $S_{CC}(k)$.

Focusing on the properties in real space through the analysis of the pair correlation functions $g_{\alpha\beta}(r)$, compared to experiments [6] in figure 6.3, unequal performances are recorded for $g_{GeGe}(r)$, $g_{GeSe}(r)$ and $g_{SeSe}(r)$. These are quantified through the positions of the main peaks and the number of neighbors as given in

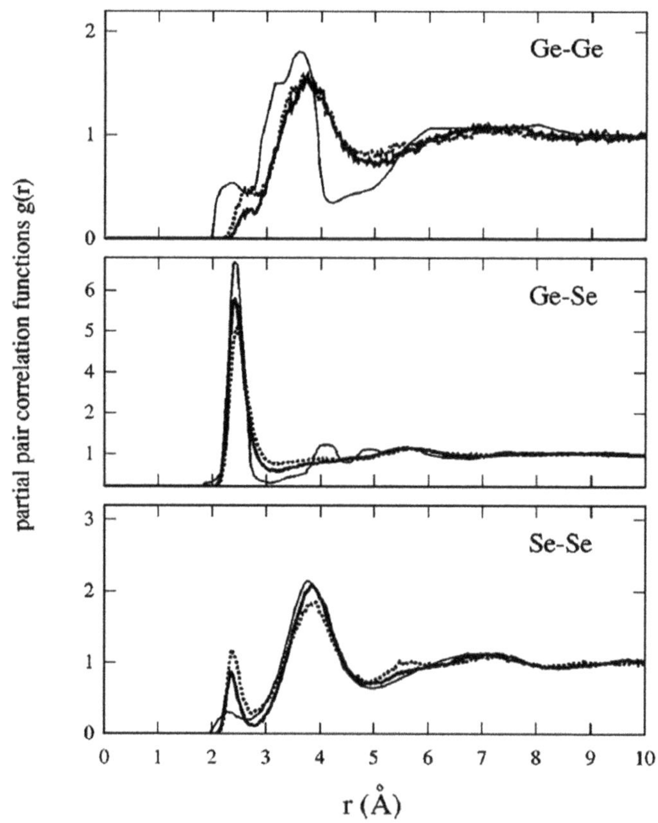

Figure 6.3. Partial pair correlation functions for liquid $GeSe_2$: PW-GGA calculations: thick line; LDA calculations: dotted line; experiments: solid line, [6]. Reprinted figure with permission from [7]. Copyright (2001) by the American Physical Society.

table 6.1. Partial pair correlations $g_{\mathrm{GeSe}}(r)$ and $g_{\mathrm{SeSe}}(r)$ are consistent with experiments, with close values for the first neighbors coordination shells numbers $n_{\mathrm{Ge}}^{\mathrm{Se}}$ and $n_{\mathrm{Se}}^{\mathrm{Se}}$. Larger differences are found in the case of $g_{\mathrm{GeGe}}(r)$. Experiments exhibit a clear distinction between shells of neighbors, with first maximum and minimum. Theory appears to smooth out the relevant peaks and minima, leading to a description of the Ge–Ge environment virtually resembling a behavior at higher temperatures.

There are fewer homopolar bonds than in experiments and, additionally, they are found at larger distances as indicated by $n_{\mathrm{Ge}}^{\mathrm{Ge}}$ in table 6.1. In [7] it was pointed out that longer interatomic Ge–Ge distances and less structured Ge–Ge pair correlation functions were found in liquid GeSe within the same theoretical framework [12]. Since such longer Ge–Ge bond lengths are typical of the metallic liquid Ge, we obtain a strong indication of the overestimate of the metallic bonding in liquid $GeSe_2$. Overall, results in reciprocal and real space have a twofold character and can be commented upon by underlying both positive and less satisfactory features. On the positive side, FPMD with Perdew–Wang GGA proved able to achieve a very good agreement with neutron scattering experiments, including the presence of homopolar bonds (not accessible to any interatomic potential at that temperature) and the appearance of the FSDP. This can be considered as a success of our approach, validating a detailed analysis of the structural motifs contained in the network. Less favorable is the description of Ge–Ge correlations, both at short and intermediate range distances. Ever since the production of these results, this shortcoming has been a strong motivation to improve the atomic-scale description in two directions, namely the consideration of other XC schemes and the study of size effects. While the Perdew–Wang GGA scheme remains a valuable choice, the overestimate of Ge–Ge bond lengths together with the flattened shape of $g_{\mathrm{GeGe}}(r)$ were strong indication of a still insufficient account of the ionic character of bonding. Looking for a size effect was also a quite natural option, while falling into the category of effect quite hard to control in an exhaustive manner, macroscopic dimension being in any case inaccessible. Further details on these issues will be given elsewhere in this book (see chapter 9).

Table 6.1. First (FPP) and second (SPP) peak positions in experimental [6] and theoretical $g_{\alpha\beta}(r)$ [7]. The integration ranges corresponding to the coordination numbers n_{α}^{β} and m_{α}^{β} are 0–2.6 Å, 2.6–4.2 Å for $g_{\mathrm{GeGe}}(r)$, 0–3.1 Å, 3.1–4.5 Å for $g_{\mathrm{GeSe}}(r)$ and 0–2.7 Å, 2.7–4.8 Å for $g_{\mathrm{SeSe}}(r)$. Error bars are the standard deviations from the mean for subaverages of 2 ps. Reprinted table with permission from [7]. Copyright (2001) by the American Physical Society.

$g_{\alpha\beta}(r)$	FPP (Å)	n_{α}^{β}	SPP (Å)	m_{α}^{β}
$g_{\mathrm{GeGe}}(r)$	2.7 ± 0.1	0.04 ± 0.01	3.74 ± 0.05	2.74 ± 0.06
$g_{\mathrm{GeGe}}^{\mathrm{exp}}(r)$	2.33 ± 0.03	0.25 ± 0.10	3.59 ± 0.02	2.9 ± 0.3
$g_{\mathrm{GeSe}}(r)$	2.41 ± 0.10	3.76 ± 0.01	5.60 ± 0.01	3.72 ± 0.03
$g_{\mathrm{GeSe}}^{\mathrm{exp}}(r)$	2.42 ± 0.02	3.5 ± 0.2	4.15 ± 0.10	4.0 ± 0.3
$g_{\mathrm{SeSe}}(r)$	2.34 ± 0.02	0.37 ± 0.01	3.84 ± 0.02	9.28 ± 0.04
$g_{\mathrm{SeSe}}^{\mathrm{exp}}(r)$	2.30 ± 0.02	0.23 ± 0.05	3.75 ± 0.02	9.6 ± 0.3

By taking advantage of $n_\alpha(l)$, defined as the average number of atoms of species α l-fold coordinated we can calculate the percentage of Ge and Se atoms coordinated to various units composed of l neighbors having a well defined chemical nature. This analysis reveals that fourfold coordinated atoms are predominant but limited to 61%. Chemical disorder is quantified by the presence of threefold coordinated Ge atoms (22.4%), and fivefold coordinated Ge atoms (10.8%) as well as 16.5% Ge atoms forming homopolar bonds. Similarly Se atoms are found in twofold configurations (70.3%), but 25% of them are threefold coordinated and 32% form homopolar bonds. The picture arising from these numbers corresponds to that of a network in which the tetrahedral units are by far the most frequent form of coordination among Ge and Se atoms (the $GeSe_4$ tetrahedra). However, there is a conspicuous number of miscoordinations under the form of homopolar bonds or other than fourfold (for Ge) and twofold (for Se) connections, marking the difference between a realistic FPMD model and descriptions based on point-charge interatomic potentials [2–4, 13].

6.2.3 Glassy $GeSe_2$

By following the same line of presentation than in the previous section, we consider first properties in reciprocal space. The calculated total neutron structure factor $S_T(k)$ (for which all considerations relative to its representation via $S_{NN}(k)$ hold as in the liquid case) is compared to the experimental data [14] in figure 6.4.

Once again, the very good agreement over the entire range of values is affected by the behavior for $k < 2$ Å^{-1}, with no clear minimum between the first two peaks and the FSDP of lower intensity. The origin of this disagreement can be traced back to the partial structure factor $S_{GeGe}(k)$ shown in figure 6.5 together with $S_{GeSe}(k)$ and $S_{SeSe}(k)$, for which the level of comparison is quite outstanding. Differences between theory and experiments in $S_{GeGe}(k)$ are noticeable for $k < 3$ Å^{-1}, following the same

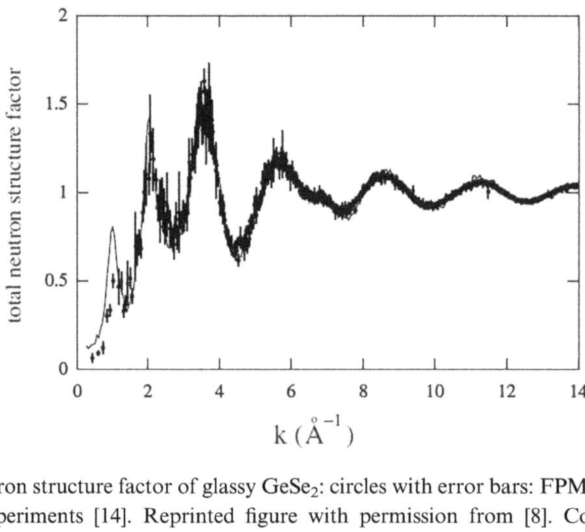

Figure 6.4. Total neutron structure factor of glassy $GeSe_2$: circles with error bars: FPMD results [8]; solid line: neutron scattering experiments [14]. Reprinted figure with permission from [8]. Copyright (2008) by the American Physical Society.

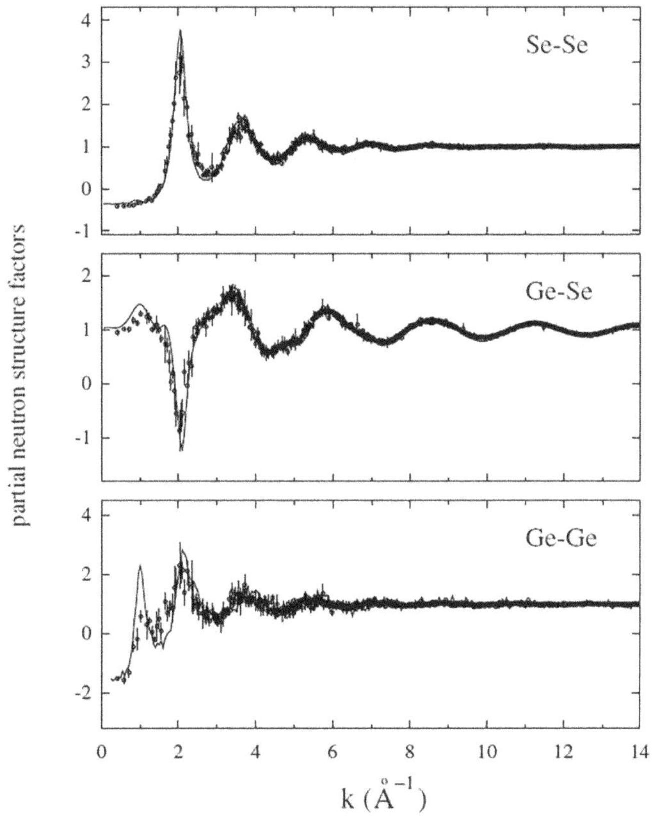

Figure 6.5. Partial structure factors of glassy $GeSe_2$: circles with error bars: FPMD results [8]; solid line: neutron scattering experiments [5]. Reprinted figure with permission from [8]. Copyright (2008) by the American Physical Society.

behavior observed for $S_T(k)$. Therefore, glassy $GeSe_2$ reproduces in reciprocal space the pattern already observed for the liquid, with a broad agreement on the intermediate range properties when it comes to Ge–Se and Se–Se correlations, this agreement worsening in the Ge–Ge case, resulting in a shape not adequately reproduced in the FSDP region.

Similar considerations hold for the case of $S_{CC}(k)$, that agrees with its experimental counterpart in the range $k \geqslant 2$ Å^{-1}, while at lower k values, the FSDP takes a shoulder shape for 1 Å^{-1} $< k <$ 1.7 Å^{-1} (figure 6.6). However, the disagreement between theory and experiments at low values of k is less severe in the glassy than in the liquid case.

When addressing the structural properties of glassy $GeSe_2$, it is worthwhile to begin our review of the results of [8] by examining $g_{GeGe}(r)$ since for $g_{SeSe}(r)$ and, to a larger extent for $g_{GeSe}(r)$, our level of theory is able to compare adequately to experiments, the few differences found in $g_{SeSe}(r)$ being almost accounted for by the error bars (see figure 6.7). The comparison between experiments and theory for $g_{GeGe}(r)$ can be based on three distinct marks in the interval 2 Å $< r <$ 4 Å. These

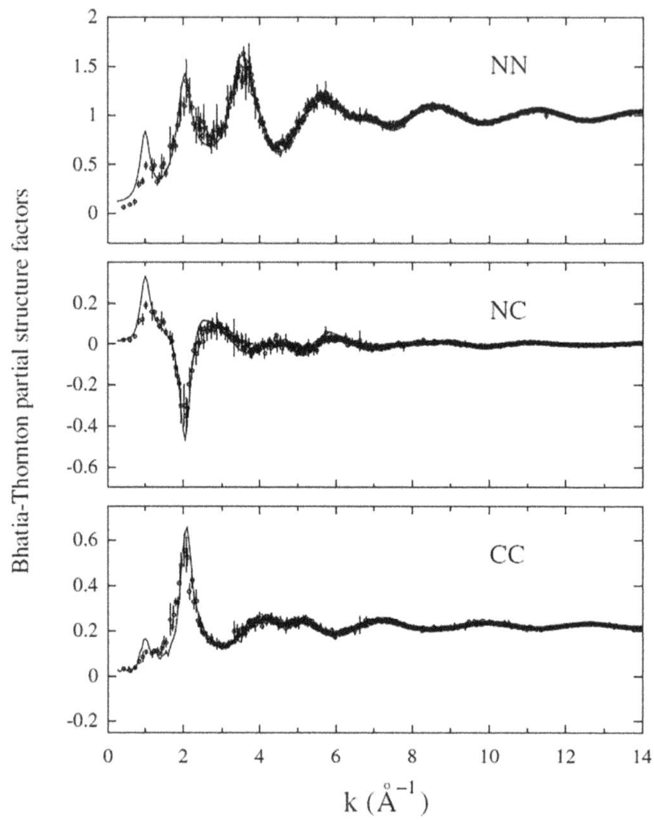

Figure 6.6. Bhatia–Thornton partial structure factors of glassy $GeSe_2$: circles with error bars: FPMD results [8]; solid line: neutron scattering experiments [5]. Reprinted figure with permission from [8]. Copyright (2008) by the American Physical Society.

correspond to homopolar Ge–Ge bonds, Ge atoms forming edge-sharing connections, and Ge atoms forming corner-sharing connections, respectively. Worthy of note is the fact that the error bars is as large as 50% in the interval 2.3 Å<r<2.7 Å around the first maximum of $g_{\mathrm{GeGe}}(r)$. We are faced here with a pattern recalling the less structured profile already observed in the case of the liquid, by confirming the limits inherent in the description of Ge–Ge correlations by using the GGA-PW scheme. A first, striking consequence is the underestimate in the value of the coordination number $n_{\mathrm{Ge}}^{\mathrm{Ge}}$ (0.07 ± 0.04 against 0.25 [5]).

The description of the short-range structure contained in table 6.2 is indicative of a larger number of both fourfold coordinated Ge atoms and twofold coordinated Se atoms than in the liquid ((75 ± 6)% vs 61% and (93 ± 6)% vs 70%, respectively). This means that the chemical order is partially restored upon cooling. In addition to the predominant tetrahedral coordination, Ge atoms are also found in first-neighbor shells made of of GeSe, $GeSe_2$ and $GeSe_3$. Also, homopolar Ge connections concern two Ge–Se_3 units that merge to form a Se_3–Ge–Ge–Se_3 ethane-like group. Table 6.2 confirms the coexistence of tetrahedra and chemical disorder as a peculiar feature of

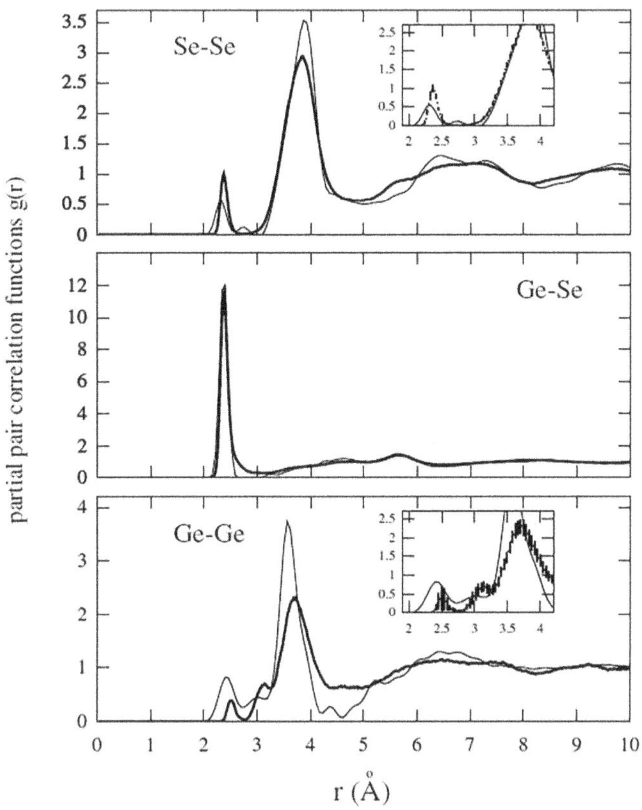

Figure 6.7. Pair correlation functions of glassy $GeSe_2$. Thick line: FPMD results, reference [8]. Thin line: neutron scattering experiments, reference [5]. In the case of $g_{GeGe}(r)$ and $g_{SeSe}(r)$, the interval of distances between 1.9 Å and 4.2 Å has been magnified in the inset and the associated error bars are shown. Reprinted figure with permission from [8]. Copyright (2008) by the American Physical Society.

chalcogenide disordered networks. Once again, from the methodological point of view, the success in reproducing most of the experimental quantities issued from structural measurements is moderately undermined by inaccuracies in the description of the environment pertaining to Ge atoms.

6.2.4 Modeling these two systems: some thoughts

When trying to summarize the main achievements contained in the investigations of glassy $GeSe_2$, the following considerations are in order. Both liquid and glassy $GeSe_2$ can be described as systems in which the tetrahedron is the main short-range structure, coexisting with homopolar bonds and miscoordinations. In the case of amorphous $GeSe_2$, several temporal trajectories have been produced, via quenches from uncorrelated configurations of the liquid. The deviation from chemical order found in the liquid (the number of defects, coordinations other than four for Ge and other than two for Se) can be reduced as a result of the cooling process, showing that high quench rates do not prevent structural change driven by the temperature

Table 6.2. Average number $n_\alpha(l)$ (expressed as a percentage) of Ge and Se atoms l-fold coordinated at a distance of 2.7 Å. For each value of $n_\alpha(l)$, we give the identity and the number of the Ge and Se neighbors. $n_\alpha(l)$ has been defined in section 2.4.1. We also compare calculated and experimental values (in percentage) for the number of Ge atoms forming edge-sharing connections, N_{Ge}(ES), the number of Ge atoms forming corner-sharing connections, N_{Ge}(CS), the number of Ge atoms involved in homopolar bonds, N_{Ge-Ge}, and the number of Se atoms involved in homopolar bonds, N_{Se-Se}. Experimental values are taken from [15]. Reprinted table with permission from [8]. Copyright (2008) by the American Physical Society.

Ge	$l = 1$		$l = 2$	
	Se	5 ± 1	Se_2	12 ± 1
	$l = 3$		$l = 4$	
	Se_3	8 ± 1	$GeSe_3$	5 ± 2
			Se_4	70 ± 4
Se	$l = 1$		$l = 2$	
	Ge	4 ± 1	Se_2	3 ± 1
			SeGe	21 ± 1
			Ge_2	69 ± 1
	$l = 3$			
	Ge_3	4 ± 1		
	N_{Ge}(ES)	N_{Ge}(CS)	N_{Ge-Ge}	N_{Se-Se}
Reference [8]	45 ± 4	50 ± 6	5 ± 2	24 ± 2
Experiment	34	41	25	20

reduction. FPMD is reliable and quite realistic throughout. However, its weak point resides in the treatment of Ge–Ge correlations for both liquid and glassy $GeSe_2$, as revealed by the shape of the pair correlation functions $g_{GeGe}(r)$. Knowing that the use of the GGA (in the Perdew–Wang form) was already crucial to improve improved the structural description with respect to LDA, this appeared as a good starting point in the search of alternative XC recipes capable of bringing theory in better agreement with experiments. Also, consideration of larger size could not be excluded, despite our repeated efforts to demonstrate the consistency of calculations with $N_{at} = 120$. Achievements aimed at going beyond the scheme employed for liquid and glassy $GeSe_2$ during the period 1998–2007 (PW XC functional and $N_{at} = 120$) will be the topic of section 6.7.

6.3 The origin of the first-sharp diffraction peak

6.3.1 FSDP in the total structure factor

The presence of the FSDP in the total structure factor and in some of the partial structure factors of disordered network-forming materials has stimulated intense work aimed at understanding its atomic-scale origin. In the previous sections we have provided examples of the occurrence of this feature in liquid and glassy $GeSe_2$. By leaving aside for a moment the specific case of these systems and the review of structural properties of other chalcogenides, it is of interest to focus on a series of attempts that were made to link the FSDP to specific structural motifs.

As a first step, we concentrate on the analysis of some interpretations proposed in the literature, with the intent of assessing their general character. The team of Vashishta [4] was the first to demonstrate that the FSDP can be observed even in binary AX_2 model systems made of charged hard spheres, provided one selects appropriate values for the charges and the relative dimensions of the hard spheres, resulting in the right combination of steric and charge effects. To put it in simple, intuitive terms, this means that the FSPD comes along when the tetrahedra take over, while it tends to disappear when this main structural motif is no more predominant. This is exactly what happens in [4] when the system is neutral, approaching the random close packing, at least for Se atoms. Therefore, the appearance of the FSDP is not related to the details of the potential or, better, to details of chemical bonding, the feature being a fingerprint of a predominant structural motif. While this statement says little about the microscopic origin (since it determines merely under which condition the FSDP appears), it is useful to approach this issue from the opposite viewpoint, namely by looking for specific conditions causing the FSDP to disappear.

6.3.1.1 Learning from liquid $GeSe_2$ at high temperatures

In [16] we have compared the total neutron structure factors of liquid $GeSe_2$ at $T = 1050$ K (as shown in section 6.2.2) and at $T = 1373$ K, for which experimental data were also available. The strong decrease of the FSDP visible in figure 6.8 can be correlated to lower percentages of Ge forming $GeSe_4$ (or defective) tetrahedra, lowering to 37%. Also, the short-range chemical disorder is enhanced at higher

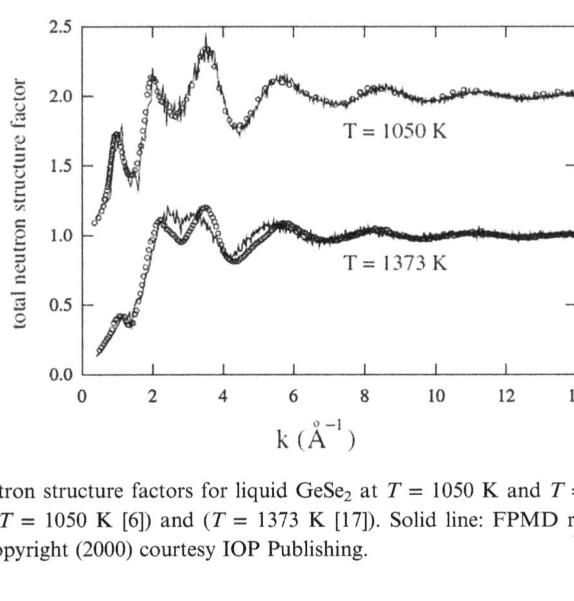

Figure 6.8. Total neutron structure factors for liquid $GeSe_2$ at T = 1050 K and T = 1373 K. Open circles: experimental results (T = 1050 K [6]) and (T = 1373 K [17]). Solid line: FPMD results (T = 1050 K [7]) (T = 1373 K [16]). Copyright (2000) courtesy IOP Publishing.

temperatures, since a larger number of coordination different than fourfold (for Ge) and twofold (for Se) are present in the network. Incidentally, it is worth mentioning that the notion of chemical order in AX_2 systems is particularly straightforward to check, since perfect chemical order corresponds to the highest number of atoms forming heterogeneous (non-homopolar) bonds.

Based on these results we are in a position to state that the FSDP (at the level of the total structure factor) shows up when there is a predominant unit accounting for more that ~50% of all structural motifs describing the coordination of atomic species. This structural motif is the tetrahedron in the case of disordered $GeSe_2$, the impact of changes in the composition being also a crucial factor to determine the structure of Ge_xSe_{1-x} networks, as it will be shown in chapter 8.

6.3.1.2 State of the art and available interpretations

Having in mind the atomic-scale origins of the FSDP, it also appropriate to check previous hypotheses advanced to rationalize the existence of correlations on length scales going beyond nearest-neighbor bond distance [18]. About 20 years ago, we started from the observation that a wide range of interpretation schemes had been made available to link the FSDP to some specific structural motifs [1–4, 18–26], by coming to the conclusion that FPMD could be instrumental to revisit two of them. The first set of interpretations can be labeled as the *layers* conjecture, since it views the FSDP as linked to some reminiscent crystalline order that manifests itself through the presence of layers. These are separated by a distance prompting the appearance of the FSDP in reciprocal space [26, 27]. While there is no reason *a priori* to look for signatures of the crystalline state in an amorphous structure beyond nearest neighbors, it should be recalled that some authors were somewhat attracted into this hypothesis by the presence of a layered structure in the corresponding crystals [22–24]. A second interpretation (termed hereafter *cluster–voids*) is based on the existence of regions having a peculiar low-density inner parts embedded in 'voids', related by distances typical of intermediate range order [18, 21].

6.3.1.3 Checking the different hypotheses

We have taken advantage of trajectories previously obtained by FPMD for liquid SiO_2 [28] and liquid $GeSe_2$ (as detailed above in section 6.2.2) at 3500 K and 1050 K, respectively. The case of liquid SiO_2 is worth commenting on briefly, since it corresponds to a disordered network chemically ordered with very few defects, unlike liquid and glassy $GeSe_2$ for which we have seen that homopolar bonds and miscoordinations are an important structural feature. The occurrence of layered regions, at the very heart of the *layers* conjecture, is based on the evaluation of the **k**-vector dependent structure factor $A(\mathbf{k})$ and its spherical average $S(k)$:

$$S(k) = \frac{1}{N_k} \sum_{|\mathbf{k}|=k} A(\mathbf{k}), \tag{6.1}$$

where N_k are the N vectors $\mathbf{k}$ for which $|\mathbf{k}| = k$, the available $\mathbf{k}$ vectors being discretized. Deviations from the average $S(k)$ are given by the second moment:

$$M^{(2)}(k) = \frac{1}{N_k} \sum_{|\mathbf{k}|=k} [\langle A(\mathbf{k})^2 \rangle - \langle S(k) \rangle^2] / \langle S(k) \rangle^2, \tag{6.2}$$

where $\langle\rangle$ means an equilibrium average. A value of 1 for $M^{(2)}(k)$ corresponds to Gaussian statistics [27], while larger $M^{(2)}(k)$ indicate the formation of planes.

In the *cluster–voids* approach, a prepeak is found in the Bhatia–Thornton concentration–concentration partial structure factor $S_{CC}(k)$ [10] for a system made of 'clusters' and 'voids' particles [18, 21]. While we took Si or Ge to define the coordinates of the clusters, the voids are determined via a Voronoi analysis [29, 30], with the vertices of the Voronoi polyhedra as positions of the void particles. From the positions of clusters (M = Si, Ge) and voids (V) the partial structure factors $S_{VV}(k)$, $S_{VM}(k)$, and $S_{MM}(k)$ are derived, with the corresponding Bhatia–Thornton $S_{CC}(k)$ taking the form

$$S_{CC}(k) = c_V\, c_M\, [c_M\, S_{VV}(k) + c_V\, S_{MM}(k) - 2(c_V c_M)^{1/2}\, S_{VM}(k)], \tag{6.3}$$

where c_M is the concentration of cluster particles and c_V the concentration of voids.

The results shown in figures 6.9 and 6.10 for liquid SiO_2 and liquid $GeSe_2$, reveal that $M^{(2)}(k)$ is smaller than 1 for any k, this means that crystalline-like layers are absent in our systems. Therefore, we have obtained a clear indication that the appearance of a FSDP does not correspond to the existence of layers. By turning to the *cluster–voids* approach, it turns out that for both liquids, the Bhatia–Thornton $S_{CC}(k)$ calculated on the system made of clusters and voids is characterized by a peak at the location of the FSDP in the total structure factor (figures 6.9 and 6.10). Therefore, at a first sight, one could conclude that the cluster–void scheme is validated by the behavior of liquid SiO_2 and liquid $GeSe_2$, in line with early indications collected for vitreous silica [21]. However, before concluding on the general validity of this model, one has to check whether or not the peak is also present when the FSDP is absent. To work on this issue, we resorted to the LDA model for liquid $GeSe_2$ referred to in section 3.6.5 since in this case the FSDP in the total structure factor is absent. As it can be seen in figure 6.11, both $M^{(2)}(k)$ and $S_{CC}(k)$ are close to those obtained for the GGA model of liquid $GeSe_2$ in figure 6.10, that is to say the one featuring a FSDP in the total structure factor. In particular, the observation of a peak at the FSDP location in $S_{CC}(k)$ obtained within the 'cluster–voids' hypothesis [21] reveals this criterion cannot be associated with the occurrence of a FSDP in the total structure factor since the peak is there even when the FSDP does not manifest itself in the total structure factor.

In summary, we were able to demonstrate that the FSDP is not correlated to crystalline-like layers, while cluster–void correlations are common to all disordered networks examined, regardless of their featuring or not a FSDP in the total structure factor. Despite the interest of the conjectures advanced to account for the FSDP, one can conclude that the only criterion standing up to criticism with regard to its

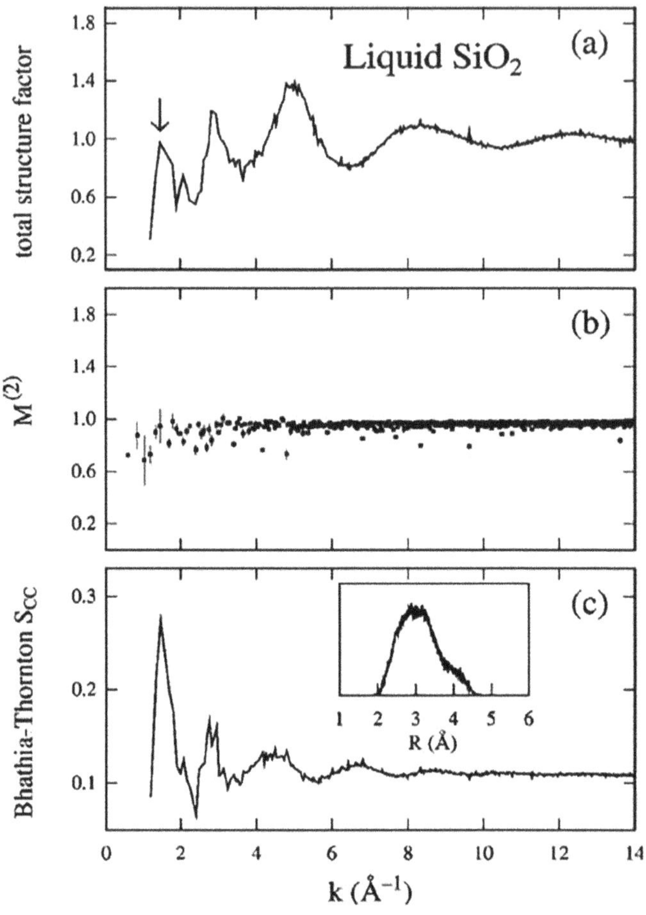

Figure 6.9. An analysis of the atomic configurations of liquid SiO_2 obtained by first-principles molecular dynamics at $T = 3500$ K [28, 31]. (A) Neutron total structure factor; (B) second moment $M^{(2)}(k)$ (see equation (6.2)); (C) Bhatia–Thornton $S_{CC}(k)$ for the cluster–void system (see equation (6.3)). The inset shows the distribution of radii corresponding to the void particles. The arrow indicates the position of the first-sharp diffraction peak. Reprinted figure with permission from [49]. Copyright (2001) by AIP Publishing.

general validity is the one linking the FSDP to the presence of a predominant structural unit, as detailed at the beginning of section 6.3.1. There we had invoked the comparative study carried out at two different temperatures for liquid $GeSe_2$ [16] to point out the different structural properties and the different intensities of the FSDP found as a function of the temperature.

6.3.2 FSDP in the concentration–concentration partial structure factor

We have seen that the establishment of a correlation between the presence of the FSDP and structural features at the atomic scale can be obtained by invoking a predominant motif, as the tetrahedron for Ge_xSe_{1-x} systems. This motif is indeed

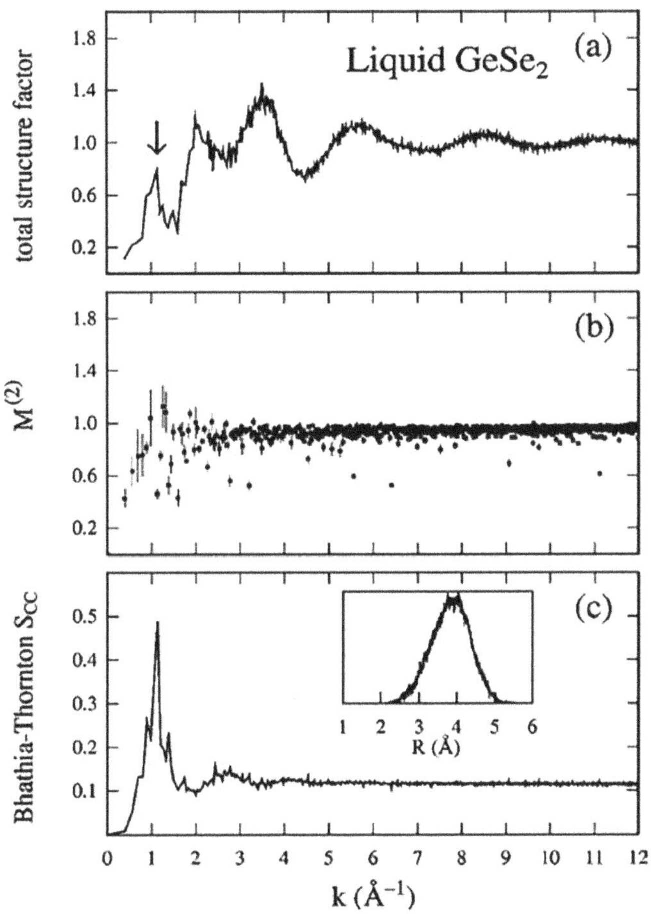

Figure 6.10. The same as in figure 6.9, but for liquid $GeSe_2$ at T = 1050 K [32]. Reprinted figure with permission from [49]. Copyright (2001) by AIP Publishing.

discernible in disordered network-forming systems and can even be recovered in simple models of charge hard spheres [4]. For the sake of clarity, let us call $FSDP_{tot}$ the FSDP in the total (neutron) structure factor we have considered so far. More subtle is to rationalize the existence of the FSDP in the partial structure factors and, in particular, in the concentration–concentration structure factor $S_{CC}(k)$, a very elusive feature since related to fluctuations of concentration on intermediate range order scales. We shall term hereafter this feature $FSDP_{cc}$. To begin our search of a microscopic (atomic) origin of $FSDP_{cc}$, we can refer to the behavior of the FSDP height (FSDP-*h* in the following) in liquid $GeSe_2$ (section 6.2.2 and [7]). FSDP-*h* was found highly dependent on time, as proved by the associated error bars (20%). This observation prompted us to reanalyze the trajectories of liquid $GeSe_2$ by considering the changes in time of the height of $FSDP_{cc}$, that takes values in between a minimum of 0.04 and a maximum of 0.36. We have separated the trajectories in two different subsets, depending on the values of the height of $FSDP_{cc}$ that can be higher or

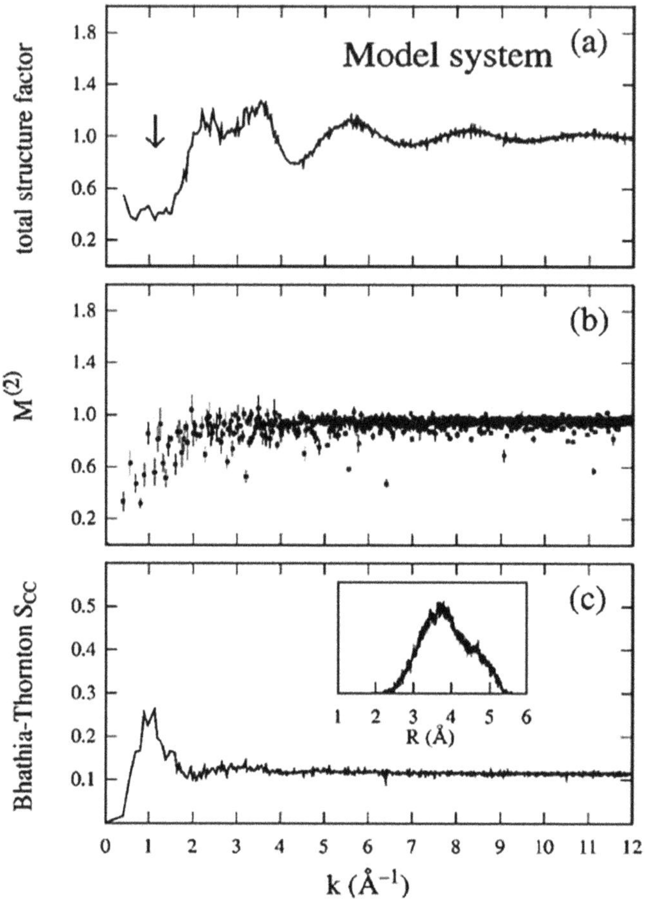

Figure 6.11. The same as in figure 6.10, but for a model of liquid $GeSe_2$ which does not show a FSDP in the total structure factor (see figure 3.6 and references therein). The arrow indicates the position of the first-sharp diffraction peak for liquid $GeSe_2$ at $T = 1050$ K (see figure 6.10). Reprinted figure with permission from [49]. Copyright (2001) by AIP Publishing.

smaller than the average 0.136. These two subsets are referred to as $\mathcal{E}_{\mathrm{low}}$ (FSDP-h in $S_{\mathrm{CC}}(k) \leqslant 0.136$) and $\mathcal{E}_{\mathrm{high}}$ (FSDP-h in $S_{\mathrm{CC}}(k) > 0.136$).

By taking advantage of this definition we introduce a new strategy for the identification of structural features that correspond to vanishing or sizeable values for the intensity of $\mathrm{FSDP_{cc}}$ [33]. As a first indication (see figure 6.13), we can notice that $S_{\mathrm{GeGe}}(k)$ averaged over $\mathcal{E}_{\mathrm{high}}$ is in much better agreement with experiments than its counterpart calculated on the full trajectory and than the same property calculated over $\mathcal{E}_{\mathrm{low}}$.

This improvement is also found in S_{CC} and $S_{\mathrm{NC}}(k)$, which are in better agreement with experiments when calculated by considering the trajectory $\mathcal{E}_{\mathrm{high}}$ (figure 6.14). Therefore, it appears that our search of a correlation between the height of $\mathrm{FSDP_{cc}}$ and specific structural feature is fully legitimate and can bring new information on

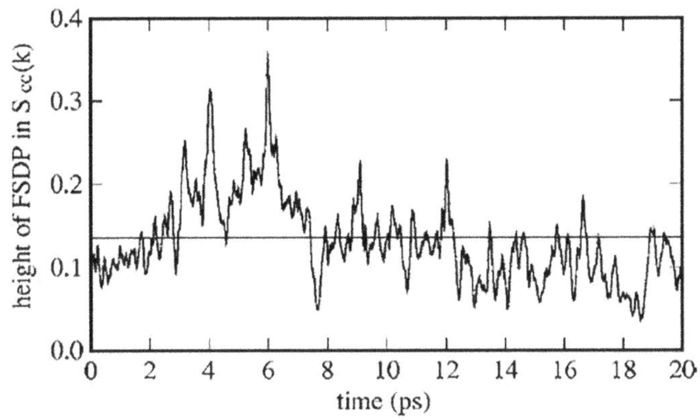

Figure 6.12. Time evolution of the height of the FSDP in $S_{CC}(k)$ for liquid $GeSe_2$. The average value of 0.136 is indicated by a horizontal line and is used to distinguish between the two sets of instantaneous configurations $\mathcal{E}_{low}$ and $\mathcal{E}_{high}$. Reprinted figure with permission from [33]. Copyright (2007) by the American Physical Society.

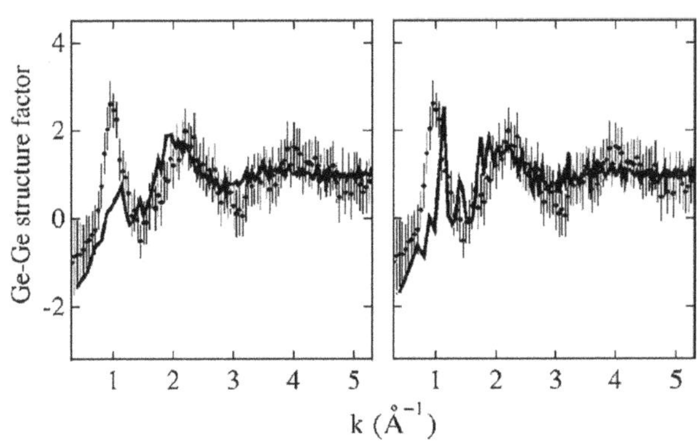

Figure 6.13. Ge–Ge partial structure factor for liquid $GeSe_2$. Experiment: dots with error bars [6] compared to the FPMD result obtained by averaging separately over all configurations in $\mathcal{E}_{low}$ (left) and $\mathcal{E}_{high}$ (right). Reprinted figure with permission from [33]. Copyright (2007) by the American Physical Society.

the atomic-scale origins of $FSDP_{cc}$. The idea is to perform the analyses employed for the studies of liquid and glassy $GeSe_2$ but on the two separate trajectories. Since we obtained close values for the coordination numbers, the bond-angle distributions, the pair correlation functions and the ring statistics on these two trajectories, we focused our attention on the number of Ge atoms belonging to two, one and zero fourfold rings (termed Ge(n) with $n = 2$, 1, and 0, respectively (see figure 6.15)).

The tetrahedra having in common only one Se atom are those contributing to the case $n = 0$, while those sharing two Se atoms ($n = 1, 2$) are those corresponding to edge-sharing connections. We found that the only notable difference between $\mathcal{E}_{low}$ and $\mathcal{E}_{high}$ in this respect is the larger value of Ge(2) in $\mathcal{E}_{high}$ (11% for $\mathcal{E}_{high}$ and 7.5% for $\mathcal{E}_{low}$) of the total number of Ge atoms. A typical configuration having a Ge(2) atom is made of two fourfold rings in a chain fashion and sharing a Ge(2) atom only. An example is given in figure 6.15 where one can observe a Ge(1)–Ge(2)–Ge(1) chain.

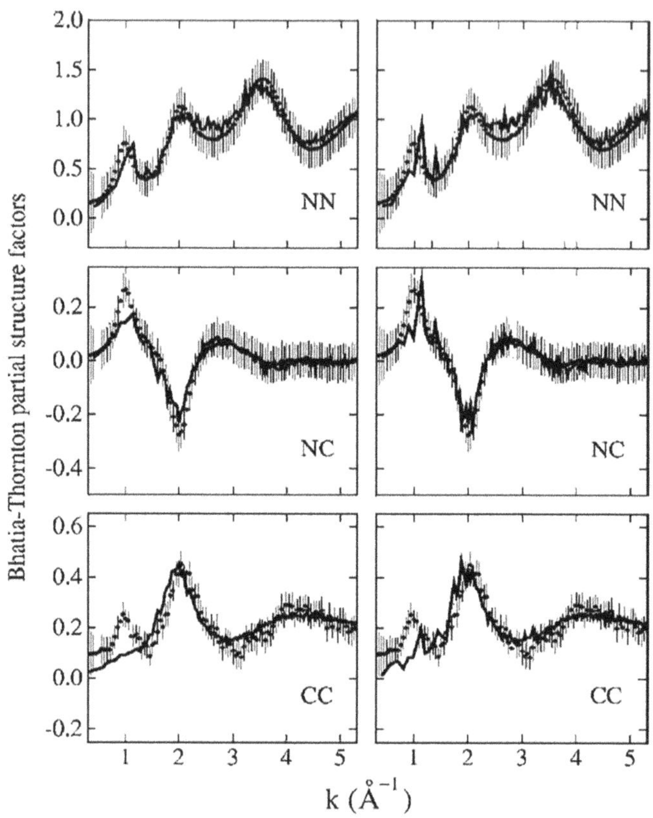

Figure 6.14. The same as in figure 6.13, but for the Bhatia–Thornton structure factors. Reprinted figure with permission from [33]. Copyright (2007) by the American Physical Society.

We found useful to select these configuration and name them Ge*. Interestingly, the number of Ge* in $\mathcal{E}_{\mathrm{high}}$ is higher than in $\mathcal{E}_{\mathrm{low}}$ (8.5% vs 7.5%, respectively).

It is important to realize that not all Ge(2) atoms form Ge* subunits[2], since some of them share not only a Ge(2) atom but in addition a Se atom threefold coordinated. This is exemplified in figure 6.15 where we have highlighted the difference between a Ge(2) atom [Ge(2)^a] being part of a Ge* unit and another Ge(2) atom [Ge(2)^b], corresponding to a configuration that features also a connection to a Se atom. In other words, ring 2 and ring 3 do not form a Ge* unit as opposed to ring 1 and ring 2. This classification allows pushing further the search of correlations between the height of $\mathrm{FSDP_{cc}}$ and the number of Ge(2) atoms and, in particular, of Ge* units.

To this end, we have recorded for each instantaneous configuration the height of $\mathrm{FSDP_{cc}}$ by obtaining its average value and error bar, to be linked to the different values of Ge(n) that are also extracted from the same configurations. No correlation is found between the amount of Ge(n) (n = 0,1) and the FSDP-h in the partial structure factor $S_{\mathrm{GeGe}}(k)$. Similarly, no correlation with FSDP-h is found for $S_{\mathrm{GeSe}}(k)$ and

[2] Ge* are defined in this section both as 'units' or 'subunits'.

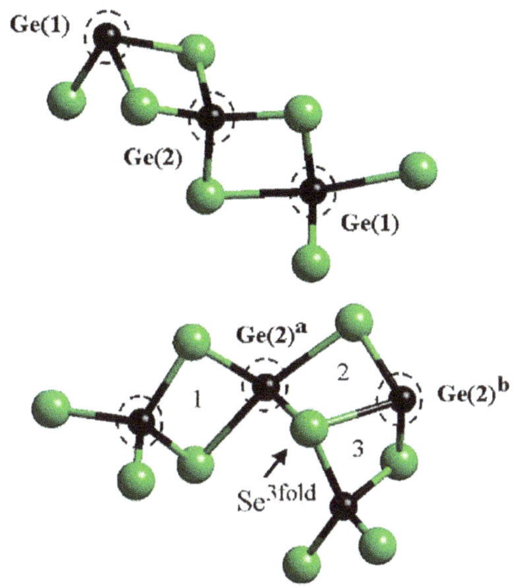

Figure 6.15. Snapshots of structural subunits found in our model of liquid $GeSe_2$. Ge atoms are black and Se atoms are green. Bonds are drawn when two atoms are separated by less than 3 Å, the first minimum in the Ge–Se pair correlation function. Ge atoms forming a Ge* subunit are surrounded by a dashed line. Reprinted figure with permission from [33]. Copyright (2007) by the American Physical Society.

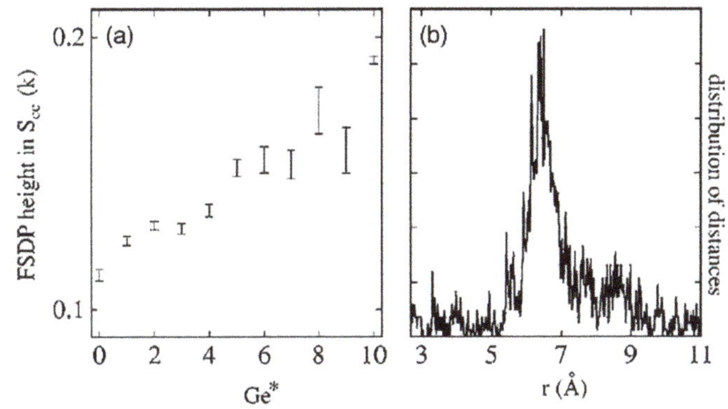

Figure 6.16. Left: height of the FSDP in $S_{CC}(k)$ as a function of the number of Ge* subunits. Right: distribution of distances between pairs of Ge atoms at the opposite ends of Ge* subunits. Reprinted figure with permission from [33]. Copyright (2007) by the American Physical Society.

$S_{SeSe}(k)$ for all values of n, this result being due to their little sensitivity to the different kind of trajectories, namely $\mathcal{E}_{low}$ or $\mathcal{E}_{high}$. While the height of the FSDP in $S_{GeGe}(k)$ is not sensitive to the occurrence of Ge(0) or Ge(1), a linear growth is observed as a function of the number of Ge* units, the height of FSDP $_{cc}$ increasing with the number of Ge*, as shown in figure 6.16(a). *Accordingly, we are in a position to label Ge* as the main structural motif accounting for the existence of the FSDP in* $S_{CC}(k)$, *thereby providing a clear link between the occurrence of fluctuations of concentration on* *intermediate range scales and the presence of Ge(1)–Ge(2)–Ge(1) chains.* As a final step of this analysis, it is important to find out whether there are distances related to the Ge* units that can be taken as corresponding to intermediate range order. This is substantiated by the frequency of occurrence of distances between pairs of Ge atoms at the ends of the chain in the Ge* units (figure 6.16(b)). The peak noticeable at ∼6.5 Å in figure 6.16(b) is consistent with the relationship between the value of r in real space (for a given peak) and the value of k of a corresponding peak in reciprocal space, $k \cdot r \approx 7.7$. We have obtained this relationship by accounting for the first maximum of the spherical Bessel function and by knowing that the FSDP position for liquid $GeSe_2$ is at 1.13 $Å^{-1}$. These results provide a prototypical example of the possibilities offered by molecular dynamics (and, in particular, by FPMD as a method to describe realistically chemical bonding at finite temperatures) to establish a correspondence between a measurable quantity and a specific structural detail. Each instantaneous configurations contains a lot of precious information that are quite often overlooked or not sufficiently exploited since one has as primary goal the calculation of average properties. In this case, little would have been learned by limiting our analysis to taking averages on the whole trajectory. On the contrary, by taking advantage of the mere observation that the FSDP strongly fluctuates in time, we were able to derive a methodology that highlights two contrasting behaviors, related to different amounts of Ge(1)–Ge(2)–Ge(1) chains of connected Ge atoms.These are at the very origin of the FSDP in the concentration–concentration structure factor $S_{CC}(k)$.

6.4 FSDP in disordered network: some considerations before to go on

So far we have seen that glassy (and liquid) $GeSe_2$ have an intriguing behavior in terms of partial and total structure factors, driving many questions on the nature of intermediate range order that go beyond the simple interest for a given system. The relevant pieces of evidence can be summarized as follows:

1. The presence of the FSDP in the total structure factor can be associated to a predominant structural motif. While this simple statement does not undermine all models developed in the search of an alternative origin, we can safely assume that such a simple explanation has a general validity, substantiated by MD approaches of an increasing complexity (from charges hard spheres to first-principles calculations).
2. While the FPMD approach based on the PW-GGA scheme is able to provide an excellent agreement between theory and experiment for the total neutron structure factor, there is striking disagreement at the level of the partial structure factors, in particular for $S_{GeGe}(k)$ and the concentration–concentration partial structure factor $S_{CC}(k)$. The presence of the FSDP in this latter, clearly visible in experiments, and meaning fluctuations of concentration on intermediate range distances, is not confirmed by our level of theory, since the FSDP in $S_{CC}(k)$ is simply absent in the liquid and is barely discernible in the glass.
3. The situation becomes quite paradoxical if one considers that the absence of the FSDP in the early models of disordered $GeSe_2$ was proposed as a general feature of AX_2 networks. On a purely intuitive basis, this cannot be taken as

a demonstration that models not accounting explicitly for the electronic structure are as performing as FPMD ones, since one expects to find a difference between $S_{CC}(k)$ and $S_{zz}(k)$ for a realistic system, while they differ simply by a constant factor in a point-charged one.

4. By looking at many instantaneous configurations within a trajectory, we have realized that the FSDP in $S_{CC}(k)$ is due to specific structural units, namely the chains of edge-sharing tetrahedra. This means that the FSDP can be observed provided certain microscopic conditions are met.

The above observations call for a series of calculations to be performed in order to clarify all these issues and move ahead into a full understanding of intermediate range order in disordered network-forming materials.

1. Experiments provided evidence for the presence of the FSDP in $S_{CC}(k)$ for a set of systems. It is important to determine whether the FPMD methodology can reproduce some of these results, thereby proving that liquid $GeSe_2$ is somewhat a special case deserving unprecedented levels of accuracy in the theoretical treatment.
2. The existence of different cases differing by the presence of the FSDP in $S_{CC}(k)$ is a motivation to classify them with respect to their structural details. Are all of the systems with no FSDP in $S_{CC}(k)$ sharing the same topology? Does the absence or the presence of this feature have an impact on (or is a consequence of) the atomic structure? Is it possible to label these networks in a comparative fashion by considering their intermediate range properties as expressed by $S_{CC}(k)$?
3. It would also be quite instructive to focus, within a first-principles scheme, on a comparison between $S_{CC}(k)$ and $S_{zz}(k)$ for some of these systems. The FSDP is absent in $S_{zz}(k)$ for a model potential of point charges. How does it behave when the electronic structure is taken into account? Is there something we can learn by considering on the same footing fluctuations of concentration ($S_{CC}(k)$) and fluctuations of charge ($S_{zz}(k)$)?

All these issues are addressed in the two sections that follow, the first in section 6.5, the second and the third in section 6.6.

6.5 Evidence of FSDP in $S_{CC}(k)$: examples

One would like to establish whether the FPMD methodology is able to capture the presence of the FSDP in the concentration–concentration for some systems. To this purpose, we have selected liquid $GeSe_4$ and liquid $SiSe_2$ [34] that were also considered in the context of our calculations on disordered network-forming materials, despite their being somewhat less representative due to the absence of experimental information on the partial structure factors[3]. These two systems were found to exhibit a higher degree of chemical order with respect to liquid and glassy $GeSe_2$, i.e. a lower number of coordinations differing from four (for Ge) and two

[3] This statement was valid at the time the calculations we are referring to were performed. Recent experimental results on the partial structure factors of glassy $GeSe_4$ can be found in [35].

(for Se) and a smaller number for homopolar bonds (this applies in particular when comparing networks of the same concentration, $GeSe_2$ and $SiSe_2$). The case of liquid $GeSe_4$ had been the object of calculations reproducing accurately the total neutron structure factor, clearly proving its character of excellent prototype of a chemically ordered network (CON) [36]. The case of liquid $SiSe_2$ was considered to ensure the availability of a model for the glassy phase [37], by concluding that the chemical order was higher than in liquid $GeSe_2$, despite the same value for the electronegativity of Ge and Si (1.8) [38]. Having invoked several times the notion of chemical order as a key tool to compare the topologies of different networks, it is instructive, before moving ahead to the explicit calculations of experimental probes relative to the structure, to have a global view of the behavior of these three disordered systems ($GeSe_4$, $SiSe_2$ and $GeSe_2$, the three in their liquid state). Table 6.3 contains data elucidating the nearest-neighbor structure of these networks, by exploiting the quantities defined in section 6.3. The same table refers also to two phenomenological models for the network structure, the random covalent network (RCN) and the CON. Both are consistent with the so-called 8−N rule, which gives the coordination number of an atom in terms of its position (column *N*) in the periodic table. The structure of a liquid at a given composition follows the RCN model if neither homopolar or heteropolar bonds are preferred. On the contrary, CON holds when the number of heteropolar bonds is the highest accessible for that composition, knowing that in any case for a given composition, the total coordination numbers for the two models are the same. With no loss of validity, we refer in this analysis to the sets of data available when the paper we refer to was published [34]. While some of these data were revisited and somewhat improved later, all of the conclusions drawn thereof remain fully valid. In particular, neither the larger systems nor alternative choices for the exchange–correlation functional changed the essence of the results.

One notices that liquid $GeSe_2$ cannot be considered as chemically ordered since n_{Se}^{Se} is far from the CON value. This same model is much more realistic that the RCN one,

Table 6.3. Values for the coordination numbers n_{Ge}, n_{Se}, n_{Si} and the total coordination number n_{tot} (also called n_{ave} in section 2.4.1) of liquids $GeSe_2$, $GeSe_4$ and $SiSe_2$. The definitions are given in section 2.4.1. Copyright (2003) courtesy IOP Publishing.

$GeSe_2$	n_{Ge}^{Ge}	n_{Ge}^{Se}	n_{Ge}	n_{Se}^{Se}	n_{Se}^{Ge}	n_{Se}	n_{tot}
FPMD	0.04	3.76	3.80	0.37	1.88	2.25	2.77
RCN	2	2	4	1	1	2	2.67
CON	0	4	4	0	2	2	2.67
$GeSe_4$	n_{Ge}^{Ge}	n_{Ge}^{Se}	n_{Ge}	n_{Se}^{Se}	n_{Se}^{Ge}	n_{Se}	n_{tot}
FPMD	0.06	3.87	3.93	1.04	0.97	2.01	2.39
RCN	1.33	2.67	4	1.33	0.67	2	2.4
CON	0	4	4	1	1	2	2.4
$SiSe_2$	n_{Si}^{Si}	n_{Si}^{Se}	n_{Si}	n_{Se}^{Se}	n_{Se}^{Si}	n_{Se}	n_{tot}
FPMD	0.02	3.90	3.92	0.13	1.95	2.08	2.69
RCN	2	2	4	1	1	2	2.67
CON	0	4	4	0	2	2	2.67

definitely unable to describe the network structure, as confirmed by the fact the coordination numbers were found to be acceptably close to experiments. Comparison with liquid $GeSe_4$ and liquid $SiSe_2$ reveals that these two systems are closer to a CON network. In particular, for liquid $GeSe_4$ this means that the majority of Ge atoms form $GeSe_4$ tetrahedra, while some of the Se atoms form homopolar bonds since the concentration is far from the stoichiometric one. Therefore, these data indicate that a transition from a network departing (moderately) from a CON (as $GeSe_2$, not featuring the FSDP in $S_{CC}(k)$) to a network more chemically ordered (as $GeSe_4$, featuring some form of FSDP in $S_{CC}(k)$, see below) occurs while increasing the content of Se atoms within the family of Ge_xSe_{1-x} disordered networks. Similarly (see below) the data on liquid $SiSe_2$ confirm the correlation between some extent of chemical order and the appearance of the FSDP in $S_{CC}(k)$.

To substantiate the previous assertions we focus on the behavior of $S_{CC}(k)$ for the systems under consideration. The case for liquid $GeSe_2$ has been already treated in section 6.2.2 and, in particular, via figure 6.2. Concerning $GeSe_4$, a new set of data were produced by starting from those published in 1998 [36][4]. As visible in figure 6.17, one obtains only a small peak in the FSDP region with the ratio between the heights of the first two peaks equal to $0.12/0.43 = 0.28$, half the experimental value of 0.56. In liquid $GeSe_4$ the FSDP-like signature at 1 $Å^{-1}$ gives a ratio between the first two peaks equal to 0.45, much higher than for the $GeSe_2$ case. In the case of liquid $SiSe_2$ a distinct bump exists at the FSDP location in $S_{CC}(k)$. Despite the higher level of chemical order found in this system when compared to liquid $GeSe_2$, this is not reflected by the ratio between the intensities of the first two peaks (0.22 versus 0.28 for liquid $GeSe_2$). Nevertheless, the FSDP-like feature is clearly more discernible in the case of $SiSe_2$, by pointing out a higher amount of fluctuations of concentration than in the corresponding $GeSe_2$ network.

The above results can be commented on by invoking a correlation between the presence of the FSDP in $S_{CC}(k)$ and the amount of chemical disorder. So far, we have been able to identify two categories, the first corresponding to the absence of the FSDP in $S_{CC}(k)$ for a system (like $GeSe_2$ when modeled within FPMD at the PW-GGA level) with a sizeable deviation from chemical order while the second features a network closer to the CON model and the appearance of a discernible feature in $S_{CC}(k)$ at the FSDP location. This is also the case of the experimental results for liquid $GeSe_2$ [6]. To complete this first classification and pave the way for further studies, one needs some indication on the behavior of essentially perfect networks (no homopolar bonds and no deviations to the regular tetrahedral arrangement) as encountered in the case of amorphous GeO_2 and SiO_2, where the difference of electronegativity between the species favors ionicity. It appears that the FSDP is absent in the measured $S_{CC}(k)$ of amorphous GeO_2 [39], a result confirmed by available FPMD data on the analogous SiO_2 networks [31]. The existence of this third category of networks (perfect chemical order and no FSDP in $S_{CC}(k)$) calls for a well established classification, involving also any possible correlation existing

[4] More on liquid $GeSe_4$ and, more generally, on this specific composition can be found in chapter 8.

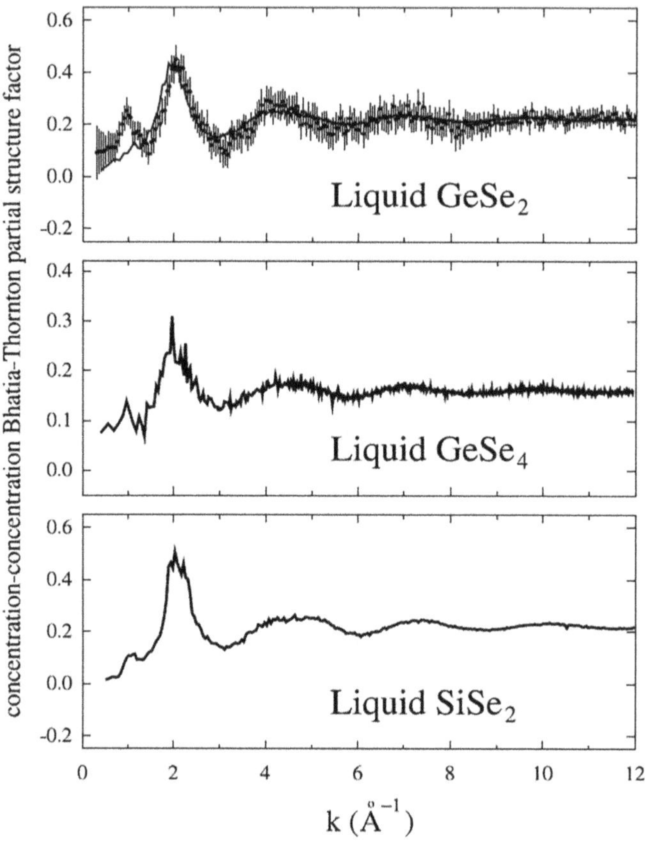

Figure 6.17. Concentration–concentration partial structure factor $S_{CC}(k)$ for liquid $GeSe_2$ [7], liquid $GeSe_4$ and liquid $SiSe_2$. The experimental data for liquid $GeSe_2$ (dots with error bars) are those of [6]. Copyright (2003) courtesy IOP Publishing.

between fluctuations of concentrations and fluctuations of charge. The idea is to establish some precise links between the atomic structures and some measurable properties, by providing information on the conditions under which the networks tend to arrange in some specific manner at the atomic scale.

6.6 What to learn from $S_{CC}(k)$ vs $S_{zz}(k)$

Before describing the behavior of $S_{CC}(k)$ and $S_{ZZ}(k)$ for different systems, classified according to the intensity recorded for the FSDP in $S_{CC}(k)$, it is useful to recall some definitions necessary to highlight similarities and differences among them. Let's consider a binary system made of α and β atoms in concentrations c_α and c_β and the Bhatia–Thornton concentration–concentration partial structure factor $S_{CC}(k)$ as defined via equation (2.6)

$$S_{CC}(k) = c_\alpha c_\beta \{1 + c_\alpha c_\beta [(S_{\alpha\alpha}(k) - S_{\alpha\beta}(k)) + (S_{\beta\beta}(k) - S_{\alpha\beta}(k))]\} \quad (6.4)$$

where $S_{\alpha\alpha}(k)$, $S_{\alpha\beta}(k)$ and $S_{\alpha\beta}(k)$ are the partial structure factors defined with respect to the atomic species.

Based on both calculations and experiments, we were also able to identify three different classes of systems, introduced in the previous section. In the first class (class I) there are networks with a perfect chemical order and no FSDP in $S_{CC}(k)$. For instance, we have systems like SiO_2 and GeO_2, known experimentally or by simulation [31, 39]. To the second class belong network systems with a distinct FSDP in $S_{CC}(k)$, characterized by small departures from chemical order, as we have seen for of liquid $GeSe_4$ and liquid $SiSe_2$ (section 6.5). The representative structure of class III is the one we obtained via FPMD for liquid $GeSe_2$. In this case, like for class I, the FSDP is absent in $S_{CC}(k)$. However, the network is far from being chemically ordered since there are a lot of structural motifs in the nearest coordination shells (section 6.2.2). In what follows, our analysis of fluctuations of concentration and fluctuations of charges on intermediate range scales is substantiated by the calculation of charge–charge structure factor $S_{zz}(k)$ for three AX_2 networks: liquid SiO_2 (class I), glassy $SiSe_2$ (class II) and liquid $GeSe_2$ (class III).

We anticipate that no FSDP shows up in any charge–charge structure factors $S_{zz}(k)$, demonstrating that no charge ordering takes place at intermediate range scales, and this regardless of the occurrence of concomitant fluctuations of concentration.

6.6.1 Calculating $S_{zz}(k)$

To obtain $S_{zz}(k)$ in the DFT–FPMD framework, one has to consider the valence electron density built in the total charge density made of ionic and electronic parts, $\rho_t(\mathbf{r}) = \sum_i z_i \delta(\mathbf{r} - \mathbf{r}_i) + \rho_e(\mathbf{r})$:

$$S_{zz}(k) = N^{-1}\langle z_v^2 \rangle^{-1} \int d\mathbf{r} d\mathbf{r}' \rho_t(\mathbf{r}) \rho_t(\mathbf{r}') e^{i\mathbf{k}\cdot(\mathbf{r}-\mathbf{r}')}. \tag{6.5}$$

Since in the pseudopotential approach the valence electrons are the only ones accounted for in $\rho_e(\mathbf{r})$, the ionic charges z_i are $z_A = +4$ and $z_X = +6$ for the case of the systems treated hereafter. In equation (6.5) we assume the spherical average over the orientations of $\mathbf{k}$, where N is the number of atoms and $\langle z_v^2 \rangle$ is a normalization factor, $\langle z_v^2 \rangle = z_{vA}^2 c_A + z_{vX}^2 c_X$. In the equation for $\langle z_v^2 \rangle$, $z_{vA} = +4$ and $z_{vX} = -2$ are the charges assigned to A and X atoms when a pointlike charge model (PLC) is selected. In the limit $k \to \infty$, $S_{zz}(\infty) = \langle z^2 \rangle / \langle z_v^2 \rangle$, with $\langle z^2 \rangle = z_A^2 c_A + z_X^2 c_X$. For AX_2 systems this gives $S_{zz}(\infty) = 3.66$.

When the PLC model holds, the total charge density takes the form $\rho_t(\mathbf{r}) = \sum_i z_{vi} \delta(\mathbf{r} - \mathbf{r}_i)$. The charge–charge structure factor $S_{zz}(k)$ becomes proportional to $S_{CC}(k)$:

$$\begin{aligned} S_{zz}^{\mathrm{PLC}}(k) &= N^{-1}\langle z_v^2 \rangle^{-1} \sum_{ij} z_{vi} z_{vj} e^{i\mathbf{k}\cdot(\mathbf{r}_i - \mathbf{r}_j)} \\ &= (c_A c_X)^{-1} S_{CC}(k). \end{aligned} \tag{6.6}$$

6.6.2 Comparing $S_{zz}(k)$ and $S_{CC}(k)$ for the three classes of networks

As a first example, we consider a system of class I, liquid SiO_2, for which we have reproduced and extended FPMD simulations published in [28, 31]. In this system, prototype of 'perfect' chemical order, the Si atoms are linked within tetrahedra to O atoms via corner-sharing connections. In figure 6.18, the structure factors $S_{zz}(k)$ and $S_{CC}(k)$ for liquid SiO_2 are compared, providing evidence that $S_{zz}(k)$ is very much different from the concentration–concentration structure factor for which we employ the definition of equation (6.6), that is $S_{CC}(k) \sim S_{zz}^{PLC}(k)$.

Also, the absence of the FSDP in $S_{zz}(k)$ (small signatures in the FSDP region not being statistically significant) are indicative of no charge fluctuations on IRO scales. It is of interest to observe that this system might be conjectured as the best candidate to be described by a pointlike charge model, due to the high ionicity of the bonds between Si and O atoms. However, this is not the case since there are noticeable differences between $S_{CC}(k)$ and $S_{zz}(k)$, illustrating the fact that structural order and

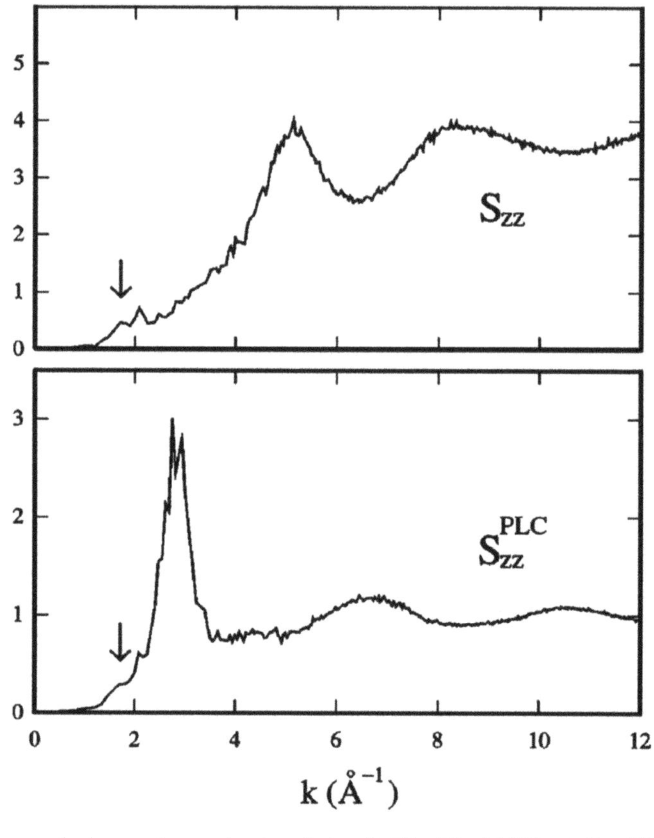

Figure 6.18. Upper panel: charge–charge structure factor $S_{zz}(k)$ of liquid SiO_2 at $T = 3500$ K. Lower panel: $S_{CC}(k)$ of liquid SiO_2 at $T = 3500$ K. $S_{CC}(k)$ is normalized as in equation (6.6). The arrows indicate the location of the FSDP in the total neutron structure factor. Reprinted figure with permission from [42]. Copyright (2004) by the American Physical Society.

charge order express two different physical properties, as a consequence of the delocalized and spatially distributed nature of the electron charge.

As a second example, turning to class II, we consider amorphous $SiSe_2$ that is characterized, as its liquid counterpart, by substantial chemical order and a few structural defects. FPMD helped to show that its atomic structure is made of both corner-sharing and (predominantly) edge-sharing connections [37] with only a few homopolar bonds [37, 40, 41], in a way consistent with coordination numbers typical of a CON-like structure, $n_{Si} = 3.95$, $n_{Se} = 2.04$, $n_{Si}^{CON} = 4$, $n_{Se}^{CON} = 2$. In figure 6.19 a peak is visible at the FSDP location in $S_{zz}^{PLC}(k)$, to be attributed mostly to Si–Si correlations. However, as for the class I system seen before, the charge–charge structure factor $S_{zz}(k)$ does not show any FSDP peak or any signature at the FSDP location (figure 6.19). Remarkably, $S_{zz}(k)$ features a bump at the main peak k value in $S_{zz}^{PLC}(k)$, $k_M \sim 2$ Å^{-1}, to indicate that both fluctuations of charge and fluctuations of concentration are present at short range. However, the peak at $k_{FSDP} = 1$ Å^{-1} proves that fluctuations of concentration are still noticeable for larger distances. We

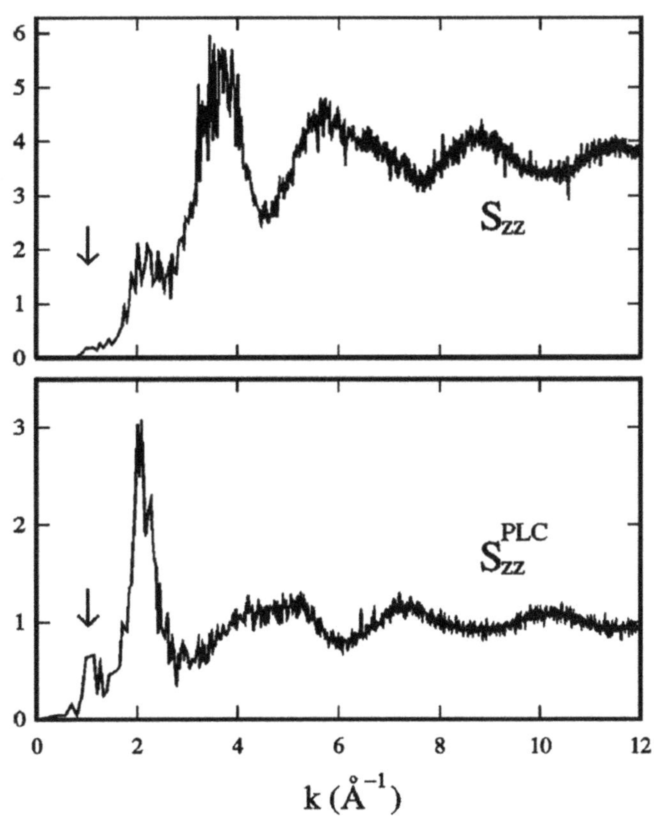

Figure 6.19. Upper panel: charge–charge structure factor $S_{zz}(k)$ of amorphous $SiSe_2$ at $T = 300$ K. Lower panel: $S_{CC}(k)$ of amorphous $SiSe_2$ at $T = 300$ K. $S_{CC}(k)$ is normalized as in equation (6.6). The arrows indicate the location of the FSDP in the total neutron structure factor. Reprinted figure with permission from [42]. Copyright (2004) by the American Physical Society.

concluded that in glassy $SiSe_2$ fluctuations of concentration over IRO length scales are correlated to a small deviation from chemical order, leading to the appearance of a FSDP in the $S_{CC}(k)$, otherwise absent in a network belonging to class I.

As we have mentioned before, to class II belongs also liquid $GeSe_2$ as measured using the method of isotopic substitution in neutron diffraction [6] since a FSDP characterizes $S_{CC}(k)$, as due to correlations involving mostly Ge–Ge interactions. On the other hand, as already underlined in section 6.2.2, the FSDP is absent in $S_{CC}(k)$ as calculated by FPMD, indicating that the absence of this feature can have two distinct origins, the first being the establishment of perfect chemical order (class I) or a high level of structural disorder (as in liquid $GeSe_2$ via FPMD class III). The charge–charge structure factor $S_{zz}(k)$ of such class III system behaves similarly to the case of glassy $SiSe_2$ (see figure 6.20). *Therefore, the lack of any feature at the FSDP location confirms that no charge ordering occurs at IRO scales in disordered network-forming materials. This conclusion holds regardless of the presence of fluctuations of concentration on the same length scale.*

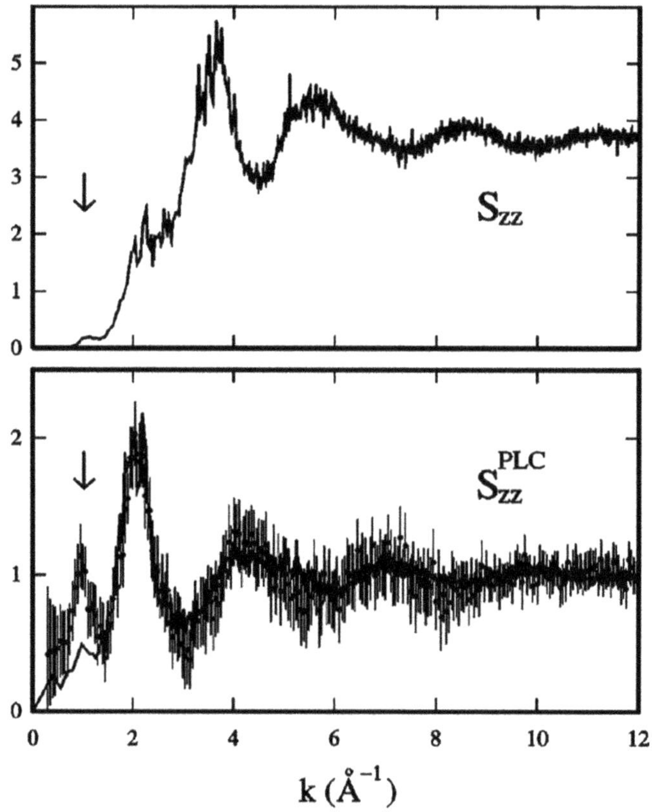

Figure 6.20. Upper panel: charge–charge structure factor $S_{zz}(k)$ of liquid $GeSe_2$ at $T = 1050$ K. Lower panel: $S_{CC}(k)$ of liquid $GeSe_2$ at $T = 1050$ K. $S_{CC}(k)$ is normalized as in equation (6.6). Full line: FPMD [42], dots with error bars: experimental results [6]. The arrows indicate the location of the FSDP in the total neutron structure factor. Reprinted figure with permission from [42]. Copyright (2004) by the American Physical Society.

To summarize and rationalize the results presented here, one can postulate that the appearance of fluctuations of concentration on IRO distances is due the constraint of charge neutrality. For a perfect network, no local variations of the average concentration are required since all atoms of the same species have the same valence. As a result, no FSDP exists in the concentration–concentration structure factor $S_{CC}(k)$. The situation is different when there is a non-negligible amount of chemical disorder via the occurrence of different valence states. In order to have charge neutrality over intermediate distances, the existence of valence states of different nature has to correspond to local variations in the network. This has the effect of producing deviations in the local concentration from the average value, giving rise to fluctuations of concentration at intermediate range distances and to the appearance of a FSDP in $S_{CC}(k)$.

6.7 Improving the description of chemical bonding

6.7.1 Contours of the GGA issue for chalcogenides

The previous sections of this chapter have focused on results for disordered chalcogenides obtained within the FPMD methodology. We have stressed the importance of going beyond the LDA approximation by using GGA schemes, as discussed and exemplified in section 3.6.5 and in section 4.2.1. The GGA scheme selected over our first period of involvement with FPMD simulations of chalcogenides (1995–2008 if taken from the very start) is the one devised by Perdew and Wang [43, 44]. In this section, we shall rationalize our choice of seeking and using alternative GGA recipes with the intention of improving the comparison with experiments for the structural properties of our prototype networks, namely liquid and glassy $GeSe_2$. We recall that the PW choice led to a very satisfactory agreement with experiments for the total neutron structure factor over the entire range of k values as shown in section 6.2.2. This was due to an improved account of the *ionic character of bonding*, as shown in section 3.6.5. However, the careful comparison of each partial pair correlation function showed, especially in the Ge–Ge case, the existence of a residual disagreement between our calculations and the experimental counterpart. The previous section has underscored as the most elusive aspect of this disagreement the one concerning the first-sharp diffraction peak in the concentration–concentration structure factor. In which direction do we have to search when looking for improvements?

6.7.2 Why BLYP?

A first observation concerns the shape of the Ge–Ge correlation function, the experimental counterpart being more structured (figure 6.3). Also the Ge–Ge distances are 15% longer than the measured values (table 6.1). These features can be ascribed, at least tentatively, to an overestimate of the metallic character of bonding. Therefore, when choosing an alternative GGA scheme, it would be desirable to select a scheme capable of optimizing the distribution of the valence-charge densities along the bonds. This means we are interested in GGA functionals promoting a more localized distribution of the valence electrons, delocalization effects being typical of schemes reminiscent of the uniform electron-gas model, as the PW one. Within this context, this section reconsiders liquid and glassy $GeSe_2$ as obtained by using a GGA scheme based on the so-called BLYP XC functional [45, 46] (B stands for the exchange energy [45] by Becke, and LYP by the correlation energy by Lee, Yang and Parr [46]). For the correlation energy, BLYP does not contain any assumption on its uniform electron-gas character, motivating its selection as an appropriate XC functional to perform better than the PW scheme. Indeed, this scheme is expected to increase the localized behavior of the electron density by lowering electronic delocalization that favors the metallic character. BLYP is thought to perform better than PW when the difference of electronegativity between the components Δ_{el} is moderate. In the opposite case (large Δ_{el}) the ionic contribution is large enough to promote charge transfer regardless of the specific XC functional, as it occurs for disordered SiO_2 ($\Delta_{el} = 1.54$), for which LDA proved fairly adequate, at least for structural properties. When one lowers Δ_{el}, the valence-charge density along the bonds is more important and, as a consequence, the relative weight of the covalent and ionic character is harder to appreciate. We have already encountered this case with the behavior of liquid $GeSe_2$, for which $\Delta_{el} = 0.54$. The choice of BLYP corresponds to the attempt of minimizing electronic delocalization effects that are unavoidable and yet undesirable when there is a competition between ionic and covalent contributions. It should be kept in mind, however, that the use of BLYP or of any other GGA recipe favoring electronic localization cannot be based on the simple analysis of the Δ_{el} value. Other factors should be included, such as the bonding changes with thermodynamic parameters as pressure and temperature. There is a direct and unambiguous way of checking whether the use of BLYP can lower the metallic character of bonding: the calculation of the electronic density of states. As shown in figure 6.21, at $T = 1050$ K the electronic densities of states calculated via the BLYP approach is characterized, around the Fermi level, by a deeper pseudogap when compared to the PW one.

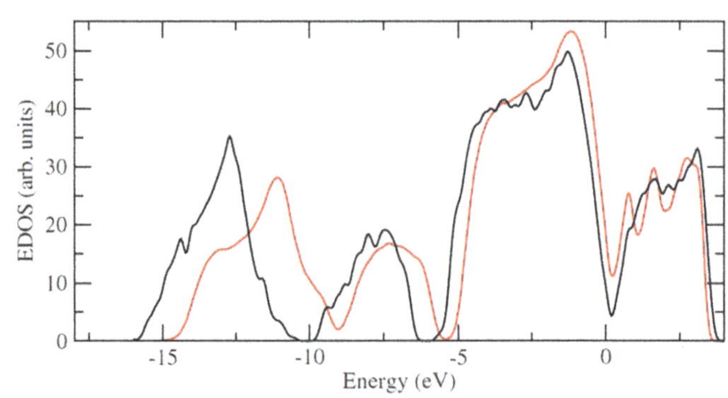

Figure 6.21. Electronic density of states (Kohn–Sham eigenvalues) of liquid $GeSe_2$: black line: BLYP results [47], red line: PW results [7]. A Gaussian broadening of 0.1 eV has been employed. Reprinted figure with permission from [47]. Copyright (2009) by the American Physical Society.

6.7.3 Liquid $GeSe_2$: BLYP vs PW, direct space and short-range properties

The analysis of the partial pair-correlation functions $g_{\alpha\beta}^{\mathrm{exp}}(\mathrm{r})$, $g_{\alpha\beta}^{\mathrm{PW}}(\mathrm{r})$ and $g_{\alpha\beta}^{\mathrm{BLYP}}(\mathrm{r})$ (figure 6.22) is an essential tool to understand whether or not the BLYP recipe brings substantial changes in the structure with respect to PW and at which (short or intermediate) range. Hints can be obtained by looking first at the numbers of neighbors and peak positions given in tables 6.4 and 6.5. Se–Se correlations obtained with the BLYP scheme are very close to those obtained when using PW, as indicated by the values for the first-coordination shells in table 6.4 and by visual inspection of $g_{\mathrm{SeSe}}^{\mathrm{BLYP}}(\mathrm{r})$ and $g_{\mathrm{SeSe}}^{\mathrm{PW}}(\mathrm{r})$ (figure 6.22). This is a typical feature of all calculations performed on chalcogenides with the intent of comparing among different XC functionals, the various approaches having little impact on the Se environment, not too dissimilar in terms of structural features to experiments. The behavior of Ge–Ge correlations is drastically different. The BLYP scheme performs better than the PW one by producing a clear first maximum in $g_{\mathrm{GeGe}}^{\mathrm{BLYP}}(\mathrm{r})$ due to homopolar Ge–Ge bonds, and a pronounced first minimum, thereby better

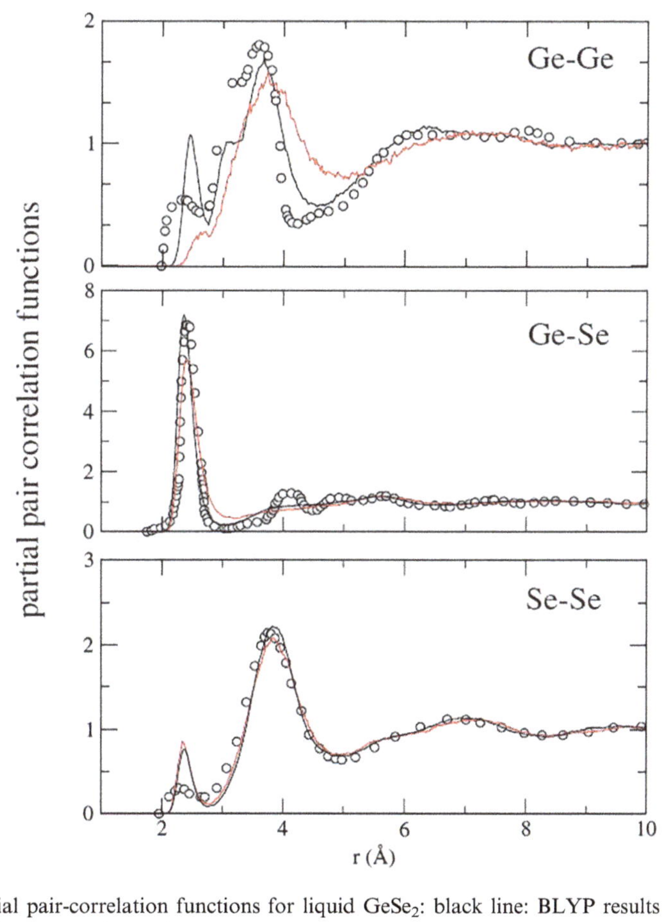

Figure 6.22. Partial pair-correlation functions for liquid $GeSe_2$: black line: BLYP results [47], red line: PW results [7], open circles: experimental results [6]. Reprinted figure with permission from [47]. Copyright (2009) by the American Physical Society.

Table 6.4. First peak position FPP and second peak position SPP in experimental [6] and calculated pair correlation functions within the PW and BLYP exchange–correlation functional. The integration ranges for the coordination numbers n_{α}^{β} and m_{α}^{β} are 0-2.6 Å, 2.6-4.2 Å for $g_{\text{GeGe}}^{\text{exp}}(\text{r})$, 0-3.1 Å, 3.1-4.5 Å for $g_{\text{GeSe}}^{\text{exp}}(\text{r})$ and 0-2.7 Å, 2.7-4.8 Å for $g_{\text{SeSe}}^{\text{exp}}(\text{r})$. Error bars are the standard deviations from the mean for subaverages of 2 ps. Reprinted table with permission from [47]. Copyright (2009) by the American Physical Society.

$g_{\alpha\beta}(\text{r})$	FPP (Å)	n_{α}^{β}	SPP (Å)	m_{α}^{β}
$g_{\text{GeGe}}^{\text{exp}}(\text{r})$	2.33 ± 0.03	0.25 ± 0.10	3.59 ± 0.02	2.9 ± 0.3
$g_{\text{GeGe}}^{\text{BLYP}}(\text{r})$	2.45 ± 0.10	0.22 ± 0.01	3.67 ± 0.10	2.70 ± 0.06
$g_{\text{GeGe}}^{\text{PW}}(\text{r})$	2.70 ± 0.10	0.04 ± 0.01	3.74 ± 0.05	2.74 ± 0.06
$g_{\text{SeSe}}^{\text{exp}}(\text{r})$	2.30 ± 0.02	0.23 ± 0.05	3.75 ± 0.02	9.6 ± 0.3
$g_{\text{SeSe}}^{\text{BLYP}}(\text{r})$	2.38 ± 0.02	0.33 ± 0.01	3.83 ± 0.02	8.9 ± 0.6
$g_{\text{SeSe}}^{\text{PW}}(\text{r})$	2.34 ± 0.02	0.37 ± 0.01	3.84 ± 0.02	9.28 ± 0.04
$g_{\text{GeSe}}^{\text{exp}}(\text{r})$	2.42 ± 0.02	3.5 ± 0.2	4.15 ± 0.02	2.9 ± 0.3
$g_{\text{GeSe}}^{\text{BLYP}}(\text{r})$	2.36 ± 0.10	3.55 ± 0.01	5.67 ± 0.10	3.85 ± 0.06
$g_{\text{GeSe}}^{\text{PW}}(\text{r})$	2.41 ± 0.10	3.76 ± 0.01	5.60 ± 0.01	3.72 ± 0.03

Table 6.5. Coordination numbers as defined in section 2.4.1. Reprinted table with permission from [47]. Copyright (2009) by the American Physical Society.

	n_{Ge}	n_{Se}	n_{ave}
BLYP	3.77 ± 0.02	2.11 ± 0.02	2.66 ± 0.02
PW	3.80 ± 0.02	2.25 ± 0.02	2.77 ± 0.02
exp [6]	3.75 ± 0.3	1.98 ± 0.15	2.57 ± 0.2

approaching $g_{\text{GeGe}}^{exp}(\text{r})$ than $g_{\text{GeGe}}^{\text{PW}}(\text{r})$. An interesting feature proving better agreement with experiments is the shoulder in the main peak of $g_{\text{GeGe}}^{\text{BLYP}}(\text{r})$, found at 3.1 Å, reflecting the presence of edge-sharing connections among tetrahedra. Turning to the Ge–Se pair correlation functions, we notice that the BLYP scheme is capable of reproducing the height of the first peak and the sharp decay down to very small values. There is a marked improvement on the value of neighbors in the first shell of coordination (3.55, BLYP; 3.50, experiments, 3.76, PW, see table 6.4).

Our analysis can be continued by looking in table 6.5 at the coordination numbers defined in section 2.4.1. In the case of the PW data, the underestimated value of $n_{\text{Ge}}^{\text{Ge}}$ was compensated by the overestimate of $n_{\text{Ge}}^{\text{Se}}$ producing a somewhat artificial agreement for n_{Ge}. In the BLYP case, one has close values for the three contributions $n_{\text{Ge}}^{\text{Ge}}$, $n_{\text{Se}}^{\text{Se}}$ and $n_{\text{Ge}}^{\text{Se}}$. As a result, the calculated and experimental average coordination numbers n_{ave} are within 3.5 %. To conclude this analysis, it appears that BLYP improves upon PW by providing a better short-range structure for liquid $GeSe_2$, as illustrated by a more structured first shell of Ge neighbors and, globally, $g_{\text{GeGe}}(\text{r})$ in much better agreement with neutron scattering data.

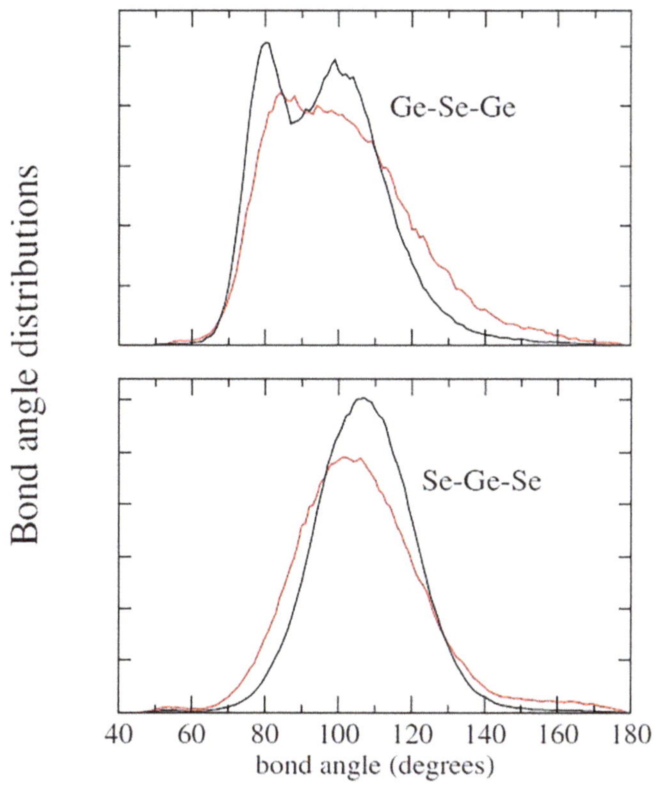

Figure 6.23. Bond-angle distributions Ge–Se–Ge (top) and Se–Ge–Se (bottom). Black line: BLYP results [47], red line: PW results [7]. Reprinted figure with permission from [47]. Copyright (2009) by the American Physical Society.

The same ideas can be further appreciated by showing the distribution θ_{SeGeSe} and θ_{GeSeGe} of the Se–Ge–Se and Ge–Se–Ge bond angles, respectively. θ_{SeGeSe} and θ_{GeSeGe} includes neighbors within less than 3 Å (figure 6.23). An improved tetrahedral order manifests itself via the shape of θ_{SeGeSe}(BLYP) that becomes symmetric around 109° with a higher intensity. Also, focusing on θ_{GeSeGe}(BLYP) for values in between 80° and 100°, there are now two discernible peaks instead of a flat maximum. This means that θ_{GeSeGe}(BLYP) results from two contributions, due to edge-sharing connections between tetrahedra (around 80°) and corner-sharing connections between tetrahedra (around 100°). These features are indications of enhanced tetrahedral organization by confirming all previous conclusions drawn on the basis of the pair correlation functions and the coordination numbers.

6.7.4 Liquid $GeSe_2$: BLYP vs PW, reciprocal space and intermediate range properties

We have seen that the BLYP approach for the XC functional is able to improve the short-range properties of liquid $GeSe_2$ by confirming that proper localization of valence electrons is a crucial property to achieve improved agreement with experiments. While applying first this 'localization' recipe (from LDA to GGA, see section 3.6.5), we made substantial progresses for both short and intermediate range properties. Is this also the case when BLYP is employed instead of the first GGA adopted, namely PW? Figure 6.24 provide some hints in this context, at least if one is willing to rely on BLYP to obtain a marked step forward in terms of intermediate range properties, such as the FSDP intensity in $S_{CC}(k)$. One notices that the performances of BLYP and PW are overall similar for $S_{NN}(k)$ and $S_{NC}(k)$, some differences being hard to ascribe to the choice of the model since minimal and comparable to the extent of statistical errors. The same holds for $S_{CC}(k)$, in which only small shoulder appears at the FSDP location, proving that the experimental peak is underestimated in both PW and BLYP approaches.

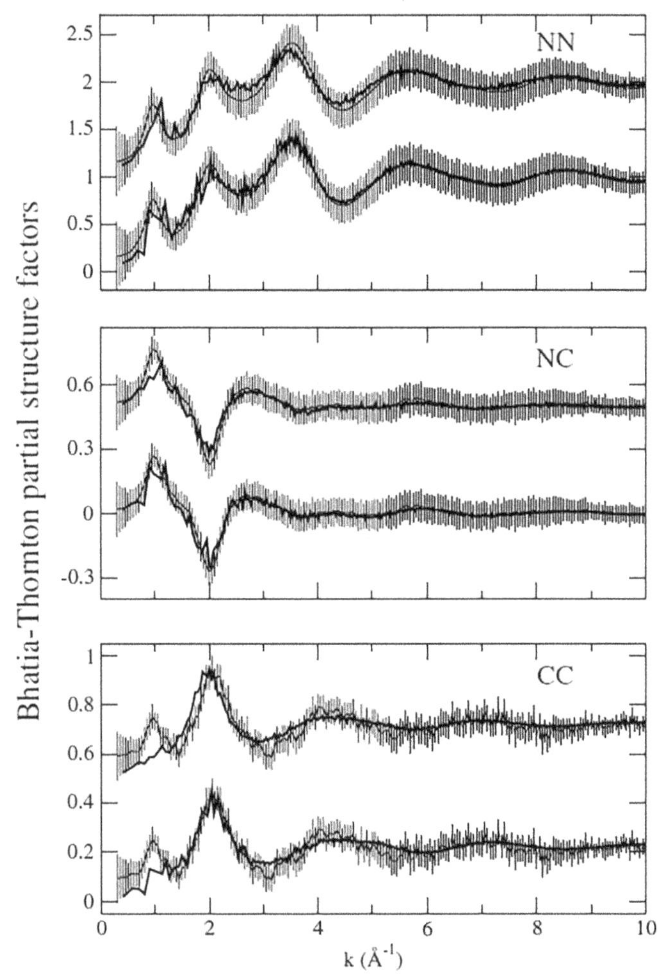

Figure 6.24. Bhatia–Thornton partial structure factors for liquid $GeSe_2$. In each panel, the experimental results of [6] are compared with the BLYP calculations [47] (bottom part) and with the PW calculations [7] (top part). $S_{NN}(k)$, $S_{NC}(k)$ and $S_{CC}(k)$ obtained within the PW scheme have been shifted up by 1, 0.6, and 0.5, respectively. Reprinted figure with permission from [47]. Copyright (2009) by the American Physical Society.

Therefore, we can conclude that the BLYP generalized gradient approximation has an undeniable effect on the short-range properties but it is much less impactful on the side of the intermediate range ones. When these results where obtained (2009) we were quite confident that the use of other XC functionals might have brought the desired improvements on the intermediate range properties. In view of what has been obtained since then, we have to admit that this kind of search has not ended yet, partially because other valuable goals have been identified leading us to postpone or reconsider the strategy needed to settle this open issue.

6.7.5 Liquid $GeSe_2$: BLYP vs PW, dynamical properties

Another way of monitoring the sensitivity of structural properties to the details of the atomic-scale model consists in calculating the diffusion coefficients. On an intuitive basis, the tetrahedra are very much stable in a network that is chemically ordered since there is a little mobility when atoms are arranged in structural units consistent with the optimal atomic bonding character (the valence) or, in terms of charges, with their positive or negative state. On the other hand, if there are departures from chemical order (as it can occurs in a AX_2 network with homopolar bonds and miscoordinations), mobility increases since the network rearranges in the search of more favorable arrangements through changes in the bonding coordination. In section 3.3 we had mentioned an example of correlation between the values of the diffusion coefficients and the atomic structure worth referring to again here. A comparison carried out between structures of liquid $GeSe_2$ obtained via FPMD and a polarizable ionic model (PIM) (see [13]) based on point charges was instrumental to highlight the drastic differences between networks allowing (FPMD) or hampering (PIM), due to stringent physical reasons, departures from chemical order. Within the PIM model, the diffusion coefficient is as low as 10^{-6} cm^2 s^{-1} at T = 3000 K, while they are much larger even at T = 1050 K for the FPMD model, as made explicit in [13] (see section 3.3). Since we have seen that the tetrahedral order is larger in the BLYP model than in the PW one for liquid $GeSe_2$ at T = 1050 K, the question arises on whether this corresponds to diffusion coefficients larger in PW than in BLYP, for which chemical order is higher. The comparison between the calculated statistical average of the mean-square displacement displacement

$$\langle r_\alpha^2(t)\rangle = \frac{1}{N_\alpha}\left\langle \sum_{i=1}^{N_\alpha} |\mathbf{r}_{i\alpha}(t) - \mathbf{r}_{i\alpha}(0)|^2 \right\rangle \tag{6.7}$$

for both species, Ge and Se, obtained within the PW and the BLYP schemes, is shown in figure 6.25, with the diffusion coefficients obtained in the infinite time limit as

$$D_\alpha = \frac{\langle r_\alpha^2(t)\rangle}{6t}. \tag{6.8}$$

One obtains values for the BLYP calculations equal to 2×10^{-6} cm^2 s^{-1} for both species. These values are definitely smaller than the PW values (2.2×10^{-5} cm^2 s^{-1} for

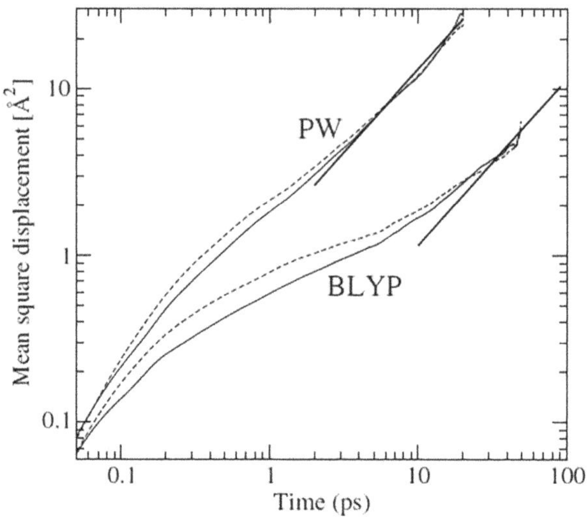

Figure 6.25. Average mean-square displacements for Ge (dashed line) and Se atoms (full line of liquid $GeSe_2$). The infinite time behavior corresponding to a slope equal to 1 in the log–log plot of $<r_\alpha^2(t)>$ vs t is also shown. Reprinted figure with permission from [47]. Copyright (2009) by the American Physical Society.

both Ge and Se) and in much better agreement with the experimental estimate (4.5×10^{-7} cm^2 s^{-1}, see [47] for details on the physical foundations of this comparison). Therefore, atomic mobility reflects well the changes in the atomic structure that remains, most likely, moderately too disordered in terms of chemical bonding. This was expected in view of the difference persisting at the level of Ge–Ge correlations in both the short and, to a larger extent, the intermediate range properties.

6.7.6 Glassy $GeSe_2$: BLYP vs PW and further thoughts

In a short paper appeared in 2010 [48], we tried to extend the above rationale to the case of glassy $GeSe_2$, by assuming that the same considerations developed for the liquid hold for a system obtained from it via rapid quench. While we found some improvements in the structural properties when compared to experiments (in particular, for the number of Ge–Ge homopolar bonds), nothing striking came up in terms of a clear-cut set of changes approaching the BLYP result to experiments. The comparison was based on a single trajectory and could not be considered conclusive. Despite the limits affecting that analysis, we obtained a further indication of the tendency toward more structured shells of Ge neighbors inherent in BLYP while the main peak remained quite small in intensity when compared to experiments. Ever since the discovery of the improved structural features obtained when using BLYP on $GeSe_2$ disordered systems, this scheme has been adopted to model other chalcogenides, even though other comparison with different XC choices were made in some cases, as glassy $GeSe_4$. We shall focus on some of these results in the next chapters.

References

[1] Salmon P S 1994 *Proc. R. Soc. Lond.* A **445** 351
[2] Vashishta P, Kalia R K, Antonio G A and Ebbsjö I 1989 *Phys. Rev. Lett.* **62** 1651–4
[3] Vashishta P, Kalia R K and Ebbsjö I 1989 *Phys. Rev.* B **39** 6034–47
[4] Iyetomi H, Vashishta P and Kalia R K 1991 *Phys. Rev.* B **43** 1726–34
[5] Petri I, Salmon P S and Fischer H E 2000 *Phys. Rev. Lett.* **84** 2413–6
[6] Penfold I T and Salmon P S 1991 *Phys. Rev. Lett.* **67** 97–100
[7] Massobrio C, Pasquarello A and Car R 2001 *Phys. Rev.* B **64** 144205
[8] Massobrio C and Pasquarello A 2008 *Phys. Rev.* B **77** 144207
[9] Waseda Y 1980 The structure of non-crystalline materials *Liquids and Amorphous Solids* (New York: McGraw-Hill)
[10] Bhatia A B and Thornton D E 1970 *Phys. Rev.* B **2** 3004–12
[11] Salmon P S 1992 *Proc. R. Soc. Lond.* A **437** 591
[12] van Roon F H M, Massobrio C, de Wolff E and de Leeuw S W 2000 *J. Chem. Phys.* **113** 5425–31
[13] Wilson M, Sharma B K and Massobrio C 2008 *J. Chem. Phys.* **128** 244505
[14] Susman S, Volin K J, Montague D G and Price D L 1990 *J. Non-Cryst. Solids* **125** 168–80
[15] Salmon P S and Petri I 2003 *J. Phys.: Condens. Matter* **15** S1509–15
[16] Massobrio C, van Roon F H M, Pasquarello A and De Leeuw S W 2000 *J. Phys.: Condens. Matter* **12** L697
[17] Petri I, Salmon P S and Spencer Howells W 1999 *J. Phys.: Condens. Matter* **11** 10219
[18] Elliott S R 1991 *Phys. Rev. Lett.* **67** 711–4
[19] Moss S C and Price D L 1985 Random packing of structural units and the first sharp diffraction peak in glasses *Physics of Disordered Materials. Institute for Amorphous Studies Series* ed D Adler, H Fritzsche and S R Ovshinsky (Boston, MA: Springer)
[20] Price D L, Moss S C, Reijers R, Saboungi M L and S Susman 1988 *J. Phys. C: Solid State Phys* **21** L1069–72
[21] Elliott S R 1992 *J. Phys.: Condens. Matter* **4** 7661–76
[22] Busse L E and Nagel S R 1981 *Phys. Rev. Lett.* **47** 1848–51
[23] Bridenbaugh P M, Espinosa G P, Griffiths J E, Phillips J C and Remeika J P 1979 *Phys. Rev.* B **20** 4140–4
[24] Phillips J C 1981 *J. Non-Cryst. Solids* **43** 37–77
[25] Wilson M and Madden P A 1994 *Phys. Rev. Lett.* **72** 3033–6
[26] Gaskell P H and Wallis D J 1996 *Phys. Rev. Lett.* **76** 66–9
[27] Wilson M and Madden P A 1998 *Phys. Rev. Lett.* **80** 532–5
[28] Sarnthein J, Pasquarello A and Car R 1995 *Phys. Rev.* B **52** 12690–5
[29] Luchnikov V A, Medvedev N N, Appelhagen A and Geiger A 1996 *Mol. Phys.* **88** 1337–48
[30] Corti D S, Debenedetti P G, Sastry S and Stillinger F H 1997 *Phys. Rev.* E **55** 5522–34
[31] Sarnthein J, Pasquarello A and Car R 1995 *Phys. Rev. Lett.* **74** 4682–5
[32] Massobrio C, Pasquarello A and Car R 1998 *Phys. Rev. Lett.* **80** 2342–5
[33] Massobrio C and Pasquarello A 2007 *Phys. Rev.* B **75** 014206
[34] Massobrio C, Celino M and Pasquarello A 2003 *J. Phys.: Condens. Matter* **15** S1537
[35] Rowlands R F, Zeidler A, Fischer H E and Salmon P S 2019 *Front. Mater* **6**
[36] Haye M J, Massobrio C, Pasquarello A, De Vita A, De Leeuw S W and Car R 1998 *Phys. Rev.* B **58** R14661–4
[37] Celino M and Massobrio C 2003 *Phys. Rev. Lett.* **90** 125502

[38] Celino M and Massobrio C 2002 *Comput. Mater. Sci.* **24** 28–32
[39] Price D L, Saboungi M-L and Barnes A C 1998 *Phys. Rev. Lett.* **81** 3207–10
[40] Boolchand P and Bresser W J 2000 *Philos. Mag.* B **80** 1757–72
[41] Johnson R W, Price D L, Susman S, Arai M, Morrison T I and Shenoy G K 1986 *J. Non-Cryst. Solids* **83** 251–71
[42] Massobrio C, Celino M and Pasquarello A 2004 *Phys. Rev.* B **70** 174202
[43] Perdew J P and Wang Y 1992 *Phys. Rev.* B **45** 13244–9
[44] Perdew J P, Chevary J A, Vosko S H, Jackson K A, Pederson M R, Singh D J and Fiolhais C 1992 *Phys. Rev.* B **46** 6671–87
[45] Becke A D 1988 *Phys. Rev. A* **38** 3098–100
[46] Lee C, Yang W and Parr R G 1988 *Phys. Rev.* B **37** 785–9
[47] Micoulaut M, Vuilleumier R and Massobrio C 2009 *Phys. Rev.* B **79** 214205
[48] Massobrio C, Micoulaut M and Salmon P S 2010 *Solid State Sci.* **12** 199–203
[49] Massobrio C and Pasquarello A 2001 *J. Chem. Phys.* **114** 7976

IOP Publishing

The Structure of Amorphous Materials using Molecular Dynamics

Carlo Massobrio

Chapter 7

The effect of pressure on the structure of glassy $GeSe_2$ and $GeSe_4$

This chapter deals with the structural transitions observed in glassy $GeSe_2$ and glassy $GeSe_4$ under the effect of an external pressure acting on the system. This is obtained by reducing the volume (increasing the density) within NVT (constant volume, constant temperature) first-principles molecular dynamics (FPMD) simulations or, in a limited number of cases, by using the constant pressure NPT (constant pressure, constant temperature) approach. The calculations are carried out in close connection with neutron scattering experiments performed by the team of P S Salmon in Bath (UK). As a prerequisite, we have also considered the issue of residual pressures affecting, in principle, simulation models implemented by taking the experimental density as the reference value for the volume. To this purpose, an appropriate volume change has been implemented for a new FPMD trajectory of glassy $GeSe_2$, leading to improved performances in terms of comparison with experimental data. We show that the structural transitions observed for glassy $GeSe_2$ and glassy $GeSe_4$ are very much different as a result of their different topologies. This can be understood by relying on the calculation of structural fingerprints introduced in section 2.3 and, in particular, the coordination numbers.

7.1 Is there any pressure left?

What we have seen so far on the behavior of $GeSe_2$ disordered systems can be summarized by invoking a predominant structural unit (the $GeSe_4$ tetrahedron), coexisting with homopolar bonds and defective Ge–Se coordinations, leading to a sizeable departure from chemical order. Pressure, or, equivalently, an increase of the density, can modify the structure of the network as reflected by the details of its topology and chemical bonding. Research efforts in this direction started in 2011 via a joined study combining neutron diffraction and FPMD, and were aimed at two systems, glassy $GeSe_2$ and glassy $GeSe_4$ [1, 2]. In what follows we shall describe these

doi:10.1088/978-0-7503-2436-6ch7

investigations, by highlighting mostly the simulation part and limiting the experimental details to what is needed to follow, without too much difficulty, the entire story. A word of caution is in order since glassy $GeSe_4$ has not been described so far in this monograph (we have taken advantage of its liquid counterpart to develop a rationale on the concentration–concentration structure factor in section 6.5). We shall devote an extensive section to its equilibrium properties later, when considering the structural changes occurring within the Ge_xSe_{1-x} family of glasses for various compositions (see chapter 8). For this reason, should the reader feel uncomfortable with glassy $GeSe_4$ under pressure before learning how it behaves under ordinary conditions, we suggest to skip section 7.3 of this chapter and read chapters 8 and 9 first.

As a prerequisite to any study of structural changes under pressure, one has to make sure that models described at the experimental density do correspond to a pressure as close as possible to the ambient one. For any practical purpose, ambient pressure can be approximated by a 'zero' value when compared to finite ones achieved in experiments (in the range of GPa) in view of the very moderate changes induced in the experimental volume for pressures smaller than, say, 0.1 GPa. The results presented so far for glassy $GeSe_2$ (within the PW or BLYP exchange–correlation (XC) functionals) contain no information on the actual value of the pressure, expected to be zero under the hypothesis that the model follows exactly the same equation of state of the experimental sample. However, this is quite often not the case, providing a solid motivation for the assessment of the pressure and a refinement of the simulations conditions prior to the introduction of changes in the density, i.e. before starting to monitor the structural changes caused by an increase of pressure. This has been accomplished in [3] by calculating first the pressure associated with the density employed in sections 6.2.3 and 6.7.6 ($\rho^{\mathbf{a}}$ = 0.034 Å^{-3} (side length 15.16 Å), slightly larger than the experimental value ($\rho^{\mathbf{exp}}$ = 0.0334 Å^{-3}) [4, 5]).

A configuration produced within the BLYP scheme (see section 6.7.6) was employed as the starting set of coordinates. The residual pressure is by far not negligible (1 GPa), to demonstrate that any assumption on the density of the model to obtain zero pressure conditions has to be checked carefully. Pressure has been minimized via expansion of the side length to 15.44 Å (density $\rho^{\mathbf{b}}$ = 0.0326 Å^{-3}) and relaxation of the whole structure, so as to lower the pressure to ~0.1 GPa. The relaxation has consisted of a total of 250 ps for temperatures in between $T = 300$ K and $T = 900$ K allowing for substantial diffusion followed by a cooling process lasting 400 ps with steps at temperatures $T = 600$ K and $T = 300$ K. The final part of 20 ps at room temperature is exploited for the calculation of the average properties, to be compared with those obtained in the presence of a pressure as large as 1 GPa (referred to in section 6.7.6). Results are presented in figure 7.1 by focusing on the pair correlation functions $g^{\mathrm{a}}_{\alpha\beta}(r)$ (corresponding to the density $\rho^{\mathbf{a}}$), $g^{\mathrm{b}}_{\alpha\beta}(r)$ (corresponding to the density $\rho^{\mathbf{b}}$) and $g^{\mathrm{ex}}_{\alpha\beta}(r)$ [6] (experiments carried out at 0.0334 Å^{-3}).

The figures reports also the coordination numbers $n_{\alpha\beta}$ (defined as n_{α}^{β} in sections 2.4, 2.4.1). The new data obtained at $\rho^{\mathbf{b}}$ for $n_{\alpha\beta}$ improve upon those obtained at $\rho^{\mathbf{a}}$. Quite impressive is the change that has occurred in $g_{\mathrm{GeGe}}(r)$, for which the three main peaks in between 2 Å and 4 Å, as customary for this kind of network, have to be

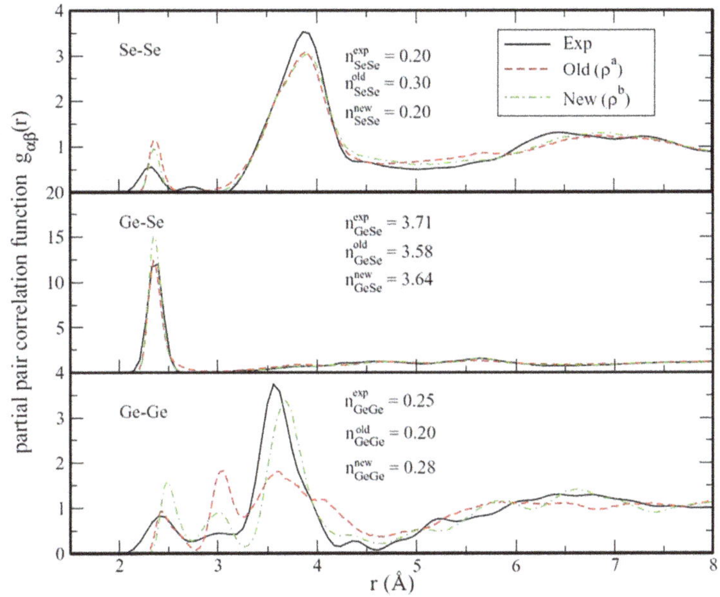

Figure 7.1. Partial pair correlation functions for glassy $GeSe_2$. The experimental results of [6] (solid curve) for the partial pair correlation functions for glassy $GeSe_2$ are compared with the results corresponding to the density $\rho^{\mathbf{a}}$ (dashed red curve) and $\rho^{\mathbf{b}}$ (dash-dotted green curve). Reprinted figure with permission from [3]. Copyright (2012) by AIP Publishing.

Table 7.1. The percentage number of Ge and Se atoms forming homopolar bonds is given by N_{Ge-Ge}, N_{Se-Se} respectively. N_{Ge}(ES), N_{Ge}(CS) are the percentage number of Ge atoms in edge-sharing connections and corner-sharing connections. Note that in [7], MD combined with a reverse Monte Carlo method was employed. Reprinted table with permission from [3]. Copyright (2012) by AIP Publishing.

	N_{Ge-Ge}	N_{Se-Se}	N_{Ge}(ES)	N_{Ge}(CS)
Reference [8]	20	30	58	22 ($\rho^{\mathbf{a}}$)
Reference [7]	17	30	38	45
Reference [15]	23	18	35	42 ($\rho^{\mathbf{b}}$)
Experiment: reference [6]	25	20	34	41

attributed for increasing distances to homopolar Ge–Ge bonds, Ge atoms linked in edge-sharing connections and corner-sharing connections, respectively. The new data reproduce much better the experimental results for the three main peaks intensities, as made clear by the observation that the second peak has become, as it should, less intense than the first one. Other improvements are worth noticing, as can be extracted from table 7.1. We found better agreement for the proportions of Ge and Se atoms involved in Ge–Ge and Se–Se homopolar bonds and for the number of Ge atoms in edge-sharing and corner-sharing connections. This latter case stems from the strong variation in the shape and intensity of the third main peak, in much better agreement with experiments in the $\rho^{\mathbf{b}}$ case.

The extent of chemical order is also modified and improved as a result of the minimization of the residual pressure. The number of Ge atoms fourfold coordinated increases from 78.1% (density $\rho^{\mathbf{a}}$) to 92% (density $\rho^{\mathbf{b}}$). If one considers those Ge atoms coordinated within a $GeSe_4$ tetrahedron the numbers change from 62.5% (density $\rho^{\mathbf{a}}$) to 78.1% (density $\rho^{\mathbf{b}}$). Enhancement of chemical order is a feature very much desirable in improved FPMD models, since the production of a glassy state from the liquid has the tendency to overestimate miscoordinations, despite the fact that the effects can be reduced by insisting as much as possible with structural relaxations via longer temporal trajectories (see section 4.4.5 and figure 4.13 for a critical discussion of this issue). Reciprocal space properties are also monitored to detect important changes in their behavior. In particular, in view of the significance attached to the concentration–concentration partial structure factor $S_{CC}(k)$ both as a quantitative test of FPMD reliability and as a quantity describing a very peculiar kind of intermediate range order, we provide the comparison of the corresponding quantities relative for $\rho^{\mathbf{a}}$ and $\rho^{\mathbf{b}}$ (figure 7.2).

The main improvement on $S_{CC}(k)$ when using the $\rho^{\mathbf{b}}$ data is the height of the main peak, at $k \sim 2$ Å^{-1}. This feature now approaches closely the experimental one, unlike the previous $\rho^{\mathbf{a}}$ data. These results mark a step forward toward more quantitative predictions of glassy structures, the need of working with realistic pressures being not specific to the case of chalcogenides. Having established first that BLYP is well suited to be employed as a reference XC functional in cases where the character of bonding contains both ionic and covalent contributions, we were able to cure an additional shortcoming of our calculations, namely the fact of working at constant *experimental* density. This sounds like a robust theoretical platform to tackle the issue of systems under pressure (next section) and consider other concentrations and a large variety of systems, as done in the next chapters.

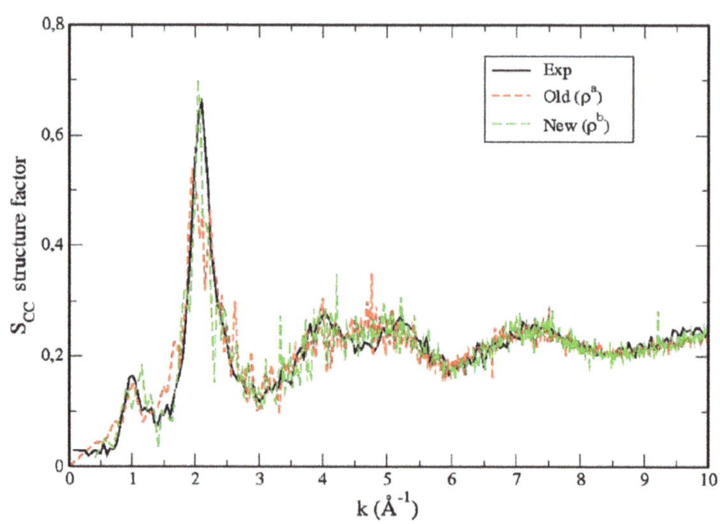

Figure 7.2. Comparison between $S_{CC}(k)$ for glassy $GeSe_2$ obtained by experiments ([6], full line), density $\rho^{\mathbf{b}}$ (green line [3]), density $\rho^{\mathbf{a}}$ (red line [8]). Reprinted figure with permission from [3]. Copyright (2012) by AIP Publishing.

7.2 $GeSe_2$ under pressure: a density-driven transition

7.2.1 Introduction: combining experiments and theory

At the origin of the combined neutron diffraction-FPMD research effort devoted to the study of structural changes in glassy $GeSe_2$ (and $GeSe_4$) under pressure, there is a basic consideration regarding the interplay between any possible form of structural change and bonding properties. This means that, when comparing systems sharing similar gross bonding features but differing in the chemical nature one should expect close behaviors or, in contrast, marked differences. Accordingly, chalcogenide glasses are expected to differ from their oxide counterparts upon densification since the stoichiometric chalcogenide $GeSe_2$ is very much different from the chemically ordered oxide counterpart SiO_2. Focusing on what was accomplished for glassy $GeSe_2$, one might argue that the combination of the two approaches (*in situ* high-pressure isotopic substitution method in neutron diffraction and FPMD carried out at increasing densities on heating cycles encompassing melting and cooling back to the glassy state) [1, 2] suffers from a slight lack of coherence. Indeed, FPMD is carried out at constant volume instead of pressure, leading to a description of the results in terms of a 'density-driven' transition instead of a pressure driven one. However, this turns out to be a minor point not affecting either the value of the comparison or the analysis of the underlying mechanisms. From the FPMD standpoint, working in the NVT ensemble turned out to be simply more practical and less expensive than producing the systems in the NPT one with the same accuracy, at least for the case of glassy $GeSe_2$. This is due to the statistical fluctuations of pressure (larger than temperature fluctuations, since independent on the system size, i.e. $P \sim N_{at}/V$, V being the volume) and the dependence of the energy cutoff in the Kohn–Sham Hamiltonian on the value of the volume, changing in time in the NPT scheme (see section 4.2.2). However, the pressure can be calculated at posterori as 1/3 of the trace of stress tensor at $T = 300$ K. We also found in the case of glassy $GeSe_2$ that the density was slow to converge, while this drawback was not encountered for glassy $GeSe_4$. In short, NVT or NPT strategies are both feasible and can be employed consistently in each case, the easiness of the practical implementations (NVT or NPT) depending on the system, respectively $GeSe_2$ and $GeSe_4$. A practical demonstration is given in a later section (section 9.4) where we shall present calculations performed for glassy $GeSe_4$ in the NPT framework.

On the FPMD technical side, the following remarks specific to this calculation can be added to the general protocol described previously (see section 3.5.2). At the beginning we employ as starting configurations those produced at vanishing stress tensor by using the BLYP XC functional, as detailed in section 7.1. Then, each simulation aimed at obtaining a new pressure (again, to be calculated at posterori since working in the NVT ensemble) was constructed via a thermal cycle based on five steps ($T = 300$ K for 40 ps, $T = 600$ K for 50 ps, $T = 900$ K for 150 ps, down to $T = 600$ K for 70 ps, and finally at $T = 300$ K for 150 ps). Statistical averages were taken at the end of a thermal cycle on the final 50 ps part of the run at $T = 300$ K. For each pressure, 2–3 different configurations were considered. The modeled

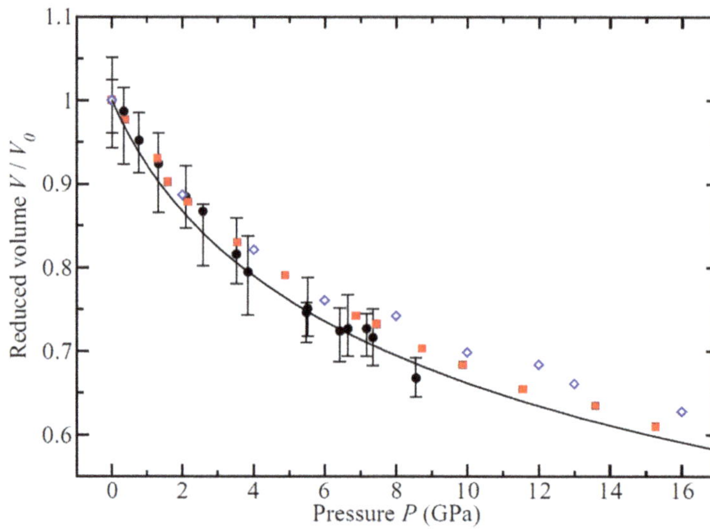

Figure 7.3. The pressure–volume equation of state for $GeSe_2$ glass under compression. The measured data are the black dots with error bars [9]. Red squares: FPMD results considered in this section [1]. Open diamonds: FPMD results obtained in [10]. The measured data are fitted to a second order equation of state (Birch–Murnaghan [11]). Reprinted figure with permission from [1]. Copyright (2014) by the American Physical Society.

equation of state is in accord with experiment [9] within the measurement error (figure 7.3).

In the present case of glassy $GeSe_2$ at increasing densities we have also resorted to hybrid functionals [12] (for a single single point at $\rho/\rho_0 = 1.529$, see figure 7.5) as mentioned in section 4.2.1. The idea was to find a better agreement with experiments at high densities. Because of the account of the exact Hartree exchange energy, these calculations are much slower (about 50 times) than those performed with the BLYP functional. For these reasons, the thermal cycle was reduced to ~20 ps divided in the following steps, $T = 300$ K for 1.2 ps, $T = 1100$ K for 3.4 ps, $T = 900$ K for 5.8 ps, and finishing at $T = 300$ K for 6.6 ps, with the results obtained after relaxation at $T = 300$ K.

7.2.2 Neutron diffraction experiments at finite pressure: the essential

As introduced in section 2.3, neutron diffraction experiments are based on the standard definitions for the total structure factor

$$F(k) = \sum_{\alpha}\sum_{\beta} c_\alpha c_\beta b_\alpha b_\beta \left[S_{\alpha\beta}(k) - 1 \right] \tag{7.1}$$

where k is the magnitude of the scattering vector, $S_{\alpha\beta}(k)$ is a partial structure factor for chemical species α and β, b_α, c_α are the coherent neutron scattering length and the atomic fraction of chemical species α.

The Fourier transformation below can provide the total pair correlation function in real space under the form

$$G(r) = \frac{1}{2\pi^2 \rho r} \int_0^\infty \mathrm{d}k \; kF(k)\sin(kr) \tag{7.2}$$

where ρ is the atomic number density. Knowing that measurements are carried out up to a maximum value $k_{\max}$, equation (7.2) holds provided $k_{\max}$ is large enough for $F(k)$ not to show any detectable patterns at high k. To compare FPMD and diffraction results consistently, the procedure sketched in section 2.3, equation (2.13) has been followed (the reciprocal-space functions obtained via FPMD simulations are Fourier transformed with $k_{\max}$ equal to the experimental value). In analyzing the results, we shall also make extensive use (see section 2.4.1) of the average nearest-neighbor coordination number $\bar{n} = c_{\mathrm{Ge}}(n_{\mathrm{Ge}}^{\mathrm{Ge}} + n_{\mathrm{Ge}}^{\mathrm{Se}}) + c_{\mathrm{Se}}(n_{\mathrm{Se}}^{\mathrm{Se}} + n_{\mathrm{Se}}^{\mathrm{Ge}})$, c_{Ge} and c_{Se} indicating the concentrations of Ge and Se atoms, respectively. Note that $\bar{n}$ is also called mean, or total and is also expressed as n_{tot} or n_{ave}, as mentioned in section 2.4.1.

Two diffractometers were employed to gather information on the pressure behavior of glassy $GeSe_2$. In the framework of isotope substitution experiments, measurements at 3.0(5), 4.7(5), 6.3(5), 7.1(5) and 8.2(5) GPa were performed with D4c [13]. The $GeSe_2$ sample of natural isotopic abundance was also considered at 1.7(5), 3.0(5), 3.9(5) and 4.7(5) GPa by using D4c, and at 8.7(5), 10.9(5), 12.8(5), 14.4(5) and 16.1(5) GPa by using the time-of-flight diffractometer PEARL [14]. Concerning the number density, data of volume vs pressure under cold compression (figure 7.3) were exploited up to 8.5 GPa, while for higher pressure the density ρ was extracted from a fit to this experimental data using a second order Birch–Murnaghan equation of state [11] $P = (3B_0/2)[(V/V_0)^{-7/3} - (V/V_0)^{-5/3}]$.

7.2.3 Understanding the structural transition: results

A representative set of results for the total structure factors measured using PEARL at higher pressures is shown in figure 7.4(a), the real-space functions being shown in figure 7.4(b). To compare the FPMD results with the diffraction results the two data sets are matched according to their number density since the equation of state issued from the calculations and experiments are not identical (figure 7.3). Focusing on the FPMD data, they appear to be in broad agreement with experiments in reciprocal space, although some noise effects due to the limited size of the simulation box ($N_{\mathrm{at}} = 120$) and the number of independent trajectories considered (2–3 for each point) are unavoidable. The level of agreement is undoubtedly better in real space since the Fourier transform applied with the same maximum cutoff value $k_{\max} = 19.55$ Å^{-1} as used for the neutron diffraction data has the effect of removing most of the noise.

The measured mean nearest-neighbor bond distance $\bar{r}$ and coordination number $\bar{n}$ found from neutron diffraction are compared to the x-ray diffraction results of Mei *et al* [9] in figures 7.5(a) and (b). In that figure, we also report the simulation results that are in agreement with experiments for pressures corresponding to a reduced number density $\rho/\rho_0 \sim 1.4$, where ρ and ρ_0 are the values at high and ambient pressure, respectively. At higher pressures, a difference stands out between theory

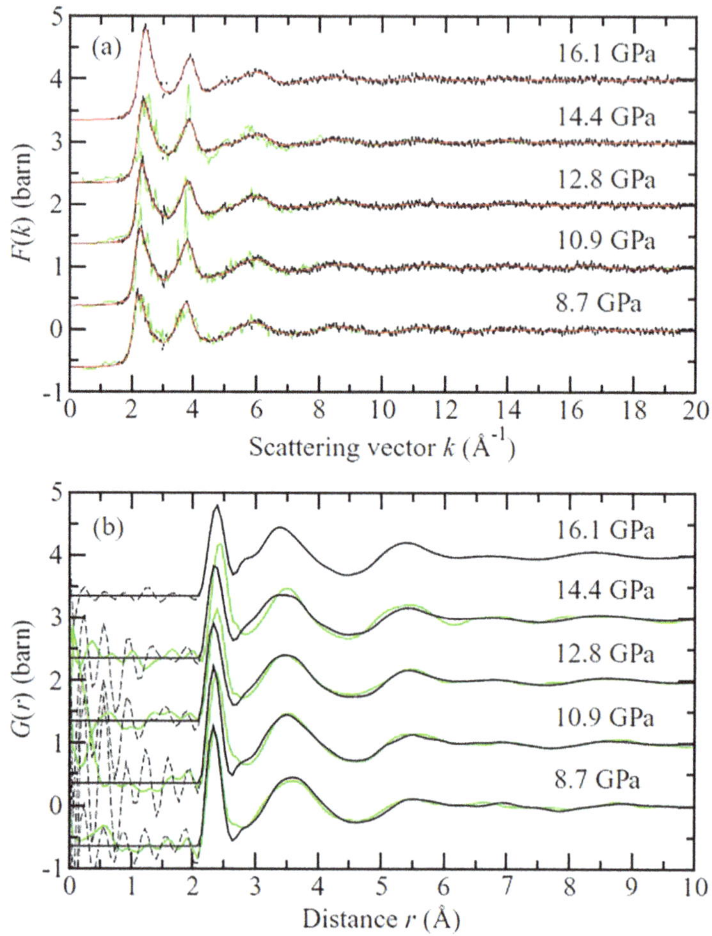

Figure 7.4. (a) The total structure factor $F(k)$ and (b) the total pair correlation function $G(r)$ for $^{N}Ge^{N}Se_2$ glass (N denotes the natural isotopic abundance) (a) PEARL diffractometer data: the points with vertical error bars. Red curves: spline fits to the measured data sets, (for $k < 1.55$ Å^{-1} fitted Lorentzian functions since this region was not accessible in the diffraction experiments (see [14]). Green curve: FPMD results for pressures of 9.87, 11.56, 13.82 and 15.27 GPa. (b) Black curve: Fourier transforms of $F(k)$ functions given in (a) (broken black curves are related to unphysical oscillations at unrealistic small r-values). Green curves: FPMD results from Fourier transforms data in (a). Reprinted figure with permission from [1]. Copyright (2014) by the American Physical Society.

and experiments since both results are characterized by an increase of $\bar{n}$. This process starts at ~8.5 GPa in FPMD (ρ/ρ_0~ 1.42) while it occurs at ~12 GPa (ρ/ρ_0 ~ 1.55) in experiments. Figure 7.5 highlights the dependence on ρ/ρ_0 of four distinct variables, extracted from FPMD trajectories, that are essential to follow the structural trend under pressure. We refer to the fraction of n-fold coordinated Ge (figure 7.5(c)) and Se (figure 7.5(d)) atoms (n = 2, 3, 4, 5 or 6) and to their expression in terms of proportions of these n-fold species featuring homopolar bonds, for Ge atoms (figure 7.5(e)) and Se atoms (figure 7.5(f)).

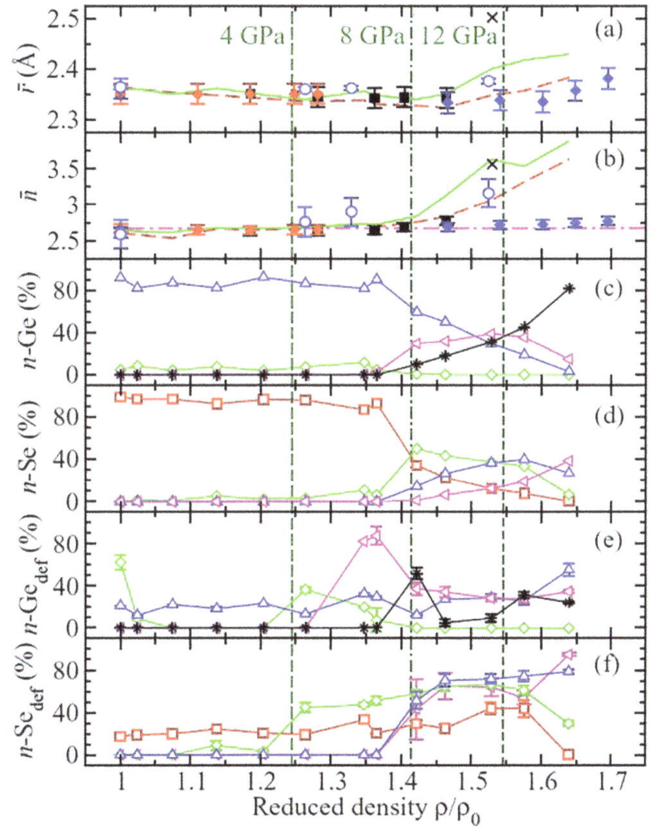

Figure 7.5. The ρ/ρ_0 dependence of the nearest-neighbor (a) bond distance $\bar{r}$ and (b) coordination number $\bar{n}$ as obtained from (i) neutron diffraction using a $^{N}Ge^{N}Se_2$ sample on either the D4c (red diamonds) or PEARL (blue diamonds) diffractometer, or using $^{70}Ge^{N}Se_2$ and $^{73}Ge^{76}Se_2$ samples and averaging the results (black squares); (ii) x-ray diffraction [9] (blue circles); (iii) FPMD [solid green curves] (results obtained before the high-temperature anneal, broken red curves) and iv) FPMD (cross) performed by using the HSE06 hybrid functional (see section 4.2.1). In (b) the horizontal chained line gives the '8 − N rule expectation of $\bar{n}$ = 2.67 [15]. Also given are the FPMD results for the fractions of n-fold coordinated (c) Ge and (d) Se atoms, and the fractions of these n-fold coordinated (e) Ge and (f) Se atoms that have homopolar bonds. In (c)–(f) the symbols denote the following coordinates species: 2-fold (red squares), 3-fold (green diamonds), 4-fold (blue triangles), 5-fold (magenta leftward triangles) or 6-fold (black stars). The vertical broken lines correspond to pressures of ~4, 8 and 12 GPa. Reprinted figure with permission from [1]. Copyright (2014) by the American Physical Society.

A further information on the structure can be obtained via the Se–Ge–Se and Ge–Se–Ge bond angle distributions defined in this case as $P(\theta_{SeGeSe}) = B(\theta_{SeGeSe})/\sin(\theta_{SeGeSe})$ and $P(\theta_{GeSeGe}) = B(\theta_{GeSeGe})/\sin(\theta_{GeSeGe})$ shown in figures 7.6(a) and (b). Here the $B(\theta)$ functions are normalized by $\sin(\theta)$ to remove the effect of a finite sampling volume [16] [1].

[1] Note that this definition differs from the non-normalized one given in section 2.4.2 and shown in other parts of this monograph.

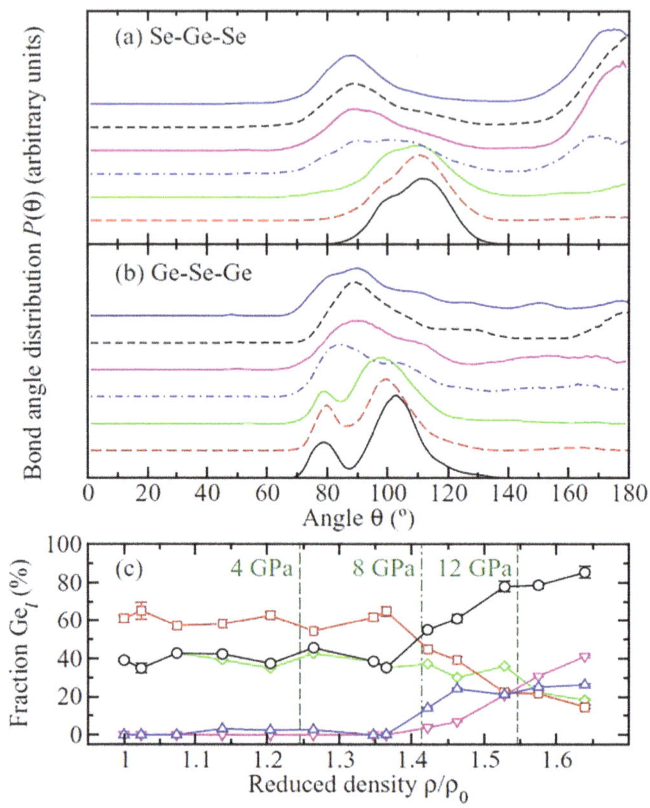

Figure 7.6. FPMD results showing the ρ/ρ_0 dependence of (a) the intra-polyhedral Se–Ge–Se and (b) the inter-polyhedral Ge–Se–Ge bond angle distributions where, from bottom to top in each panel, ρ/ρ_0 = 1, 1.204, 1.365, 1.463, 1.528, 1.575 or 1.639. In (c) the density dependence of the fractions of CS Ge atoms (Ge_0, open squares) and ES Ge atoms (open circles) is given, together with the contributions from Ge_1 (diamonds), Ge_2 (triangles) and Ge $_{3/4}$ (reversed triangles) units. Reprinted figure with permission from [1]. Copyright (2014) by the American Physical Society.

The fractions of Ge atoms that are involved either in CS or ES motifs (see figure 4.1) are given in figure 7.6(c). Corner-sharing connections are named Ge_0 while the edge-sharing connections are subdivided into Ge_ℓ (ℓ = 1, 2, 3/4), where ℓ = 1, 23/4 (3 or 4) is the number of edge-sharing connections for each Ge atoms (or, equivalently, the number of Ge belonging to ℓ fourfold rings, see for instance section 6.3.2).

7.2.4 Understanding the structural transition: rationale

In the $GeSe_2$ glass, measurements and FPMD results indicate a densification process that consists of two steps, roughly separated at $\rho/\rho_0 \sim 1.42$. Below these values, there is agreement between diffraction data and FPMD results and $\bar{n}$ remains constant up to $\rho/\rho_0 \sim 1.42$. For higher values, theory and experiments point out an increase of $\bar{n}$ that occurs with a different rate when comparing FPMD and diffraction data. FPMD data shows a clear increase of 5- and 6-fold coordinated

Ge atoms (figure 7.5(c)), as well as Se atoms with coordinations higher than 2 (figure 7.5(d)). Homopolar bonds become also detectable for ρ/ρ_0 larger than the threshold value especially in figure 7.5(e), so as to legitimate the notion of homopolar bonds as defects acting as mediators in the density-driven structural transformation. Worthy of notice is also the reduction of corner-sharing connection and the increase of edge sharing ones in figure 7.6(c). The larger increase with density of $\bar{n}$ found in calculations can be interpreted as an overestimate of the metallic behavior of bonding, experiments being close to the retention of a semi-conducting behavior. This is the reason underlying the attempt of increasing the ionic character of bonding by resorting to a scheme accounting exactly for the Hartree exchange (hybrid exchange–correlation functionals) (see points in figures 7.5(a) and (b)). Since the calculation did not bring any better agreement with the experiments (and being unable to produce a longer statistics at the time of data production) the following explanation appears as the most plausible. In experiments, the glass was cold-compressed instead of undergoing a thermal treatment as it was done via FPMD. To prove that this makes a big difference we performed simulations closely mimicking the cold-compression procedure (essentially by increasing the density at room temperature, see the red broken curve in figures 7.5(a) and (b) to be compared with the rapidly increasing green one). The rate of change of $\bar{r}$ and $\bar{n}$ with density is reduced for $\rho/\rho_0 > 1.42$. These results show that there should be an energy barrier separating two energy valleys, that forms for large pressures. This barrier can be surmounted by heating the system, at a temperature comparable to that employed (~900 K, as we did in FPMD) for achieving substantial diffusion in our production of the glassy state by cooling.

It is possible to establish a correspondence between the observed behavior of the structural parameters and a set of snapshots representing atomistic configurations extracted from the FPMD trajectories (figure 7.7). For $\rho/\rho_0 = 1$ the network is made of a large majority of corner-sharing and edge-sharing Ge atoms belonging to one fourfold ring (Ge_0 and Ge_1 atoms, respectively). The second configuration at

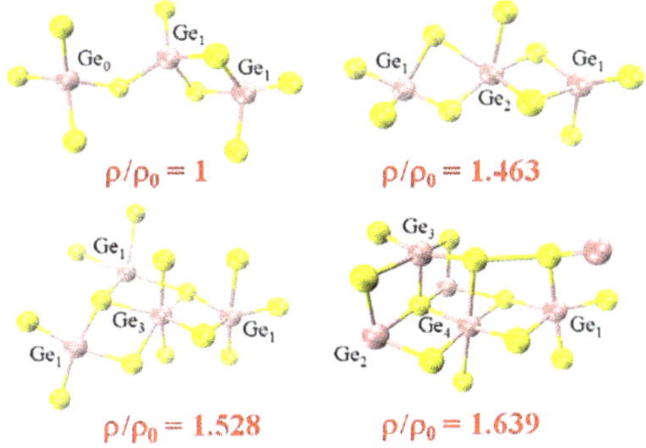

Figure 7.7. Atomistic configurations from FPMD for $GeSe_2$ glass at different reduced densities. Ge atoms are dark (purple) and Se atoms are light (yellow). Bonds are drawn when two atoms are separated by a distance $\leqslant r_{min}$ given by the position of the first minimum in $g_{GeSe}(r)$. Reprinted figure with permission from [1]. Copyright (2014) by the American Physical Society.

$\rho/\rho_0 = 1.463$ is indicative of the formation of additional bonds of Se with Ge_1 atoms, leading to 5-fold coordinated Ge_2 units. Since the impact of Ge_0 atoms is limited, the peak at ~111° in the Se–Ge–Se bond angle distribution is shifted to smaller angles, followed by the disappearance of the twin peaks structure in the Ge–Se–Ge bond angle distribution (figures 7.6(a) and (b)). Now the network topology consisting of corner- and edge-sharing units tends to be superseded by a different one with more Ge_2 and a few $Ge_{3/4}$ configurations (see $\rho/\rho_0 \simeq 1.528$), resulting in 6-fold coordinated Ge atoms. Accordingly, the Se-Ge-Se bond angle distribution features a peak to around 90° and a second peak near 180°. When the process of structural transition is completed the network can be described as a pseudo-cubic arrangement of Ge-centered units, the cubic nature manifesting itself through the peak at ~90° in the Ge–Se–Ge bond angle distribution.

Therefore, the structural transition under pressure (or increasing density) in glassy $GeSe_2$ is far from being simple in terms of structural motifs, since we have seen that homopolar bonds come also into play. For these reasons it is of interest to review its salient features and highlight the different level of agreement with available experimental measurements. The agreement between FPMD and experiments is excellent up to $P \sim 8.5$ GPa ($1 \leqslant \rho/\rho_0 \leqslant 1.42$) in terms of mean nearest-neighbor bond distance $\bar{r}$ (figure 7.5(a)) and coordination number $\bar{n}$ (figure 7.5(b)), not to mention the total pair correlation function. For larger pressures, the different trends observed in FPMD and diffraction data can be ascribed to a more pronounced semiconducting character of the real samples when compared to the tendency toward metallicity inherent in the FPMD behavior of quantities like $\bar{r}$ and $\bar{n}$. Simulations show that densification occurs via the formation of 5- and 6-fold coordinated Ge atoms and 3-, 4- and 5-fold coordinated Se atoms. These higher coordinated Ge and Se atoms have homopolar bonds, acting as mediating agents of the structural transformation. Finally, it appears that an energy barrier to structural rearrangement exists but it cannot be explored in the cold-compression diffraction experiments. However, it can be accessed via the high-temperature annealing stage in the simulations. These results stimulate further investigations of structural changes in chalcogenide glasses, the next section being devoted to the study of the $GeSe_4$ case.

7.3 $GeSe_4$ under pressure: when theory and experiments agree

For both experiments and theory, methodologies employed for glassy $GeSe_4$ are very similar to those sketched above in the case of glassy $GeSe_2$. In particular, measurements were performed on the D4c and PEARL diffractometers at ambient pressure, 3.0(5), 4.7(5), 6.3(5), 7.0(5) and 8.1(5) GPa and 8.7(5), 10.9(5), 12.8(5) and 14.4(5) GPa, respectively. To establish the volume/pressure equation of state, ρ was taken from a fit to the measured pressure dependence of the reduced volume V/V_0 for $GeSe_4$ glass under cold-compression [17] (figure 7.8), by using a third-order Birch–Murnaghan equation of state:

$$P = (3B_0/2)[(V/V_0)^{-7/3} - (V/V_0)^{-5/3}] \\ \times \{1 + (3/4)(B_0' - 4)[(V/V_0)^{-2/3} - 1]\} \tag{7.3}$$

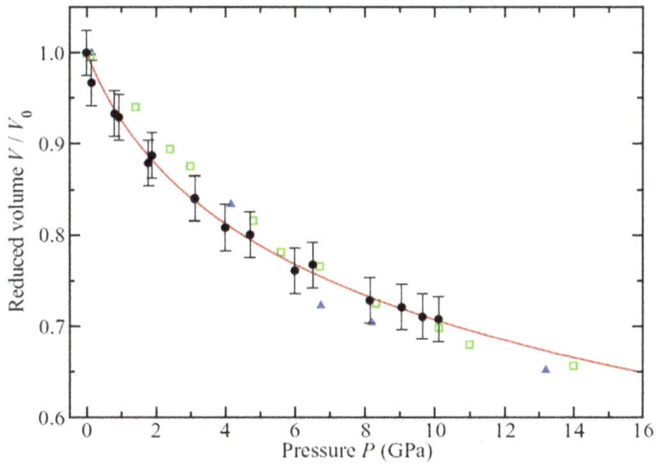

Figure 7.8. The pressure–volume equation of state for amorphous $GeSe_4$ as measured under compression in the experiments of Skinner *et al* [17] (solid (black) circles with vertical error bars) and Kalkan *et al* [18] (open (green) squares). The solid (red) curve gives a fit to the data of [17] using a third-order Birch–Murnaghan equation of state, and the solid (blue) triangles give the FPMD results from [2]. Reprinted figure with permission from [2]. Copyright (2016) by the American Physical Society.

where V is the volume at pressure P and V_0 is the volume at zero pressure. On the side of FPMD, we followed the indications of section 7.1 to prepare an initial configuration at $T = 300$ K by increasing the volume of a previous NVT model set at the experimental density so as to minimize the residual pressure to the value ~0.15 GPa. By taking advantage of affordable times of convergence for the density within the NPT ensemble when compared to the $GeSe_2$ case (see section 7.2.1), this framework was more effectively exploited by performing calculations at high pressure for the values of 4.15, 6.73, 8.18 and 13.2 GPa. At the end of a thermal cycle quantitatively similar to the $GeSe_2$ case, averages were taken along 50 ps at $T = 300$ K trajectory. The simulated points are consistent with the experimental equation of state [17, 18] within the measurement error (figure 7.8).

7.3.1 Behavior under pressure

Simulated and measured neutron total structure factors are shown in figure 7.10, with the corresponding real-space functions in figure 7.11. The agreement is quite impressive over the covered pressure range, FPMD being able to reproduce essentially all features both in terms of peak positions and intensities. In real space there are some differences at the level of the second peak, knowing that both results (experiments and theory) have been obtained from Fourier transform from reciprocal space of the total structure factors. Because of this procedure, the FPMD data in reciprocal space are much less noisy than in the case of $GeSe_2$, thereby allowing for a smoother comparison with experiments.

A first information on the structural behavior of glassy $GeSe_4$ under pressure as modeled via FPMD is given by $\bar{r}$ (figure 7.9(a)) and $\bar{n}$ (figure 7.9(b)). Calculations are

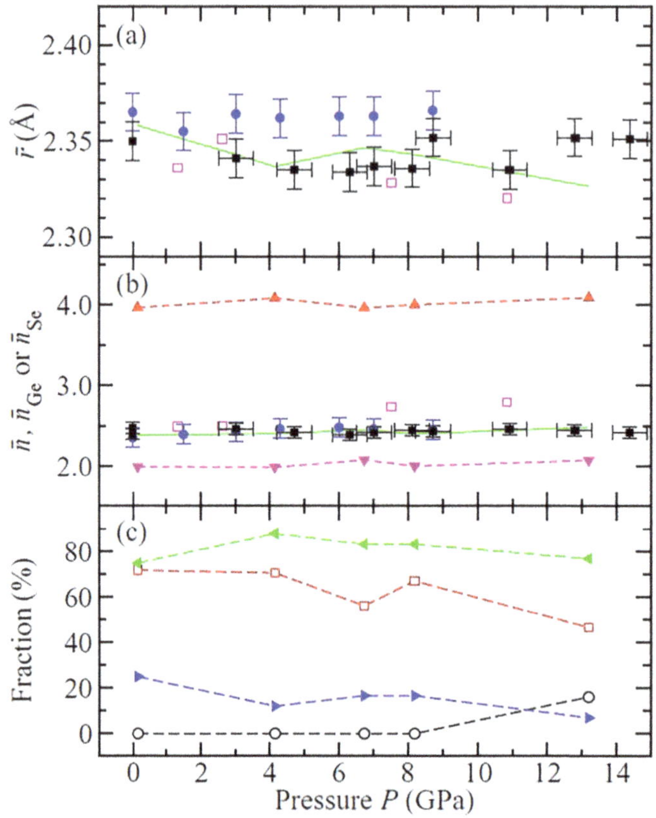

Figure 7.9. The pressure dependence of (a) the mean nearest-neighbor distance $\bar{r}$ as measured in (i) the x-ray diffraction work of Skinner *et al* [17] ((blue) solid circles with vertical error bars) and Kalkan *et al* [18] ((magenta) open squares), (ii) neutron diffraction data of [2] ((black) solid squares with error bars), and (iii) as obtained via FPMD (solid (green) curve); (b) the mean coordination number $\bar{n}$ as measured in (i) the x-ray diffraction work of Skinner *et al* [17] ((blue) solid circles with vertical error bars) and Kalkan *et al* [18] ((magenta) open squares), (ii) neutron diffraction data of [2] ((black) solid squares with error bars), and (iii) as obtained via FPMD (solid (green) curve). The latter is shown in different contributions from $\bar{n}_{Ge}$ (termed n_{Ge} in section 2.4.1) ((red) upward-pointing triangles) and $\bar{n}_{Se}$ (termed n_{Se} in section 2.4.1) ((magenta) downward-pointing triangles). We show also (c) the FPMD results for the fractions of Ge ((black) open circles) and Se ((red) open squares) atoms involved in homopolar bonds, and the fractions of Ge atoms involved in corner-sharing ((green) leftward-pointing triangles) and edge-sharing ((blue) rightward-pointing triangles) tetrahedra. Reprinted figure with permission from [2]. Copyright (2016) by the American Physical Society.

able to follow the experimental values, both for the initial pressure-induced bond reduction ($\bar{r}$) and the little changes with pressure ($\bar{n}$). We provide the evolution of the partial pair correlation functions $g_{\alpha\beta}(r)$ in figure 7.12. For pressures lower than the highest value we considered, the absence of Ge–Ge homopolar bonds in the first peak at $\simeq 2.35$ Å in $g_{GeGe}(r)$ means that Ge–Se and Se–Se correlations only are responsible for the values $n_{Ge} = 4$ and $n_{Se} = 2$ (figure 7.9(b)) leading to $\bar{n} = 2.4$. At the highest pressure, homopolar Ge–Ge bonds first contribute to the value of n_{Ge} (figure 7.9(c)), manifesting themselves in $g_{GeGe}(r)$ by a small-r feature with peaks at 2.35 and 2.58 Å

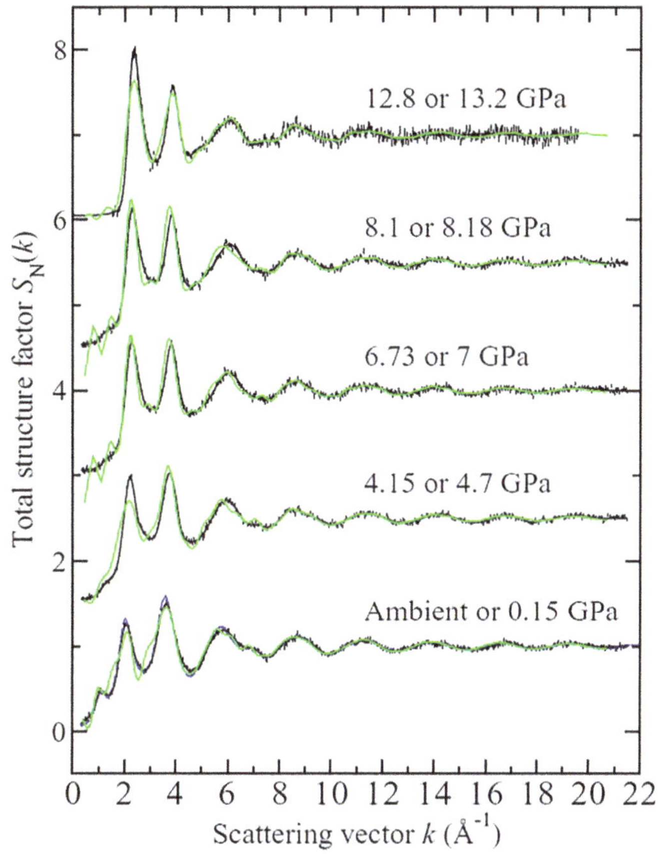

Figure 7.10. The pressure dependence of the neutron total structure factor $S_N(k)$ as measured for $GeSe_4$ glass at ambient pressure, 4.7, 7, 8.1 and 12.8 GPa (points with vertical error bars), and as calculated by Fourier transforming the FPMD real-space total pair-distribution functions at 0.15, 4.15, 6.73, 8.18 and 13.2 GPa (solid light (green) curves). Reprinted figure with permission from [2]. Copyright (2016) by the American Physical Society.

(figure 7.12(c)). However, n_{Ge} and n_{Se} do not increase appreciably ($n_{Ge} \simeq 4.1$ and $n_{Se} \simeq 2.1$ at 13.2 GPa), corresponding to a mean coordination number $\bar{n}$ in the range $\simeq$2.39 at 0.15 GPa $-\simeq$2.48 at 13.2 GPa (figure 7.9(b)). To understand why $n_{Ge} = 4$ and $n_{Se} = 2$, it has to be said that this network is (to a very large extent) made of $Ge(Se_{1/2})_4$ tetrahedra and Se–Se homopolar bonds, these two motifs being at the origin of the fourfold (for Ge) and twofold (for Se) preferential ccordination. When looking at the second peak at $\simeq$3.75 Å in $G_N(r)$ at ambient pressure, it appears that the most dominant contribution comes from $g_{SeSe}(r)$ since Se has the largest atomic fraction. With increasing pressure, one notices the growth of a small shoulder at $\simeq$3.25 Å on the second peak in $g_{SeSe}(r)$ (figure 7.12(a)). Concerning $g_{GeGe}(r)$, there are peaks at 2.97 and 3.68 Å that stem from edge- and corner-sharing Ge-centered tetrahedra, respectively (figure 7.12(c)). At 8.18 GPa the first peak is now located at 3.02 Å while the second peak broadens and shifts to 3.44 Å. At the highest pressure these two

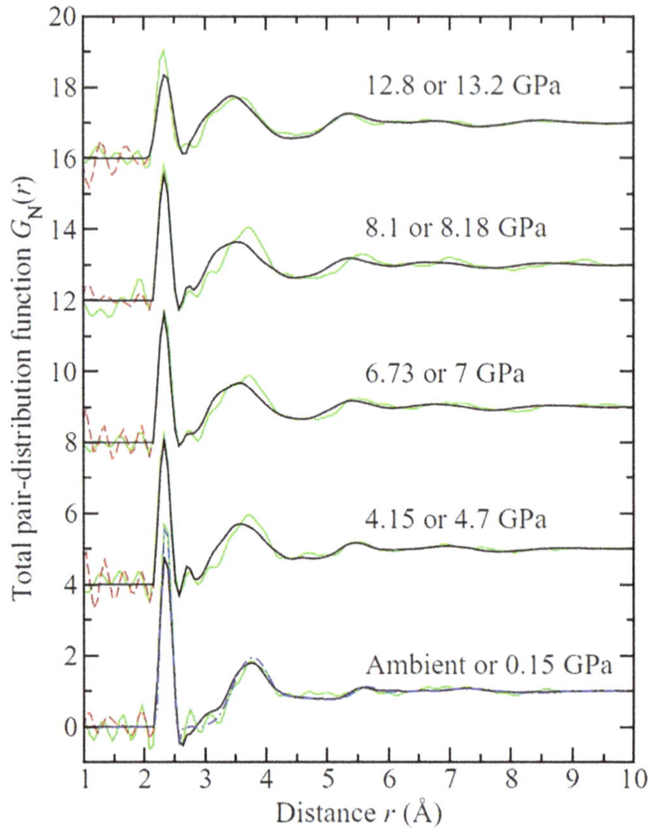

Figure 7.11. The pressure dependence of the neutron total pair-distribution function $G_N(r)$ for $GeSe_4$ glass. The solid (black) curves show the Fourier transforms of the spline-fitted measured $S_N(k)$ functions shown in figure 7.10 at ambient pressure, 4.7, 7, 8.1 and 12.8 GPa. The chained (red) curves show the Fourier transform artifacts at unphysically small r-values. The solid light (green) curves show the Fourier transforms of the FPMD $S_N(k)$ functions (as calculated directly in reciprocal space) at pressures of 0.15, 4.15, 6.73, 8.18 and 13.2 GPa using the same k_{max} values as for the corresponding measured data sets. Reprinted figure with permission from [2]. Copyright (2016) by the American Physical Society.

peaks merge to give one broad peak at ~3.42 Å. It is of interest to know the percentage of Ge atoms involved in corner- or edge-sharing connections. These are given in figure 7.9(c). At the lowest pressure, all of the Ge atoms are found in either one of these two configurations, while $\simeq$ 84% of them can be counted as corner or edge sharing at 13.2 GPa. Despite these decrease, $Ge(Se_{1/2})_4$ remains as the predominant Ge-centered structural unit in $GeSe_4$ glass in the range from ambient to ~13 GPa.

7.3.2 Behavior under pressure: rationale

It is instructive to introduce a classification of configurations in $GeSe_4$ by referring to figure 7.13 where a *twofold* coordinated B (Se) atom can be connected to either (a) two B (Se) atoms, (b) one atom A(Ge) and one atom B (Se), or (c) two A (Ge) atoms. Such

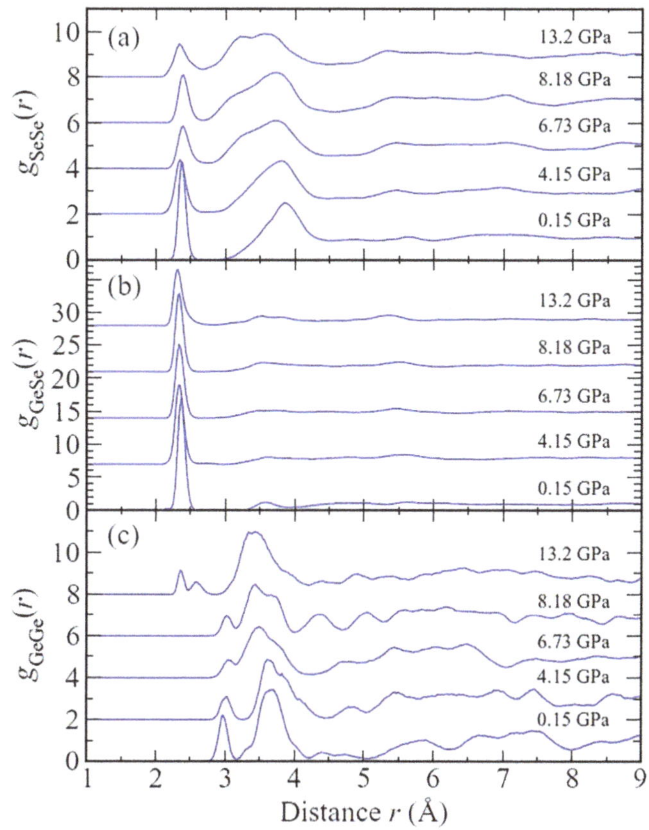

Figure 7.12. The pressure dependence of the partial pair-distribution functions (a) $g_{SeSe}(r)$, (b) $g_{GeSe}(r)$ and (c) $g_{GeGe}(r)$ obtained from the FPMD simulations. Reprinted figure with permission from [2]. Copyright (2016) by the American Physical Society.

classification provides a qualitative description of the network under the hypothesis that all Se atoms are twofold coordinated. The configurations are labeled as BB, AB, and AA, respectively and can be employed to define two extreme cases for the network organization.

In the first, the fractions of AA, BB, and AB units are 0%, 0%, and 100%, corresponding to a corner-sharing network with Ge connected by Se dimers (AB configurations). There are no BB and AA configurations since Se atoms are not available to form Se chains longer than Se_2 or to have two Ge atoms as first neighbors. A model featuring AB = 0 does correspond to a phase separation, since Se atoms are free to organize themselves in chains (BB configurations), while the rest of the system is made of $GeSe_2$-like units (AA configurations). These concepts will be reiterated and further assessed when considering the structural properties of glassy $GeSe_4$ as they were rationalized over the years through a set of FPMD investigations (see chapter 8).

The percentages of ℓ-fold coordinated Ge atoms (ℓ = 3, 4 or 5) and ℓ-fold coordinated Se atoms (ℓ = 1, 2 or 3) (FPMD results, figure 7.14(a) and (b)) show that a predominantly tetrahedral system is moderately altered with increasing pressure

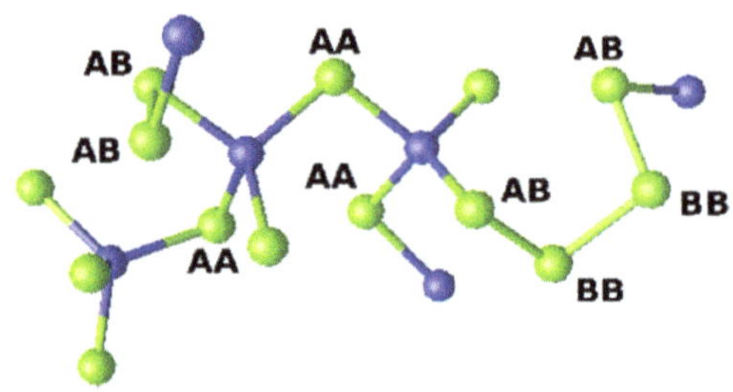

Figure 7.13. A representative subset of Ge and Se atoms in amorphous $GeSe_4$ under ambient conditions, where Ge atoms are dark (blue) and Se atoms are light (green). A Se atom in a connection pathway between two Ge atoms is labeled as AA, a Se atom between one Ge atom and one Se atom is labeled as AB, and a Se atom between two Se atoms is labeled as BB. Reprinted figure with permission from [2]. Copyright (2016) by the American Physical Society.

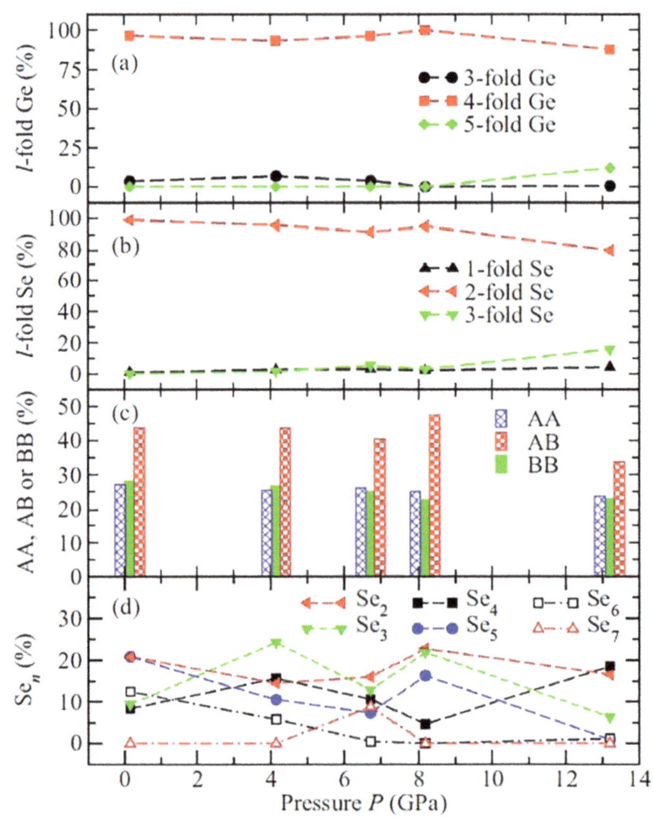

Figure 7.14. The pressure dependent fractions of (a) ℓ-fold coordinated Ge atoms (ℓ = 3, 4 or 5) and (b) ℓ-fold coordinated Se atoms (ℓ = 1, 2 or 3) obtained from the FPMD simulations. (c) The pressure dependent fractions of Se atoms that are 2-fold coordinated in either an AA, AB or BB configuration. (d) The pressure dependent fractions of Se atoms that are 2-fold coordinated in Se_n chains where n = 2, 3, 4, 5, 6 or 7. Reprinted figure with permission from [2]. Copyright (2016) by the American Physical Society.

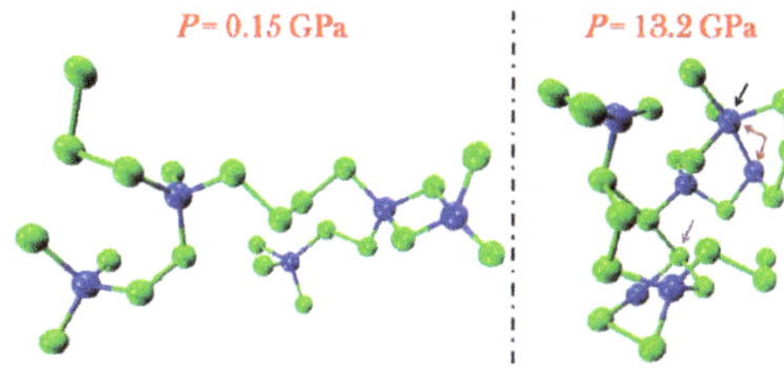

Figure 7.15. Atomistic configurations taken from the FPMD simulations of glassy $GeSe_4$ at low and high pressures, where Ge atoms are dark (blue) and Se atoms are light (green). For the high-pressure data set, the arrows point (clockwise from top) to a 5-fold coordinated Ge atom, to the pair of Ge atoms in a homopolar Ge–Ge bond, and to a 3-fold coordinated Se atom. Reprinted figure with permission from [2]. Copyright (2016) by the American Physical Society.

via the formation of 5-fold and 3-fold coordinated Ge and Se atoms, respectively. At the lowest pressure (figure 7.14(c)), AB configurations amount to ~44% and the proportions of AA and BB configurations are close, ~27% and ~28%, respectively. These percentages remain essentially unchanged at all pressures, proving that no phase separation occurs in glassy $GeSe_4$ at any value of the pressure considered. Also, the FPMD model points toward a large distribution of chain lengths as shown in figure 7.14(d). The highest pressure of 13.2 GPa marks a slight change in the topology due to an increase of threefold coordinated Se atoms, with percentages of twofold AB, AA and BB Se atoms taking the values ~34%, 24% and 23%, respectively. Therefore, despite a sizeable reduction, Ge–Se–Se (AB) connections remain the most frequent when compared to AA and BB.

The above findings can be summarized and better understood with the help of figure 7.15 where we have reproduced atomistic configurations of glassy $GeSe_4$ pertaining to the lowest and highest pressure points. In the low pressure structure one has both corner- and edge-sharing $Ge(Se_{1/2})_4$ tetrahedra (figure 7.9(c)). The tetrahedra are linked according to AA (Ge–Se–Ge), AB (Ge–Se–Se) and BB (Se–Se–Se) connections, where the AA motif is due to Ge atoms at the centers of edge-sharing tetrahedra (figure 7.14(c)). As the pressure is increased to 8.18 GPa, one notices the low-r shoulder at $\simeq 3.25$ Å on the second peak in $g_{\mathrm{SeSe}}(r)$ (figure 7.12(a)). At 13.2 GPa, some non standard coordination units begin to appear, as 5-fold coordinated Ge atoms, homopolar Ge–Ge bonds and 3-fold coordinated Se atoms (figure 7.15), leading to a small increase in $\bar{n}$. Therefore, the whole account of structural features shows that ordering in glassy $GeSe_4$ is not affected by the application of pressure, at least up to values ~8 GPa. Minimal changes (i.e. miscoordinated atoms departing from the twofold (Se) and fourfold (Fe) coordinations) are noticeable at higher pressure. This behavior preserving the basic structural motifs is due to the high flexibility of Se_n allowing for significant densification.

References

[1] Wezka K, Bouzid A and Pizzey K J *et al* 2014 *Phys. Rev.* B **90** 054206

[2] Bouzid A, Pizzey K J, Zeidler A, Ori G, Boero M, Massobrio C, Klotz S, Fischer H E, Bull C L and Salmon P S 2016 *Phys. Rev.* B **93** 014202

[3] Bouzid A and Massobrio C 2012 *J. Chem. Phys.* **137** 046101

[4] Petri I and Salmon P S 2002 *Phys. Chem. Glasses* **43** 185–90

[5] Azoulay R, Thibierge H and Brenac A 1975 *J. Non-Cryst. Solids* **18** 33–53

[6] Petri I, Salmon P S and Fischer H E 2000 *Phys. Rev. Lett.* **84** 2413–6

[7] Parthapratim Biswas D N, Tafen and Drabold D A 2005 *Phys. Rev.* B **71** 054204

[8] Massobrio C, Micoulaut M and Salmon P S 2010 *Solid State Sci.* **12** 199–203

[9] Mei Q, Benmore C J and Hart R T *et al* 2006 *Phys. Rev.* B **74** 014203

[10] Durandurdu M and Drabold D A 2002 *Phys. Rev.* B **65** 104208

[11] Birch F 1947 *Phys. Rev.* **71** 809–24

[12] Krukau A V, Vydrov O A, Izmaylov A F and Scuseria G E 2006 *J. Chem. Phys.* **125** 224106

[13] Fischer H E, Cuello G J, Palleau P, Feltin D, Barnes A C, Badyal Y S and Simonson J M 2002 *Appl. Phys.* A **74** s160–2

[14] Salmon P S, Drewitt J W E, Whittaker D A J, Zeidler A, Wezka K, Bull C L, Tucker M G, Wilding M C, Guthrie M and Marrocchelli D 2012 *J. Phys.: Condens. Matter* **24** 415102

[15] Salmon P S 2007 *J. Non-Cryst. Solids* **353** 2959–74

[16] Zeidler A, Salmon P S, Martin R A, Usuki T, Mason P E, Cuello G J, Kohara S and Fischer H E 2010 *Phys. Rev.* B **82** 104208

[17] Skinner L B, Benmore C J, Antao S, Soignard E, Amin S A, Bychkov E, Rissi E, Parise J B and Yarger J L 2012 *J. Phys. Chem.* C **116** 2212–7

[18] Kalkan B, Dias R P, Yoo C-S, Clark S M and Sen S M 2014 *J. Phys. Chem.* C **118** 5110–21

Chapter 8

Structural changes with composition in Ge_xSe_{1-x} glassy chalcogenides

The Ge_xSe_{1-x} family of disordered systems is characterized by important changes in the structural organization as a function of the composition. At $x = 0.33$, stoichiometry allows maximizing the number of tetrahedral connection and the underlying chemical order. This was one of the main messages of chapter 6. In this chapter we move away from stoichiometry by considering the Ge_xSe_{1-x} glassy networks at $x < 0.33$ ($x = 0.2$, $GeSe_4$) and $x > 0.33$ ($x = 0.4$, Ge_2Se_3). Starting from early calculations on liquid $GeSe_4$ we introduce the notion of structural variability and its implication in the definition of a boundary between different regimes of elasticity occurring around $x = 0.2$. The existence and the extent of the so-called intermediate phase are studied for glassy $GeSe_4$ and glassy $SiSe_4$. For glassy $GeSe_4$ we are also able to assess the sensitivity of the structural properties to the choice of the exchange–correlation functional via the calculation of the electronic density of states. Turning to the other side of the concentration range, for $x = 0.4$, we highlight a system (glassy Ge_2Se_3) in which Ge–Ge contacts become unavoidable to accommodate those Ge atoms not forming tetrahedra. The availability of several concentrations around and at the stoichiometric one makes possible a comparative study of these networks in terms of the main structural quantities. The chapter ends with a detailed study of the electronic localization properties pointing out the different nature of the Wannier centers identifiable at $x = 0.4$ for both the liquid and the glass.

8.1 Composition makes the difference: early calculations on liquid $GeSe_4$

Working on glassy $GeSe_2$ has allowed us to get acquainted with a set of notions covering topology (tetrahedra and their geometry/connections), intermediate range

doi:10.1088/978-0-7503-2436-6ch8

order (establishing well beyond nearest neighbors) and deviations from chemical order. Chemical order is the highest when, for a given concentration, the number of heteropolar bonds is maximized. It occurs for the stoichiometric composition in the absence of homopolar bonds and with $Ge(Se_{1/2})_4$ tetrahedra as the only structural unit throughout the network. We have also taken advantage of this composition to introduce useful concepts and tools related to fluctuations of correlations and charge, not to mention the delicate interplay between ionic and covalent character of bonding that underlies the search for the optimal exchange–correlation energy. It was also quite instructive to increase the density to realize how this kind of networks can be modified by acting on external thermodynamic variables (pressure and temperature). For instance, the reduction or increase of temperature in the disordered state has mostly the effect of impacting the percentage of structural defects. Having set the scene with tools, concepts and results for a prototypical system, our research line became naturally oriented toward other compositions, so as to enhance materials science understanding through the study of a family of systems.

Disordered $GeSe_4$ systems are at least as intriguing and instructive as $GeSe_2$ ones, due to the existence of a wealth of physical changes occurring for $\bar{n} = 2.4$, this value taking the meaning of a very special threshold that has stimulated a flurry of ideas and developments starting well before any attempt to model this system at the atomic scale [1, 2]. Interestingly, a paper appeared in 1998 [3] and devoted to first-principles molecular dynamics (FPMD) calculations on liquid $GeSe_4$ contains most of the essential information on the structural properties at this concentration. We have quoted some of the results of that work in the context of section 6.5. In what follows we refer to the results of that publication to introduce the issue of topology in disordered $GeSe_4$, by deferring to the next section a first account of structural properties for the case of glassy $GeSe_4$ (section 8.2). The partial pair correlation functions obtained in [3] are shown in figure 8.1.

Let's assume one knows very little about a Ge_xSe_{1-x} network. The simplest way of envisioning it is to take the extreme case of $x = 0$, for which one can think of an assembly of Se chains and rings, having a highly floppy and flexible structural organization. By increasing the number of Ge atoms at the expenses of Se atoms, the original Se-made structures are crosslinked so as to make the network more rigid. As pioneered in early investigations [1, 2], the number of constraints associated to the existence of bonds in the network can be enumerated as a function of the composition. This leads to an equality reached between the number of constraints and the number of degrees of freedom at a critical composition, that appears to be at $x = 0.2$ ($GeSe_4$) for Ge_xSe_{1-x} network systems. A qualitative and yet, quite instructive picture in terms of a stiffness transition was proposed marking the frontier between an under-constrained (floppy) and an over-constrained (rigid) network. The occurrence of interesting phenomena at the stiffness transition, well documented in the literature [4, 5] and the need to complement and go beyond phenomenological models was a great motivation to employ FPMD to obtain an atomic scale picture by asking as first, essential information, whether disordered $GeSe_4$ is a chemically ordered network. This is indeed the case, as we anticipated in

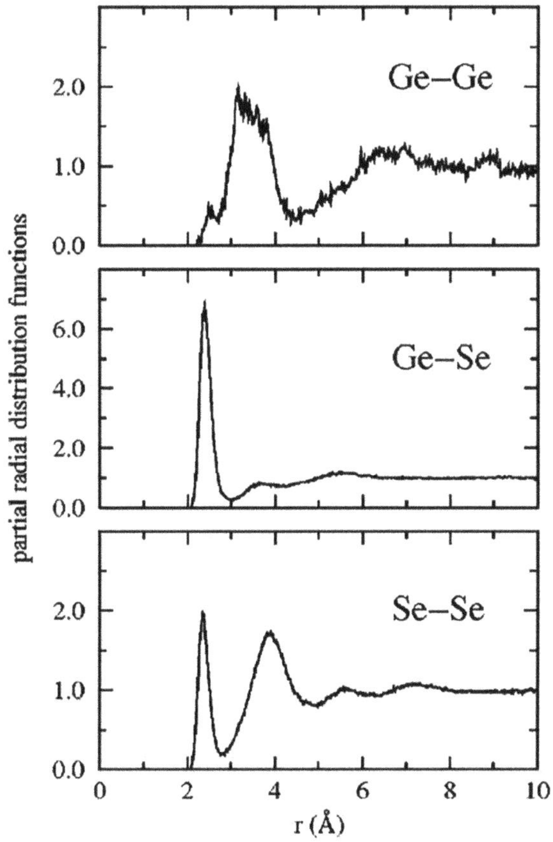

Figure 8.1. Calculated partial correlation functions for liquid $GeSe_4$ [3]. Reprinted figure with permission from [3]. Copyright (1998) by the American Physical Society.

section 6.5 in the context of the search for the atomic-scale origins for fluctuations of concentrations (the first sharp diffraction peak (FSDP) in $S_{CC}(k)$), where it was shown that liquid $GeSe_4$ is much more chemically ordered than liquid $GeSe_2$. Besides the very good agreement with experiments in term of total neutron structure factor, FPMD results on liquid $GeSe_4$ were also instrumental in producing the first estimate of the relative percentages of AA, BB and AB connections, defined in figure 7.13 and equal to 30%, 30% and 40%.

Despite the available definitions of different phenomenological networks and of how they relate to the FPMD models, leading to the data of table 6.3, some attention is required when focusing on the $GeSe_4$ topology after having considered mostly $GeSe_2$ systems. In short, homopolar bonds are not a criterion anymore to make the difference between chemical order and disorder, since they are obviously there for Se while, in principle, they should be absent for Ge (table 6.3). At this point, it is useful to make the connections with the concepts developed in section 6.5, where we introduced different levels of chemical disorder corresponding to the different intensities of the FSDP in $S_{CC}(k)$. One could argue that there is a slight contradiction between the above claim of chemical order holding in liquid $GeSe_4$ and the existence of the FSDP in $S_{CC}(k)$, that we attributed to small deviations from chemical order (see figure 6.17). In reality, these views can be reconciled by recalling that other disordered networks like SiO_2 feature strictly no deviations from the tetrahedral arrangement and a higher level of ionicity, hampering the formation of homopolar bonds unless at very high temperatures. Therefore, it makes still perfect sense not to assimilate $GeSe_4$ with SiO_2 when invoking chemical disorder, the second being much a very good representative of the notion of 'perfect' chemical disorder corresponding to no FSDP in $S_{CC}(k)$, as discussed in section 6.5.

8.2 Glassy $GeSe_4$ and glassy $SiSe_4$ and the 'structural variability'

The existence of a threshold at the mean coordination number $\bar{n} = 2.4$ [4, 5] has been long considered as a pure floppy (under-constrained) to rigid (over-constrained) transition [1, 2], well substantiated by experimental evidence and worth investigating with atomic-scale models. However, results in favor of a finite window of compositions around $x = 0.2$, separating the floppy and stressed-rigid regimes, started to appear [6–9], leading to the definition of the so-called *intermediate phase* (IP) having very peculiar physical properties in terms of self-organization, constraints and stress [10]. *It is worth making clear at this point that this monograph is not the right place to find details on these phenomenological models, that have inspired some of the calculations presented hereof but do not fall within the category of atomic-scale modeling that is our current mainstream.* The idea pursued is to rely on FPMD to complement intuition and qualitative approaches with a more quantitative assessment. The underlying questions are: is it possible to find any evidence of the intermediate phase substantiated by accurate modeling predictions? Is there anything special happening at the atomic scale when crossing the transition region? When considering two systems like $GeSe_4$ and $SiSe_4$, does it make sense to compare their intermediate phases intervals?

To move ahead and address these issues, it is convenient to refer to figure 7.13, to the introduction of the different Se coordinations done in section 7.3.2 and to the concept of *structural variability* [11]. This is very much related to the different weights of AA, BB and AB configurations within the network and it has been made explicit as follows. If we take as reference a system a network of undefective AB_4 tetrahedra connected in a corner-sharing fashion, with all A atoms linked by B_2 dimers, the fractions of AA, BB and AB units are 0%, 0% and 100%, respectively. According to mean-field ideas as those developed in [11], this is exactly a system (the 'full AB network') with no intermediate phase, or equivalently with a vanishing intermediate phase interval of concentrations. Therefore, the above percentages of AA, BB and AB units do promote a single elastic transition, while any deviation from the full AB network indicates a structural variability (different kinds of self-organization). Accordingly, the width of the intermediate phase for an A_xB_{1-x} material is related to the structural variability, a larger variability leading to a wider range of IP compositions. Quantitative models are expected to substantiate these ideas, knowing that the counting of the different Se coordinations is reliable provided the structural properties compare favorably with experiments. We shall concentrate on this issue before addressing the significance of the structural variability.

8.2.1 Structural properties

It should be said upfront that the values of AA, BB and AB configurations found for liquid $GeSe_4$ are a strong indication that an intermediate phase should exist also in the glass, since it is hard to conceive that the network could transform to a full AB network on quenching. Calculations can go beyond this obvious conclusion, by considering two systems, like $GeSe_4$ and $SiSe_4$, and comparing their behavior around the transition threshold. To this purpose glassy $GeSe_4$ was generated from the liquid phase detailed above via a cooling schedule of several tens of ps. The same strategy was applied to $SiSe_4$ for which an initial random configuration was accurately stabilized at high temperatures and quenched on time scales comparable to those of the $GeSe_4$ case [12].

The total structure factors $S_T(k)$ (figure 8.2) and the total pair correlation function $g_T(r)$ (figure 8.3) of glassy $GeSe_4$ and glassy $SiSe_4$ obtained by FPMD compare very well to experiments. These results are an example of calculation

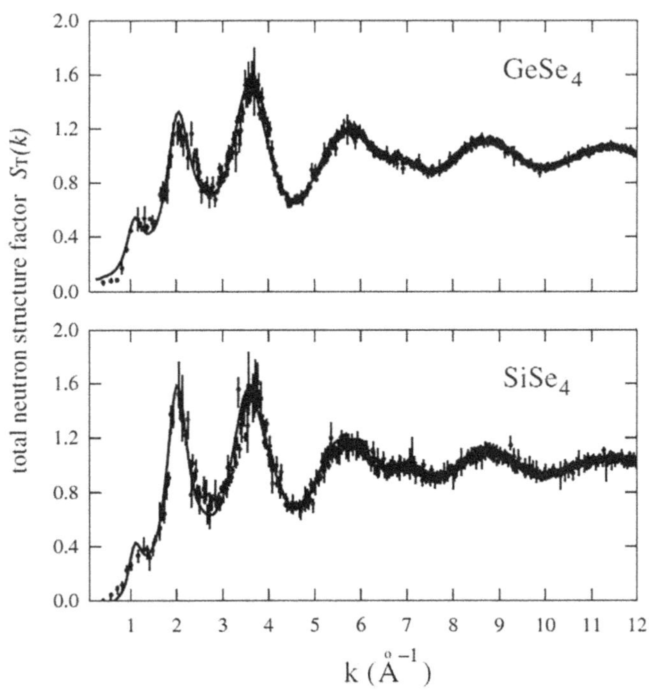

Figure 8.2. The measured (full curve) and calculated (dots with error bars) total neutron structure factor $S_T(k)$ for glassy $GeSe_4$ (upper panel) and glassy $SiSe_4$ (lower panel). Measurements and calculations methodologies are described in a joint paper, see [12]. Reprinted figure with permission from [12]. Copyright (2009) by the American Physical Society.

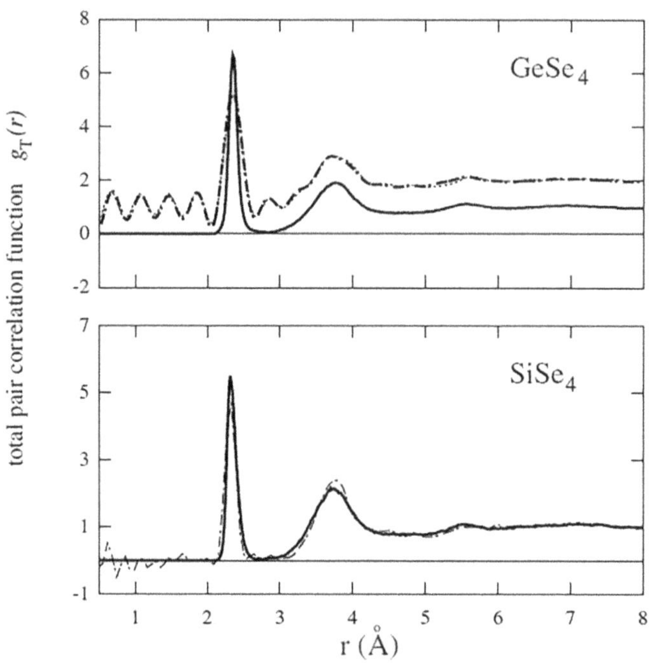

Figure 8.3. (a) Upper panel: total pair correlation function for $GeSe_4$. The dash-dotted curve gives the measured $g_T^{exp}(r)$ obtained by Fourier transforming the measured $S_T(k)$ function with a cutoff value k_{max} = 15.95 Å^{-1}, the dotted curve gives the FPMD $g_T^{th(b)}(r)$ obtained from the calculated $S_T(k)$ function by applying the same procedure, and the full curve gives the FPMD $g_T^{th}(r)$ obtained by direct calculation from the real space coordinates. (b) Lower panel: total pair correlation function for $SiSe_4$. The dash-dotted curve gives the measured $g_T^{exp}(r)$ obtained by Fourier transforming the measured $S_T(k)$ function with k_{max} = 40 Å^{-1} and the full curve gives the FPMD $g_T^{th}(r)$ obtained by direct calculation from the real space coordinates. Reprinted figure with permission from [12]. Copyright (2009) by the American Physical Society.

performed as in point **C** of section 2.3, with equations (2.11) and (2.13). In the case of $g_T(r)$ we have also included in figure 8.3 the direct calculation of the total pair correlation function as given by equation (2.10) of section 2.3.

By taking advantage of the concepts developed in section 6.5, it is of interest to consider the concentration–concentration partial structure factor $S_{CC}(k)$, for which, at the time of production of the FPMD data, no experiments were available either for $GeSe_4$ or $SiSe_4$. Given the presence of a small feature in the region of the FSDP (see figure 8.4), one can infer a small amount of chemical disorder in both systems, as it will be further demonstrated in section 8.2.2 where a small departure from a chemically ordered network model will be quantitatively described and emphasized.

Turning to real space, we display the calculated partial pair correlation functions $g_{SeSe}(r)$, $g_{ASe}(r)$ and $g_{AA}(r)$ for both glassy $GeSe_4$ and glassy $SiSe_4$ in figure 8.5. As expected, there is a first peak in $g_{SeSe}(r)$ indicative of Se–Se homopolar bonds in both systems. Focusing to $g_{GeSe}(r)$ and $g_{SiSe}(r)$, the striking feature is the sharp main peak indicative of a large majority of $GeSe_4$ and $SiSe_4$ tetrahedral units.

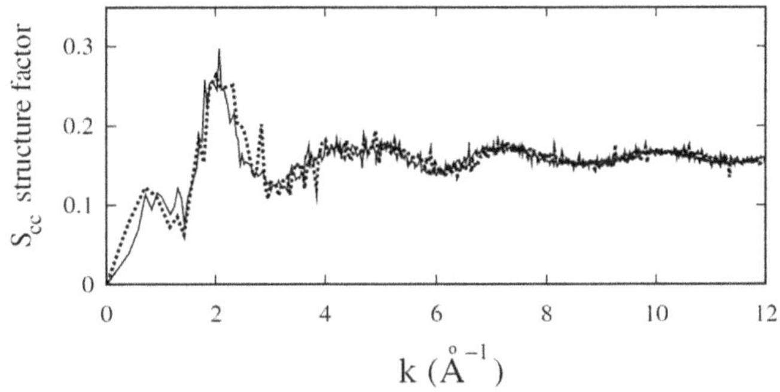

Figure 8.4. The calculated Bhatia–Thornton concentration–concentration partial structure factor $S_{CC}(k)$ for glassy $GeSe_4$ (solid curve) and glassy $SiSe_4$ (dotted curve). Reprinted figure with permission from [12]. Copyright (2009) by the American Physical Society.

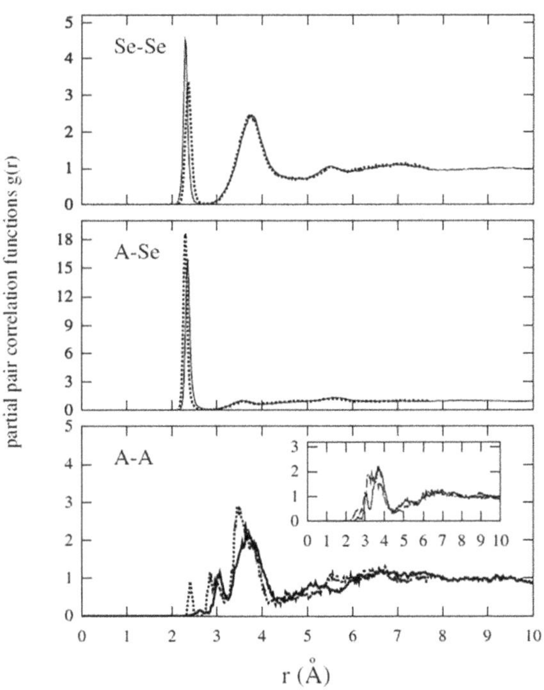

Figure 8.5. The calculated partial pair correlation functions for glassy $GeSe_4$ (solid curves) and glassy $SiSe_4$ (dotted curves). The label Se–Se indicates $g_{SeSe}(r)$, the label A–Se indicates either $g_{GeSe}(r)$ or $g_{SiSe}(r)$ and the label A–A indicates either $g_{GeGe}(r)$ or $g_{SiSi}(r)$. Inset: comparison between the calculated partial pair correlation function $g_{GeGe}(r)$ for glassy $GeSe_4$ (solid curve) and the calculated partial pair correlation function $g_{GeGe}(r)$ for liquid $GeSe_4$ (dash-dotted curve) (see figure 8.1 in section 8.1). Reprinted figure with permission from [12]. Copyright (2009) by the American Physical Society.

By considering $g_{GeGe}(r)$ and $g_{SiSi}(r)$, one finds that the pattern typical of Ge–Ge correlation in network-forming systems of this family is also reproduced when Si takes the place of Ge, with three peaks in the region comprised between 2 Å and 4 Å, as already observed for glassy $GeSe_2$ in section 6.2.3. These are due to homopolar Ge–Ge (Si–Si) bonds, Ge(Si) atoms in edge-sharing tetrahedral connections, and Ge(Si) atoms involved in corner-sharing tetrahedral connections, respectively. The availability of the results for both the liquid and the glass reveals that (see inset of figure 8.5) the shape of $g_{GeGe}(r)$ is dramatically different in the two cases, to indicate that the quench is at the origin of significant differences between the pair correlation functions of the liquid and of the glass. This argument was made explicit in section 4.4.5 were we referred to the case of relaxation in glassy $GeSe_2$ (see figure 4.13).

Another set of typical fingerprints of disordered chalcogenides, carrying useful information on the structure, is provided by the bond-angle distributions of figure 8.6 for which the two main peaks appearing in the Ge–Se–Ge and Si–Se–Si stem from edge-sharing tetrahedra, i.e. to Ge or Si centered subunits which have in common two Se atoms, and to corner-sharing tetrahedra which share only a single Se atom. In glassy $SiSe_4$, the main peaks in θ_{SiSeSi} are found at at 80° and 99° while in glassy $GeSe_4$ θ_{GeSeGe} has a peak at 80° with a higher intensity. A second, broader peak has

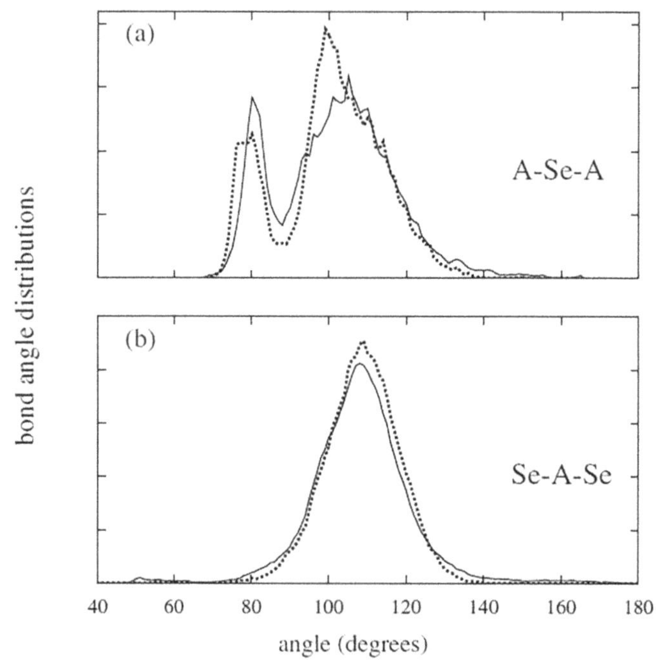

Figure 8.6. The calculated bond-angle distributions for glassy $GeSe_4$ (solid curves) and glassy $SiSe_4$ (dotted curves). The label A–Se–A indicates either the Ge–Se–Ge or Si–Se–Si bond-angle distribution while the label Se–A–Se indicates either the Se–Ge–Se or Se–Si–Se bond-angle distribution. Reprinted figure with permission from [12]. Copyright (2009) by the American Physical Society.

its maximum at 105°. The bond-angle distributions for the Se–Ge–Se and Se–Si–Se triads are symmetrical around 109°, as it occurs for a tetrahedral network highly chemically ordered and with a large majority of Ge and Si atoms fourfold coordinated.

8.2.2 Structural variability

We have mentioned above the essential ideas relative to the issues of the structural variability and of a finite width (window) of compositions (the intermediate phase, IP) for a number of properties across the $x = 0.2$ value in the case of Ge_xSe_{1-x} and Si_xSe_{1-x} systems (with $\bar{n} = 2(1 + x)$). The real nature of the IP has been demonstrated by experimental evidence by allowing a comparison between different systems [11]. For instance, it appears that the IP window is wider for glassy Si_xSe_{1-x}. Since finite widths are thought to be related to non-negligible amounts of AA and BB configurations in both systems, a larger width for the Si_xSe_{1-x} system compared to Ge_xSe_{1-x} would call for a larger structural variability. It is of interest to check whether FPMD is able to confirm this kind of experimental evidence. To this purpose, the motifs in glassy $GeSe_4$ and glassy $SiSe_4$ that are responsible for the structural variability and the finite width of the intermediate phase have been identified by using FPMD methods [12].

We found that AB units are the most frequent (47% in glassy $GeSe_4$ and 42% in glassy $SiSe_4$) but AA units and BB units are also far from being absent (with 23% of AA units in glassy $GeSe_4$ and 25% in glassy $SiSe_4$ and with 26% of BB units in glassy $GeSe_4$ and 30% in glassy $SiSe_4$). These numbers point to glassy $SiSe_4$ as the system possessing a larger structural variability, resulting from the observation that AB($GeSe_4$) > AB($SiSe_4$), BB($GeSe_4$) < BB($SiSe_4$) and AA($GeSe_4$) < AA($SiSe_4$). This behavior is consistent with the larger number of edge-sharing tetrahedra in glassy $SiSe_4$, typical examples of AA linkages in which two Se atoms have two Ge (or Si) nearest neighbors to form a fourfold Ge–Se–Ge–Se ring. Overall, FPMD and experiments agree on the larger width of the intermediate phase window for the case of silicon selenide glasses. While this agreement is very satisfactory, one should somewhat soften its impact as a stringent demonstration of the occurrence of the intermediate phase via first-principles methods. First, the notion of structural variability appears as a mere consequence of the existence of edge-sharing connections and of a departure from a full AB network. Also, no evidence exists for the absence of AA and BB connections in any calculation performed around the $x = 0.2$ transition value for A_xB_{1-x} materials. The idea of comparing two windows for the intermediate phase is also quite intriguing and highly valuable. However, strictly speaking, the evidence presented above (and in more detail in [12]) should be made more compelling by repeating the calculations on larger sizes since the differences between the two glasses are minimal. Nevertheless, the FPMD results presented in 2009 were the first to establish a connection between the intermediate phase and a realistic atomic-scale model.

Concerning glassy $GeSe_4$, the essence of these results were confirmed via a new set of calculations performed by using a different exchange–correlation functional

(PBE) [13]. This functional was employed much in the same spirit of the BLYP/PW comparison reported in section 6.7, although with a different motivation, since PBE is in principle known for being closer to PW than BLYP (see section 4.2.1). In reality, the PBE calculations were compared first to the early PW ones (instead of the BLYP ones, since the calculations were performed by different teams in different times). Since some BLYP runs were also produced shortly after the PBE ones, we were in a position to compare three different sets of data for the same system. While no drastic difference were found among the three approaches in terms of structures, this was a good opportunity to complement the information obtained on the impact of exchange–correlation functionals along the lines followed in section 6.7 for $GeSe_2$ disordered systems.

Figure 8.7 shows the electronic density of states (EDOS) obtained with the three different functionals, the first label referring to the functional employed to obtain the structure via a thermal cycle, while the second indicates which functional has been used for a further trajectory of relaxation. This has been implemented to infer the degree of sensitivity of each trajectory to changes of the exchange–correlation functional as a result of the second trajectory. Two considerations are in order. First, there seems to be no real gap in the EDOS, in agreement with the known underestimate of this quantity within DFT and the semiconductor character of these compounds. Both PBE and BLYP have a deeper pseudogap than PW, in line with our rationale on the higher metallic nature of bonding of PW when compared to BLYP (section 6.7). *Even though our calculation strategy has been based on BLYP ever since the first application to glassy $GeSe_2$* detailed in section 6.7, PBE is also well performing in the specific case of $GeSe_4$. As a second point, it appears

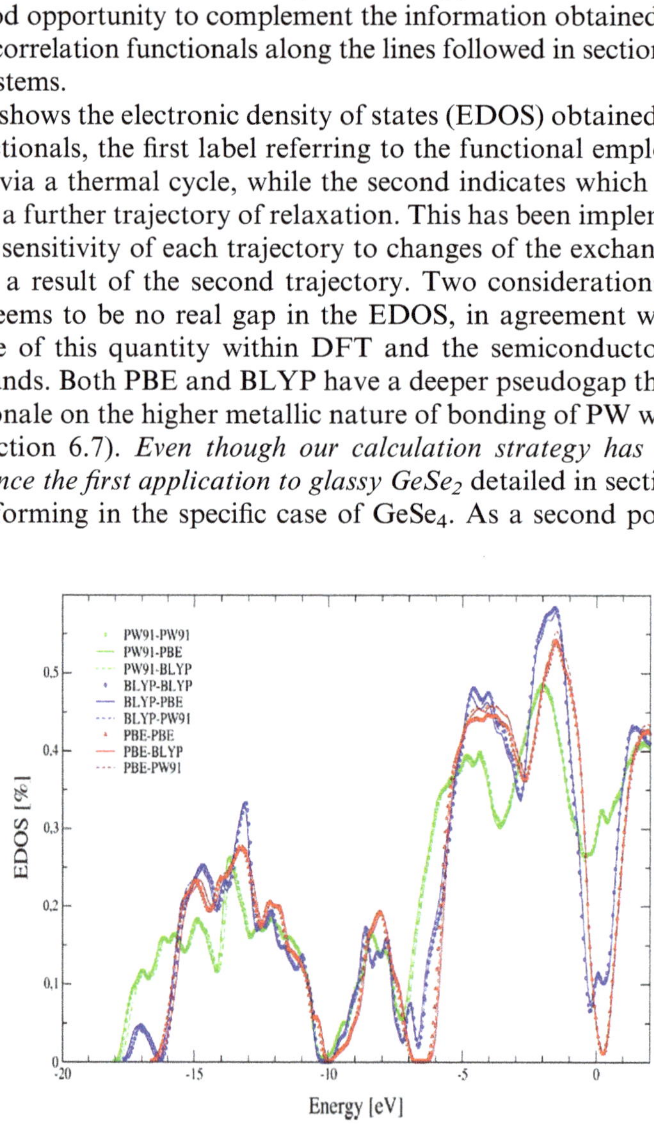

Figure 8.7. Electronic density of states of glassy $GeSe_4$ obtained with three different exchange–correlation functionals (first label on the left). For each of them, the structure obtained was relaxed by using the exchange–correlation functionals indicated by the second label on the right.

that changing to a new functional after production of a trajectory with a given one does not induce any modification (or very little) in the EDOS. This is due to the vanishing atomic mobility at low temperatures on the timescale of our calculations and, to a smaller extent, to the similar structural features of the three models, as underlined above when comparing PW and PBE. We recall that differences between the LDA and the PW structure were obtained in section 3.6.5 for the case of a liquid, a system for which mobility is substantial and allows surmounting diffusion barriers.

8.3 Altering stoichiometry by adding Ge: glassy Ge_2Se_3

After having considered $GeSe_2$ and $GeSe_4$, one can move to the other side of the Ge_xSe_{1-x} family of concentrations, for x larger than 0.33. In this region, glasses are known to form up to $x \leqslant 0.43$. These networks are characterized by Ge atoms not having enough Se atoms to accommodate in tetrahedra, the first consequence being the expected absence of Se chains [14]. When selecting Ge_2Se_3, the idea is to learn about this system by comparing its structural properties to those of glassy $GeSe_2$, glassy $GeSe_4$ and their parent liquid structures, by exploiting a number of results made available in previous studies [12, 15–24]. Our FPMD-BLYP framework (employed for consistency to treat all networks contributing to a comparison, including those for which, as glassy $GeSe_4$, published results were only available within the PW scheme) was applied to a system of N_{at} = 120 (48 Ge and 72 Se) atoms. Incidentally, it is worth noting that the work on glassy Ge_2Se_3 marks the end of calculations performed on N_{at} = 120 atoms by moving to N_{at} = 480 being carried out in parallel to an investigation on size effects on liquid $GeSe_2$ [19][1]. More on this issue will be given in later sections. In particular, comparisons between results for N_{at} = 120 and N_{at} = 480 are provided in section 9.4 for the cases of glassy $GeSe_4$ and GeS_4.

As a specific technical detail, we can mention that the glassy configurations were obtained by starting from ten decorrelated initial set of coordinates of the liquid [18] by setting first the density of the initial liquid configuration at T = 1000 K at the one of the amorphous state [14] and then by cooling the system at T = 300 K over more 100 ps. Concerning the electronic densities of states, these were calculated on 100 configurations equally distributed along the trajectory. The total neutron structure factor $S_T^{th}(k)$ is compared with its experimental counterpart $S_T^{exp}(k)$ [25] in figure 8.8, showing very good agreement over the full range of wavevectors in terms of intensities and peaks positions. The same can be said for the total pair correlation function (figure 8.9) for which, as we did in section 8.2.1 we have followed the different definitions introduced in section 2.3.

We begin the comparative study among the four systems glassy Ge_2Se_3, liquid Ge_2Se_3, glassy $GeSe_4$ and glassy $GeSe_2$ by considering the partial structure factors in

[1] At this point it is useful to reiterate that all calculations performed with N_{at} = 120 and considered so far are strictly *reliable and trustworthy*. However, especially for glasses, a switch to larger sizes appeared more appropriate to allow for an improved structural determination.

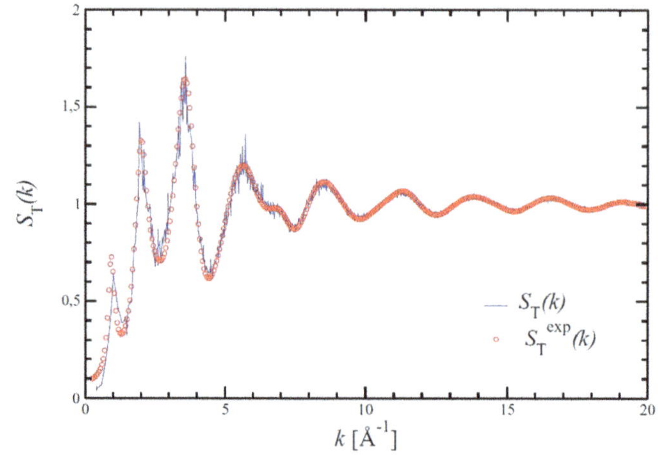

Figure 8.8. Total neutron structure factor for glassy Ge_2Se_3 at T = 300 K. The experimental result $S_T^{exp}(k)$ given in [25] (red circles) is compared to the calculated function $S_T^{th}(k)$ (solid blue curve). For sake of clarity only a representative subset of error bars are given at specific location in the reciprocal space. Reprinted figure with permission from [26]. Copyright (2012) by the American Physical Society.

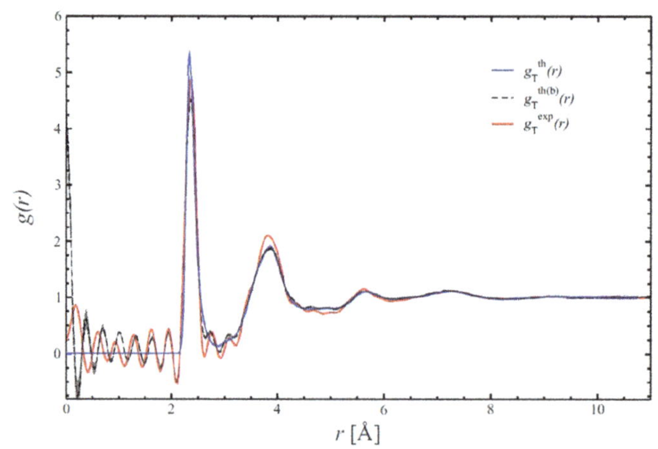

Figure 8.9. Total pair correlation function for glassy Ge_2Se_3 at T = 300 K. The experimental function $g_T^{exp}(r)$ of [25] (solid red line) was obtained by Fourier transforming the measured total structure factor $S_T^{exp}(k)$ (see figure 8.8) with a cutoff value k_{max} = 19.95 Å^{-1}. The same procedure was applied to the computed function $S_T^{th}(k)$ (see figure 8.8) to obtain $g_T^{th(b)}(r)$ (broken black curve). The total pair distribution function $g_T^{th}(r)$ (solid blue curve) is the result of a direct calculation from the real space coordinates. Reprinted figure with permission from [26]. Copyright (2012) by the American Physical Society.

figure 8.10. When producing glassy Ge_2Se_3 from liquid Ge_2Se_3, the partial structure factors $S_{SeSe}(k)$ and $S_{GeSe}(k)$ undergo an expected increase of the peak intensities. However, this is not the case for a small feature in the region ($k \sim 1$ Å^{-1}) of $S_{GeSe}(k)$, meaning that Ge–Se correlations beyond nearest neighbors have little impact on the intermediate range order.

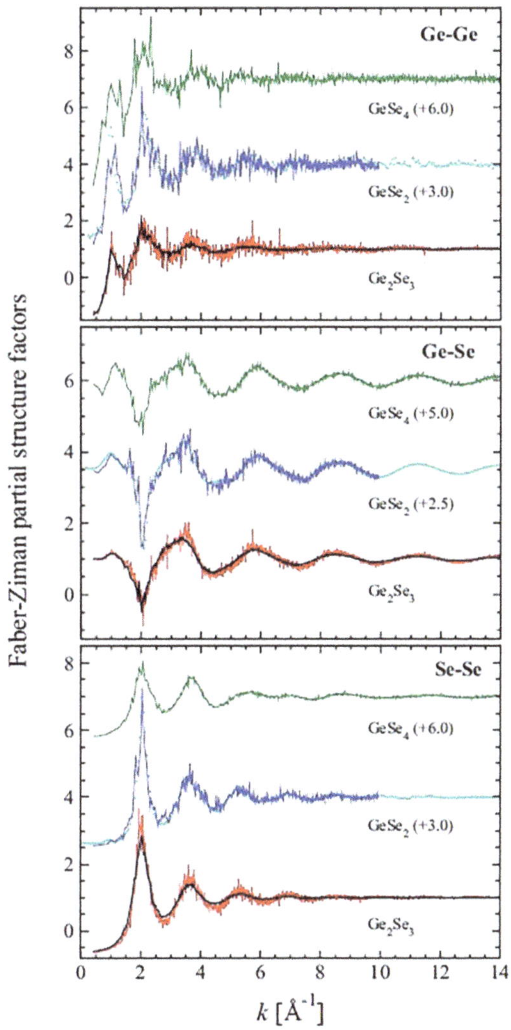

Figure 8.10. The partial structure factors $S_{GeGe}(k)$ (top panel), $S_{GeSe}(k)$ (middle panel) and $S_{SeSe}(k)$ (bottom panel) for glassy Ge_2Se_3 ([26] solid red lines), liquid Ge_2Se_3 ([18] solid black lines), glassy $GeSe_2$ ([27] solid blue lines) and glassy $GeSe_4$ ([12] solid green lines). Experimental counterpart for glassy $GeSe_2$ is also presented ([14] light blue dots). Reprinted figure with permission from [26]. Copyright (2012) by the American Physical Society.

Focusing on Ge–Ge reciprocal space correlations, the ratio $R_{FSDP/Main}$ between the intensities of the first two peaks (FSDP and the main peak) is larger in glassy Ge_2Se_3 to indicate that there are more important Ge–Ge intermediate range correlations in the glass than in the liquid. Also, one can notice the following: (a) $R_{FSDP/Main}$ is larger in $S_{GeGe}(k)$ for Ge_2Se_3 with respect to $GeSe_4$, (b) the highest main peak ($k \sim 2$ Å^{-1}) in $S_{GeSe}(k)$ is found in glassy $GeSe_2$, (c) the largest intensity at $k \sim 1$ Å^{-1} in $S_{GeSe}(k)$ is found for glassy $GeSe_4$ and (d) in the case of $S_{SeSe}(k)$, the main peak has a reduced intensity for glassy $GeSe_4$, due to the smaller number of Se atoms in $GeSe_4$ tetrahedra.

In figure 8.11 we show the Bhatia–Thornton (BT) $S_{NN}(k)$, $S_{NC}(k)$, and $S_{CC}(k)$ partial structure factors [28, 29] for glassy $GeSe_2$, glassy $GeSe_4$, glassy Ge_2Se_3 and liquid Ge_2Se_3. The comparison among the different BT for the three glasses under consideration reveals that there is an effect in the FSDP region due to the smaller concentration of Ge atoms appearing in $S_{NC}(k)$. This stems from the $c_{Ge}S_{GeGe}(k)$ term (see equation (2.7)), more important in glassy $GeSe_2$ and glassy Ge_2Se_3 than in

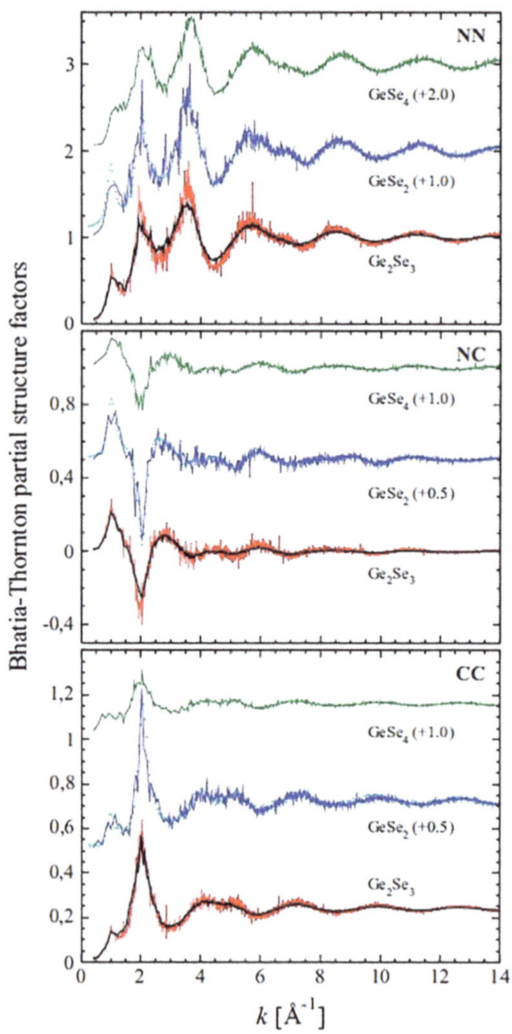

Figure 8.11. The Bhatia–Thornton partial structure factors $S_{NN}(k)$ (top panel), $S_{NC}(k)$ (middle panel) and $S_{CC}(k)$ (bottom panel) for glassy Ge_2Se_3 (solid red lines), liquid Ge_2Se_3 ([18] solid black lines), glassy $GeSe_2$ ([27] solid blue lines) and glassy $GeSe_4$ ([12] solid green lines). Experimental counterpart for glassy $GeSe_2$ is also presented (from [14] light blue dots). Reprinted figure with permission from [26]. Copyright (2012) by the American Physical Society.

glassy $GeSe_4$ because of the higher concentration of Ge atoms in these systems and the smaller intensity of the FSDP in $S_{GeGe}(k)$ for glassy $GeSe_4$. Turning to $S_{CC}(k)$, the most visible feature at short wavevectors occurs for glassy $GeSe_2$, glassy Ge_2Se_3 and liquid Ge_2Se_3. This allows establishing a connection with the results detailed in section 6.6.2 based on a correlation between the intensity of FSDP in $S_{CC}(k)$ and the chemical disorder. Accordingly, chemical disorder is expected to be more important in glassy $GeSe_2$, glassy Ge_2Se_3 and liquid Ge_2Se_3 than in glassy $GeSe_4$ [30].

Overall, we have shown that liquid Ge_2Se_3 and glassy Ge_2Se_3 share the same features of the partial structure factors. This is by far not trivial, since our methodology is able to detect differences in reciprocal space between glasses and liquids, should they exist [12, 17]. Changing the concentration of Ge and Se atoms is reflected by differences in the absolute and relative intensities of the FSDP (at $k \sim 1$ Å^{-1}) and the first main peak (at $k \sim 2$ Å^{-1}). In figure 8.12 we display the calculated partial pair correlation functions $g_{\alpha\beta}(r)$ for glassy Ge_2Se_3 together with the calculated functions for liquid Ge_2Se_3 [18], glassy $GeSe_2$ and glassy $GeSe_4$. The predominance of a Ge centered tetrahedral motif in the glassy state is indicated by the sharper shape of the main peak in $g_{GeSe}(r)$ when compared to the liquid case. The case of $g_{SeSe}(r)$ is indicative of the absence of any little shoulder at distances typical of Se–Se homopolar bonds. This is due to the disappearance of homopolar bonds that were created by thermal activation in the liquid state on vanishing intervals of time [12]. When commenting on the results obtained for liquid Ge_2Se_3 [18], the partial correlation function $g_{GeGe}(r)$ was analyzed on the basis of one peak at $r = 2.47$ Å due to Ge–Ge homopolar bonds and a large second peak in between 3 Å and 4.5 Å. By quenching from the liquid we are now in a position to disentangle the contribution of the edge-sharing and corner-sharing connections, in line with other studies on Ge_xSe_{1-x} systems, since a three peak structure is visible in glassy Ge_2Se_3 after cooling. On the other hand, Ge–Ge homopolar bonds are largely preserved, since they stem from Ge atoms in excess with respect to those necessary to form tetrahedral bonds with Se atoms. The availability of three concentrations, located respectively at the stoichiometric value ($GeSe_2$), in the Se-richer part ($GeSe_4$) and in the Ge-richer part (Ge_2Se_3), allows following the trend of $g_{\alpha\beta}(r)$ with the composition. For $g_{GeSe}(r)$, we can reiterate that the $GeSe_4$ coordination is the most frequent in the three cases as it expresses the existence of a structural motif that stands out in the three concentrations. The different shapes observed in $g_{SeSe}(r)$ at $r \sim 2.4$–2.5 Å are related to the impact of Se–Se connections, since they are not only inevitable in glassy $GeSe_4$, but they also appear in glassy $GeSe_2$ to indicate chemical disorder. Similarly, there is a three peak structure in $g_{GeGe}(r)$ due to homopolar bonds, edge-sharing and corner-sharing tetrahedra, but the relative weight of the three corresponding peaks depends on the composition. In fact, there are very few Ge–Ge homopolar bonds in glassy $GeSe_4$, but they increase in glassy $GeSe_2$. Finally, Ge–Ge connections cause the presence of an intense peak in glassy Ge_2Se_3, as it should be since there are Ge atoms that cannot accommodate within $Ge(Se_{1/2})_4$ tetrahedra.

We provide the coordination numbers n_α^β of glassy $GeSe_2$, glassy $GeSe_4$, glassy Ge_2Se_3 and liquid Ge_2Se_3 in table 8.1. Some of these numbers reflect quite explicitly

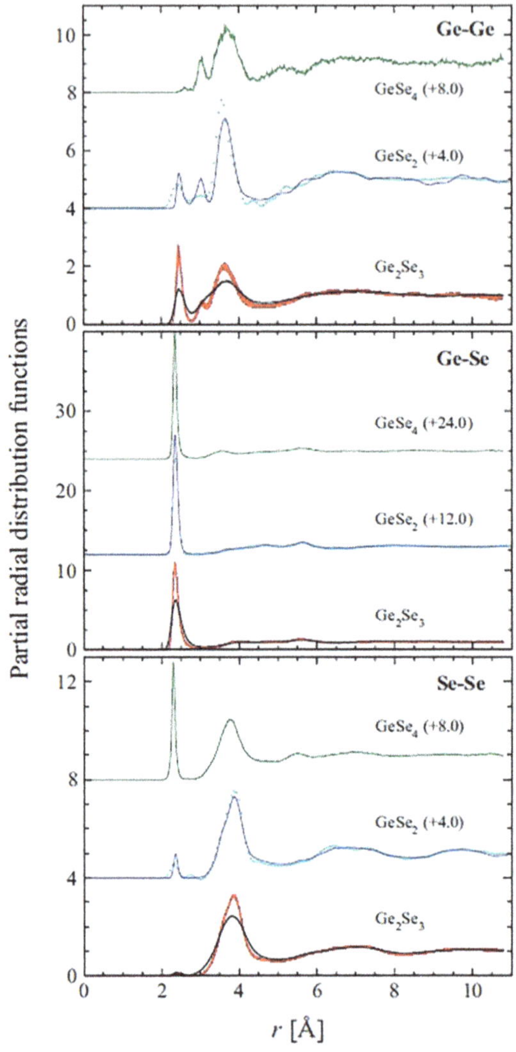

Figure 8.12. The partial pair correlation functions (here named partial radial distribution functions) $g_{\mathrm{GeGe}}(r)$ (top panel), $g_{\mathrm{GeSe}}(r)$ (middle panel) and $g_{\mathrm{SeSe}}(r)$ (bottom panel) for glassy Ge_2Se_3 ([26] solid red lines), liquid Ge_2Se_3 ([18] solid black lines), glassy $GeSe_2$ ([27] solid blue lines) and glassy $GeSe_4$ ([12] solid green lines). Experimental counterpart for glassy $GeSe_2$ is also presented ([14] light blue dots). Reprinted figure with permission from [26]. Copyright (2012) by the American Physical Society.

the above observations on the shape of the pair correlation function. For instance, $n_{\mathrm{Ge}}^{\mathrm{Ge}}$ is higher in glassy Ge_2Se_3 and in liquid Ge_2Se_3 than in glassy $GeSe_2$ and in glassy $GeSe_4$, due to Ge atoms not forming tetrahedra with Se atoms. Similarly, $n_{\mathrm{Se}}^{\mathrm{Se}}$ decreases in glassy Ge_2Se_3 with respect to liquid Ge_2Se_3 (0.01 against 0.08), since there are only very few Se chains. In table 8.1 we also give the values of n_{α}^{β} for two simple models of the network structure of disordered Ge_xSe_{1-x}, chemically ordered network (CON) and random covalent network (RCN). In the CON model, Ge–Se

Table 8.1. The first peak position (FPP) and second peak position (SPP) in $g_{\alpha\beta}(r)$ and the nearest-neighbor coordination numbers n_α^β obtained from FPMD models of glassy Ge_2Se_3, liquid Ge_2Se_3, glassy $GeSe_2$ and glassy $GeSe_4$ [12, 18, 27]. The predictions of the CON and RCN models are also listed [14]. Reprinted table with permission from [26]. Copyright (2012) by the American Physical Society.

$g_{\alpha\beta}(r)$	Model	FPP (Å)	SPP (Å)	n_α^β	$\bar{n}_\alpha^\beta$ (CON)	n_α^β (RCN)
$g_{GeGe}(r)$	Glassy $GeSe_4$	2.64	3.65	0.01	0	1.3333
	Glassy $GeSe_2$	2.47	3.67	0.28	0	2
	Glassy Ge_2Se_3	2.47	3.63	0.52	1	2.2857
	Liquid Ge_2Se_3	2.47	3.70	0.48	1	2.2857
$g_{GeSe}(r)$	Glassy $GeSe_4$	2.36	3.54	3.92	4	2.6667
	Glassy $GeSe_2$	2.34		3.84	4	2
	Glassy Ge_2Se_3	2.35	5.61	3.21	3	1.7143
	Liquid Ge_2Se_3	2.36	5.65	3.15	3	1.7143
$g_{SeGe}(r)$	Glassy $GeSe_4$	2.36	3.54	0.98	1	0.6666
	Glassy $GeSe_2$	2.34		1.82	2	1
	Glassy Ge_2Se_3	2.35	5.61	2.14	2	1.1429
	Liquid Ge_2Se_3	2.36	5.65	2.10	2	1.1429
$g_{SeSe}(r)$	Glassy $GeSe_4$	2.29	3.76	1.01	1	1.3333
	Glassy $GeSe_2$	2.34	3.76	0.2	0	1
	Glassy Ge_2Se_3	2.37	3.88	0.01	0	0.8571
	Liquid Ge_2Se_3	2.39	3.81	0.08	0	0.8571

bonds are favored and only Ge–Se and Ge–Ge bonds are allowed for $x > 0.33$ while the opposite is true (Ge–Se and Se–Se bonds allowed) for $x < 0.33$. In the RCN model bond types are distributed in a statistical manner such that Se–Se bonds are allowed for $x > 0.33$ and Ge–Ge bonds are allowed for $x < 0.33$. While the RCN model fails to describe the glasses under consideration, the CON model is quite acceptable for glassy $GeSe_4$, and less reliable for glassy $GeSe_2$ and glassy Ge_2Se_3.

The total coordination numbers for Ge and Se and the total one are collected in table 8.2 and compared to the measured values and to those expected from the '8 − N' rule where Ge atoms are fourfold coordinated and Se atoms are twofold coordinated. Table 8.2 shows that the '8 − N' rule is followed quite accurately by all systems. The network can be further described in terms of structural units as introduced in section 2.4, by recalling that for each specific unit, $n_\alpha(l)$, the result is expressed by calculating the ratio between the mean number of the $n_\alpha(l)$ events and total number of atoms of type α. The proportions of l-fold coordinated atoms and of each specific unit $n_\alpha(l)$ are presented in figure 8.13 (for glassy Ge_2Se_3 and liquid

Table 8.2. The coordination numbers for Ge, n_{Ge}, and Se, n_{Se}, in glassy Ge_2Se_3. The results are compared with those obtained from FPMD models of the liquid Ge_2Se_3, glassy $GeSe_2$ and glassy $GeSe_4$ [12, 18, 27]. The calculated average coordination number, n_{tot}, for each system is also listed and the values are compared with the experimental results n_{tot}^{exp} of [14] and with the expectations of the '8 − N' rule, n_{tot}^{8-N}. Reprinted table with permission from [26]. Copyright (2012) by the American Physical Society.

Model	n_{Ge}	n_{Se}	n_{tot}	n_{tot}^{exp}	n_{tot}^{8-N}
Glassy $GeSe_4$(BLYP)	3.96	2.01	2.40	2.44	2.4
Glassy $GeSe_4$(PW)	3.93	1.99	2.37	2.44	2.4
Glassy $GeSe_2$	3.92	2.02	2.66	2.69	2.67
Glassy Ge_2Se_3	3.73	2.15	2.78	2.81(5)	2.8
Liquid Ge_2Se_3	3.63	2.18	2.76	2.8(2)	2.8

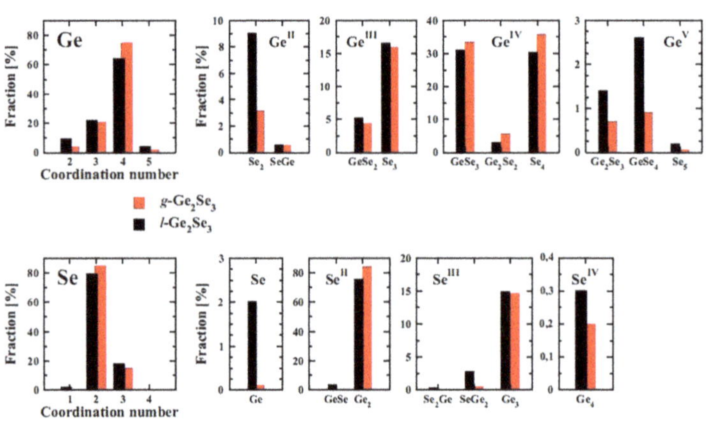

Figure 8.13. Percentage of l-fold coordinated atoms, also decomposed in term of each specific unit $n_\alpha(l)$, for both Ge (top panel) and Se (bottom panel). The systems considered are glassy Ge_2Se_3 and liquid Ge_2Se_3. Reprinted figure with permission from [26]. Copyright (2012) by the American Physical Society.

Ge_2Se_3 [18, 26]) and figure 8.14 (for glassy $GeSe_2$ and glassy $GeSe_4$ [12, 27]). We note first in figure 8.13 that glassy Ge_2Se_3 has a large number of fourfold Ge atoms and a reduced number of twofold Ge atoms. The same occurs in the Se case, with a decrease of miscoordinated threefold Se atoms in favor of an increase of twofold Se atoms in the glass. There is an increase (by 5%) in glassy Ge_2Se_3 of twofold Se atoms at the expense of (miscoordinated) threefold Se atoms. Twofold Ge atoms present in liquid Ge_2Se_3 had very reduced lifetimes [18] that vanish as a result of cooling, undermining the importance of these units. This promotes the increase of the fourfold connection for which the formation barriers to be overcome are lower. Overall, in glassy Ge_2Se_3, the number of Se atoms not twofold coordinated is lower than in the liquid Ge_2Se_3, resulting in ~85% of Se atoms twofold coordinated.

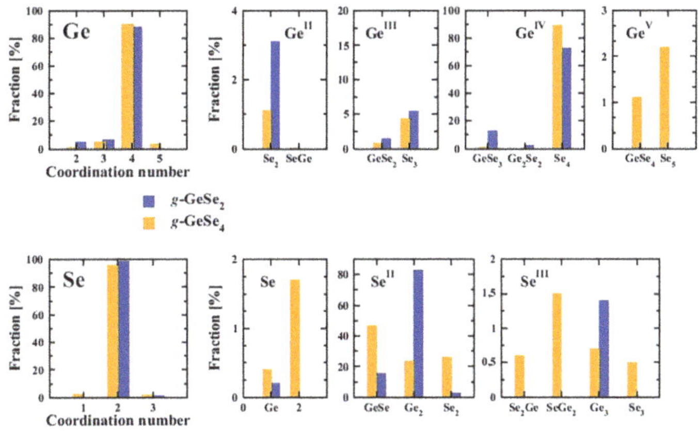

Figure 8.14. Percentage of l-fold coordinated atoms, also decomposed in term of each specific unit $n_\alpha(l)$, for both Ge (top panel) and Se (bottom panel). The systems considered are glassy $GeSe_2$ and glassy $GeSe_4$. Reprinted figure with permission from [26]. Copyright (2012) by the American Physical Society.

Figure 8.14 allows comparing the topologies of glassy $GeSe_4$ and glassy $GeSe_2$, both characterized, as shown in previous examples through the pair correlation functions, by a predominant tetrahedral coordination. The most interesting difference between the two networks resides in the proportions of AA, AB and BB connections (defined in figure 7.13). In glassy $GeSe_2$ Se–Se chains are a few, leading to a small amount of BB connections. The $GeSe_2$ network contains a majority of AA connections, due to its structure in which Se atoms are mostly bound to Ge atoms in tetrahedra. Much closer values are found in glassy $GeSe_4$, as detailed in section 8.2.2. In summary, two kinds of structural evolutions are encountered when considering a comparative fashion liquid Ge_2Se_3, glassy Ge_2Se_3, glassy $GeSe_4$ and glassy $GeSe_2$. Chemical order is enhanced when moving from the liquid to the glass, by confirming the observation that some of the bonds that establish in the liquid are short lived and disappear on quenching. Changes in composition have the effect of altering the nature of coordination motifs present in the network to interconnect the predominant tetrahedra. Among all units observed, Se_N chains are most peculiar, $N = 12$ being the largest values recorded for $GeSe_4$ in [13].

8.3.1 A glimpse on the correlation between atomic and electronic structure

The availability of four different systems obtained either by considering the liquid and the glass phases for a given system or by varying the composition opens interesting perspectives on the side of the correlation between electronic structure (bonding properties) and network topology. At the time of our first attempts, looking more deeply into this field appeared sensible since, in addition to the electronic density of states, we could benefit from other tools (like the maximally localized Wannier centers analysis, see section 2.5.2) well adapted to our search of a

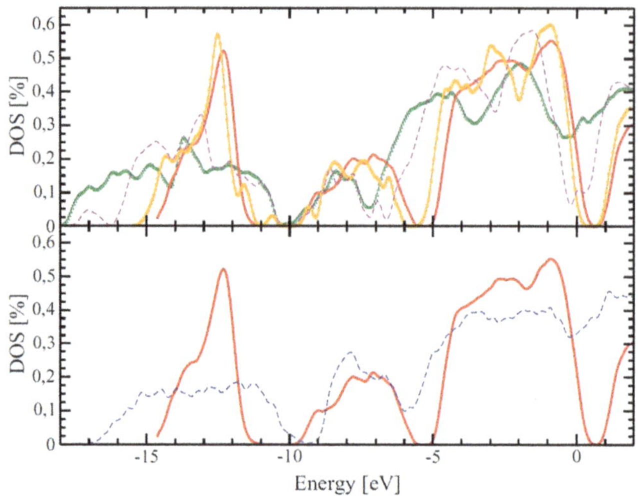

Figure 8.15. The EDOS extracted from the Kohn–Sham eigenvalues. Top panel: result for glassy Ge_2Se_3 (solid red curve) is compared to that obtained for glassy $GeSe_2$ (from [27], orange curve with circles), and glassy $GeSe_4$, PW results ([12] green curve with triangles) and BLYP results (broken pink curve). Bottom panel: result for glassy Ge_2Se_3 (solid red curve) is compared to that obtained for liquid Ge_2Se_3 ([18], broken blue curve). A Gaussian broadening of 0.1 eV has been employed. Reprinted figure with permission from [26]. Copyright (2012) by the American Physical Society.

definition for the nature of bonding[2]. Figure 8.15 provides a comprehensive view of the EDOS for glassy Ge_2Se_3, liquid Ge_2Se_3, glassy $GeSe_2$ and glassy $GeSe_4$. The EDOS of glassy Ge_2Se_3 has a deep pseudogap around the Fermi unlike its liquid counterpart, meaning that the enhanced tendency to metallicity of liquid Ge_2Se_3 coexist with an atomic structure not differing substantially from the case of the glass. This apparent contradiction can be resolved by invoking the frequent occurrence of bonding breaking and formation in the liquid, leading to a structure quite similar to the glass on a temporal average but quite different in terms of bonding behavior and, ultimately, EDOS pattern.

In the vicinity of the Fermi level, the patterns followed by the EDOS of glassy $GeSe_2$ and glassy Ge_2Se_3 are quite similar, a small gap opening up for glassy $GeSe_2$. To allow making a connection with figure 8.7, figure 8.15 contains the two results (PW and BLYP) available for $GeSe_4$. Once again, the better performances of the BLYP approach are worth underlying, since the depth of the pseudogap is consistent with experimental evidence [31, 32].

The Wannier function centers (WFCs) study starts with the observation of figure 8.16, relative to liquid Ge_2Se_3. Focusing first on Ge_1 and Ge_2 (separated by 2.69 Å, a distance larger that Ge–Ge bonds, ∼2.43 Å) we notice that a single WFC (W_4) is found at only 0.31 Å from Ge_2. This is an indication of the presence of a

[2] So far, we have reported in this monograph the cases of electronic density of states exemplified in section 2.5.1 by figure 2.8 and, in this chapter, section 8.2.2, the one shown in figure 8.7 relative to a given system studied with different exchange–correlation schemes.

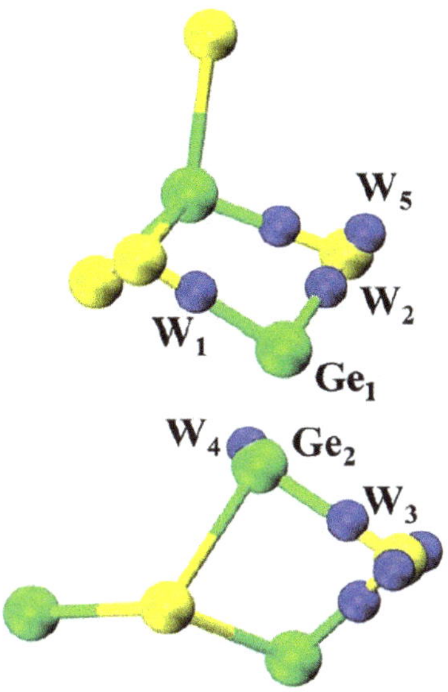

Figure 8.16. Example of bonding analysis based on the Wannier centers (WFCs) of different natures. Ge atoms appear in green, Se atoms in yellow and the WFCs in blue. Reprinted figure with permission from [26]. Copyright (2012) by the American Physical Society.

dangling bond (DB) acting as a precursor for the formation of a homopolar Ge–Ge bond. Let's now focus on the identification of the ionic character of bonding. One expects to have Wannier centers closer to Se than to Ge, since Se is more electronegative than Ge. This is indeed the case since the distances separating the pairs Ge_1–W_1, Ge_1–W_2 and Ge_2–W_3 are 1.39 ± 0.7 Å, whereas the corresponding values are 0.90 ± 0.4 Å for the case of Se atoms. Among the WFCs not participating to the bonding W_5 is located at ~0.41 Å from the closest Se atom.

We have seen that in both glassy Ge_2Se_3 and liquid Ge_2Se_3 there are Ge atoms forming homopolar connections since there are not enough Se atoms to form tetrahedral connections. A typical case is shown for glassy Ge_2Se_3 in figure 8.17, where Ge_1 and Ge_2 form a homopolar bond. The WFC labeled W_1 is located in the middle of this bond (Ge_1–W_1 = 1.19, Ge_1–W_1 = 1.22 Å). Se_1, Se_2 and Se_3 are bound to Ge_1 so as to have the WFCs W_2, W_3 and W_4 closer to their corresponding Se atoms. As in the case of the liquid, their distances are all around ~0.94 Å.

As a third example, in figure 8.18 we can spot a chain of four Ge atoms with three distinct bond distances, Ge_1–Ge_2 = 2.50 Å, Ge_2–Ge_3 = 2.54 Å and Ge_3–Ge_4 = 2.48 Å. This is due to the fact that Ge_1 is threefold coordinated to Se atoms, Ge_2 to one Se atom, Ge_3 has no Se bonded to it and Ge_4 is twofold coordinated with Se atoms. The Ge–WFCs distances are Ge_1–W_1 = 1.06 Å, Ge_2–W_1 = 1.44 Å, Ge_2–W_2 = 1.01 Å, Ge_3–W_2 = 1.53 Å, Ge_3–W_3 = 1.60 Å, Ge_4–W_3 = 0.90 Å, reflecting different degrees of ionicity and covalency along the connected Ge atoms. In this

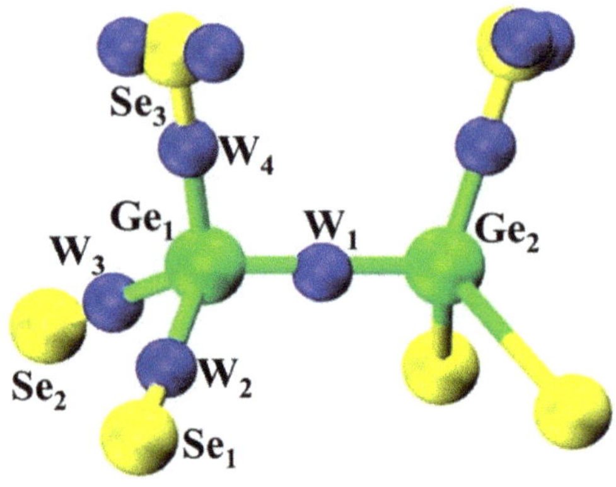

Figure 8.17. Details of homopolar bonds in the amorphous phase. We show a typical configuration with a single homopolar bond Ge–Ge. This can be taken as the final state with respect to figure 8.16. Ge in green, Se in yellow and WFCs in blue. Reprinted figure with permission from [26]. Copyright (2012) by the American Physical Society.

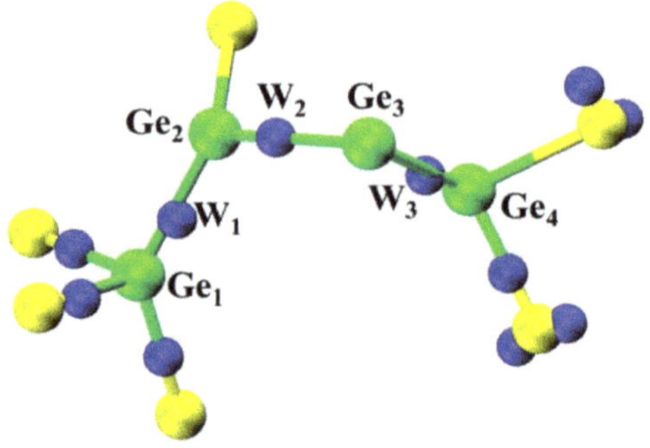

Figure 8.18. A sequence of three homopolar Ge–Ge bonds. Ge are in green, Se in yellow and WFCs in blue. Reprinted figure with permission from [26]. Copyright (2012) by the American Physical Society.

respect, Ge_3 has the largest ionic character due to the larger distances from the closer WFCs. As such, it is the best candidate to bind to available Se atoms that move in its nearest environment during the temporal evolution.

8.3.2 What to learn from glassy Ge_2Se_3

The availability and the usefulness of three different glasses around the stoichiometric composition appeared clearly in this section. One is able to compare the different structures and understand the effect of composition on the topologies of the networks. Also, comparing glass and liquid Ge_2Se_3 is quite useful to realize, once again, that FPMD is perfectly capable of capturing the differences between these systems despite the well known issue of 'unrealistic' quench rates. In reciprocal space, liquid and glassy Ge_2Se_3 differ by the intensities of the peaks, in particular for the FSDP in $S_{GeGe}(k)$. Considering glassy Ge_2Se_3, glassy $GeSe_2$ and glassy $GeSe_4$, the highest ratio between the FSDP and the main peak is found in glassy Ge_2Se_3, the three systems showing the signatures of a small departure from chemical order (a small mark in the FSDP region). Glassy $GeSe_4$ exhibits the largest degree of chemical order. The partial pair correlation functions are characterized by: (a) three peaks located at r <5 Å in $g_{GeGe}(r)$ with the relative heights depending on composition (these are due to the homopolar Ge–Ge bonds, edge-sharing and corner-sharing connections) and (b) a sharp maximum caused by Ge–Se nearest neighbors distances in $g_{GeSe}(r)$.

On the side of the electronic properties, the EDOS for Ge_2Se_3, $GeSe_2$ and $GeSe_4$ are similar, $GeSe_2$ being the system with the largest gap among the three. Finally, through the investigation of the Wannier centers one is able to identify covalent bonds (Ge–Ge homopolar bonds), dangling bonds (undercoordinated Ge atoms) and iono-covalent bonds (Ge–Se bonds in tetrahedra).

References

[1] Phillips J C 1979 *J. Non-Cryst. Solids* **34** 153–81

[2] Thorpe M F 1983 *J. Non-Cryst. Solids* **57** 355–70

[3] Haye M J, Massobrio C, Pasquarello A, De Vita A, De Leeuw S W and Car R 1998 *Phys. Rev.* B **58** R14661–4

[4] Bresser W, Boolchand P and Suranyi P 1986 *Phys. Rev. Lett.* **56** 2493–6

[5] Boolchand P 1986 *Phys. Rev. Lett.* **57** 3233

[6] Xingwei Feng W J, Bresser and Boolchand P 1997 *Phys. Rev. Lett.* **78** 4422–5

[7] Selvanathan D, Bresser W J and Boolchand P 2000 *Phys. Rev.* B **61** 15061–76

[8] Wang Y, Boolchand P and Micoulaut M 2000 *Europhys. Lett.* **52** 633–9

[9] Wang Y and Murase K 2003 *J. Non-Cryst. Solids* **326-327** 379–84

[10] Thorpe M F, Jacobs D J, Chubynsky M V and Phillips J C 2000 *J. Non-Cryst. Solids* **266-269** 859–66

[11] Sartbaeva A, Wells S A, Huerta A and Thorpe M F 2007 *Phys. Rev.* B **75** 224204

[12] Massobrio C, Celino M, Salmon P S, Martin R A, Micoulaut M and Pasquarello A 2009 *Phys. Rev.* B **79** 174201

[13] Sykina K, Furet E, Bureau B, Le Roux S and Massobrio C 2012 *Chem. Phys. Lett.* **547** 30–4

[14] Salmon P S 2007 *J. Non-Cryst. Solids* **353** 2959–74

[15] Massobrio C, Pasquarello A and Car R 2001 *Phys. Rev.* B **64** 144205

[16] Massobrio C and Pasquarello A 2008 *Phys. Rev.* B **77** 144207

[17] Micoulaut M, Vuilleumier R and Massobrio C 2009 *Phys. Rev.* B **79** 214205

[18] Le Roux S, Zeidler A, Salmon P S, Boero M, Micoulaut M and Massobrio C 2011 *Phys. Rev.* B **84** 134203

[19] Micoulaut M, Le Roux S and Massobrio C 2012 *J. Chem. Phys.* **136** 224504

[20] Zhang X and Drabold D A 2000 *Phys. Rev.* B **62** 15695–701

[21] Durandurdu M and Drabold D A 2002 *Phys. Rev.* B **65** 104208

[22] Alemany M M G and Chelikowsky J R 2003 *Phys. Rev.* B **68** 054206

[23] Tafen D N and Drabold D A 2003 *Phys. Rev.* B **68** 165208

[24] Tafen D N and Drabold D A 2005 *Phys. Rev.* B **71** 054206

[25] Salmon P S and Liu J 1994 *J. Phys.: Condens. Matter* **6** 1449–60

[26] Le Roux S, Bouzid A, Boero M and Massobrio C 2012 *Phys. Rev.* B **86** 224201

[27] Bouzid A and Massobrio C 2012 *J. Chem. Phys.* **137** 046101

[28] Bhatia A B and Thornton D E 1970 *Phys. Rev.* B **2** 3004–12

[29] Fischer H E, Barnes A C and Salmon P S 2005 *Rep. Progress Phys.* **69** 233-99

[30] Massobrio C, Celino M and Pasquarello A 2004 *Phys. Rev.* B **70** 174202

[31] Street R A, Nemanich R J and Connell G A N 1978 *Phys. Rev.* B **18** 6915–9

[32] Tanaka K 1989 *Phys. Rev.* B **39** 1270–9

Chapter 9

Moving ahead, better and bigger: GeS_2, $GeSe_9$ and $GeSe_4$ vs GeS_4

Three glasses systems (GeS_2, $GeSe_9$ and GeS_4) are described in this chapter, together with a fourth one ($GeSe_4$) already considered in chapter 8 and employed here for comparative purposes. All calculations take advantage of larger systems (~500 atoms) and improved computational resources, thereby marking a transition in terms of sizes of the periodic systems employed. *This 'new era' of first-principles molecular dynamics (FPMD) calculations strengthens the overall impact of this theoretical tool without undermining what has been obtained with smaller periodic systems.* Glassy GeS_2 allows capturing the differences between a fully self-consistent FPMD scheme and a cheaper approach based on a non self-consistent version of density functional theory (DFT), by pointing out the differences between the two corresponding structures. The study of glassy $GeSe_9$, performed in connection with neutron scattering experiments, supersedes previous results obtained on a smaller system and allows drawing instructive conclusions on the sensitivity of the results to the simulation protocol. Finally, a detailed comparison between glassy $GeSe_4$ and glassy GeS_4 highlights the importance of complementing a description of the structure with an analysis of chemical bonding in terms of localization properties. This allows concluding that glassy GeS_4 has a higher ionic character than $GeSe_4$.

9.1 Introduction

The set of calculations (the first published in 2013) contained in this chapter are representative of a methodology and of a level of analysis concerning the structural properties of glasses that has come to a full maturity. Systems sizes are larger that the early, somewhat limited (and yet irreproachable) size $N_{at} = 120$, trajectories are extended as much as possible and several structural features are considered in more detail, with a specific focus on the correlation with electronic properties. This new

doi:10.1088/978-0-7503-2436-6ch9

era of enhanced quantitative productivity comes not only from increased expertize and knowledge, quite often resulting on expected improvements, but also on the availability of better high-performance computers and increased human resources devoted to accomplish the targeted goal. Glassy GeS_2 is a revealing example of changes occurring when a fully self-consistent DFT scheme, as the one inherent in FPMD, is employed instead of an alternative, simplified total energy approach known as the Harris functional [1]. Also, comparing GeS_2 and $GeSe_2$ is quite instructive to infer their bonding character and understand how ionicity manifests itself in these systems. The case of $GeSe_9$ is useful to understand the technicalities inherent in the treatment of a network in which one of the two species is highly predominant, since for N_{at} = 120 the number of atoms of the minority species (12) becomes so small to raise questions on the overall statistical accuracy of the structural properties, especially those that are dependent on Ge atoms [2]. Finally, a study is performed for GeS_4 in comparison to $GeSe_4$, where in this case two different system sizes are employed to highlight similarities and differences and make sure that the results are minimally affected by the number of atoms contained in the periodic systems. Also, the issue of the extent of ionicity is treated in great detail, allowing to reach a detailed and insightful description of unprecedented quantitative nature.

9.2 Glassy GeS_2

Amorphous germanium disulfide is another example of system belonging to the class of AX_2 networks (A = Ge, Si; X = O, Se, S) but, at the time we began studying its properties, it appeared to have attracted much less attention than its $GeSe_2$ counterpart largely described in previous sections of this monograph (see chapters 6 and 7). The extent of chemical order has been the subject of experimental work aimed at establishing the role played by homopolar bonds [3–5]. Atomic-scale modeling on glassy GeS_2 was made available by a set of calculations employing a non self-consistent DFT approach (HFMD hereafter), based on a functional due to Harris [1] and implemented within the local density approximation, LDA [6–8]. Therefore, it appeared timely and needed to employ the fully self-consistent FPMD recipe to this case, as detailed in [9] and recalled here (see chapters 3 and 4). Within the FPMD implementation, the equilibrium trajectories were generated as follows for a system of N_{at} = 480 atoms at the experimental density. In the first three cases, the starting configurations were obtained by assigning random positions to the atoms, by avoiding bonds shorter than of 1.7 Å. A first trajectory lasting 10 ps at T = 2000 K was followed by quench at T = 1200 K during 9 ps by decreasing the temperature with intervals of 100 K. At the temperature T = 300 K the system spent 14 ps, with physical quantities calculated as averages on the last 6 ps. The corresponding quench rate is $q(1)_{\mathrm{r}}^{\mathrm{FPMD}} = 5\times10^{14}$ K s^{-1}. For a fourth trajectory we used a different quench rate, $q(2)_r^{\mathrm{FPMD}} = 3\times10^{13}$ K s^{-1} with a thermal cycle made of 10 ps at T = 300 K, 10 ps at T = 700 K and 11 ps at T = 300 K for the averages (FMPD(1) corresponds to the mean of the three first trajectories and FMPD(2) to the fourth one). The marks FPMD (FPMD(1) and FPMD(2) when needed) and HFMD will be employed whenever comparing these two sets of results.

9.2.1 Real space properties

In figure 9.1 the total pair correlation functions $g_{tot}^{FPMD}(r)$ (FPMD(1) and FPMD(2)) are compared with the HFMD results of [7] ($g_{tot}^{HFMD}(r)$) and the experimental data sets ($g_{exp}^{B}(r)$, $g_{exp}^{S}(r)$) due to the teams of Bychkov [10, 11] and Salmon [5], respectively.

Both FPMD results are able to reproduce all the experimental features, such as the number of relevant peaks and their positions. The only difference is found at the level of the third peak in between 3 Å and 4 Å, for both FPMD and HFMD, the experimental data showing higher intensities. Besides the enhanced sharpness of the first peak in $g_{tot}^{HFMD}(r)$, HFMD and FPMD tend to agree on the whole range of distances. The partial pair correlations functions are shown in figure 9.2, featuring $g_{GeGe}^{FPMD}(r)$ (FPMD(1) and FPMD(2)), $g_{GeGe}^{HFMD}(r)$, $g_{GeS}^{FPMD}(r)$ (FPMD(1) and FPMD (2)), $g_{GeS}^{HFMD}(r)$, $g_{SS}^{FPMD}(r)$ (FPMD(1) and FPMD(2)) and $g_{SS}^{HFMD}(r)$. We shall refer without loss of generality to the FPMD(1) data, very similar to the FPMD(2) ones, indicating that in this case the structural properties have little sensitivity to the quench rates. Let's begin our analysis by considering the third peak in between 3 Å and 4 Å in $g_{tot}^{FPMD}(r)$. This can be ascribed to S–S correlations since it corresponds to the position of the second peak of $g_{SS}^{FPMD(1)}(r)$. The second peak of $g_{GeGe}^{FPMD(1)}(r)$ is found at 2.91 Å, the distance between Ge atoms in edge-sharing (ES) connections, as shown in the inset of figure 9.2. Finally, the first peak in $g_{tot}^{FPMD(1)}(r)$ is due to the Ge–S bonds responsible for the very intense peak of $g_{GeS}^{FPMD(1)}(r)$.

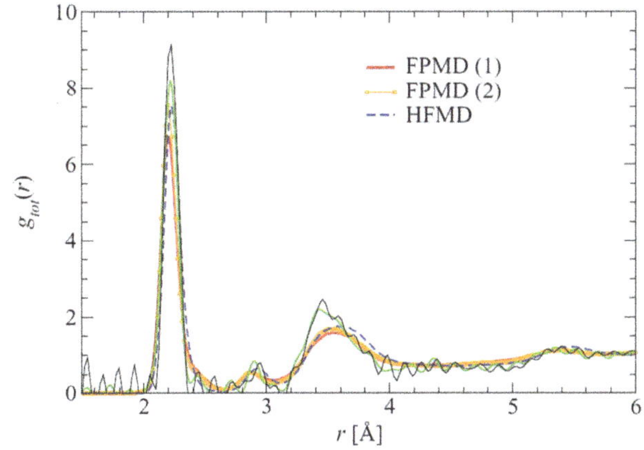

Figure 9.1. Total pair correlation function of glassy GeS_2. We compare the results from [7] obtained with the Harris functional (dashed blue line) with the FPMD calculation (thick red line, FPMD(1) and yellow line FPMD(2), see [9]) and the experimental measurements from [10, 11] (green symbols) and [5] (straight black line). Reprinted figure with permission from [9]. Copyright (2013) by the American Physical Society.

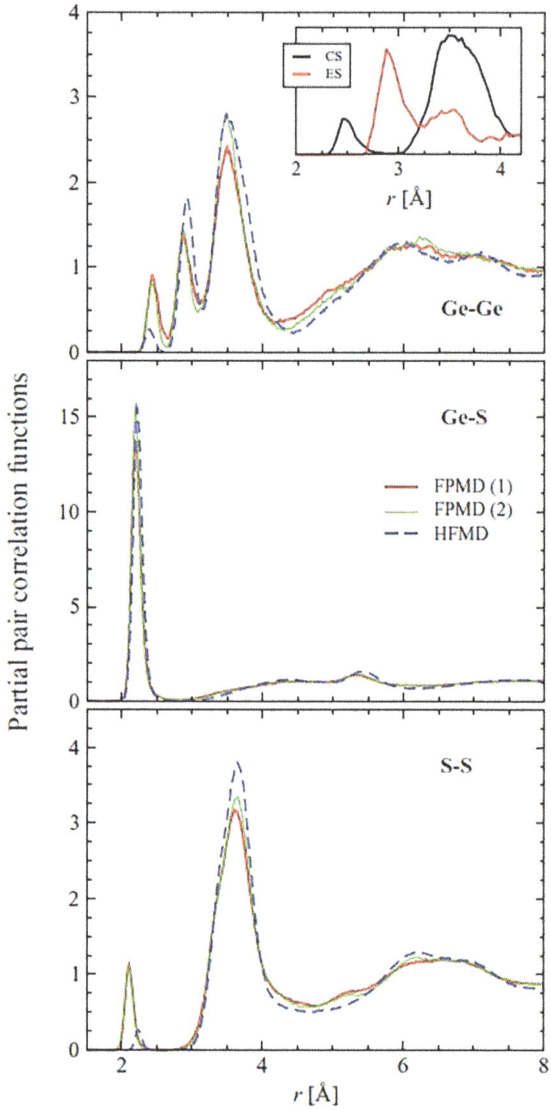

Figure 9.2. Partial pair correlation functions of glassy GeS_2. Comparison between the results from [7] obtained with the Harris functional (dashed blue line), and the FPMD calculations (thick red line and green line, see [9]). In the inset the FPMD(1) Ge–Ge correlation is decomposed to highlight the edge-sharing (ES) contributions and homopolar bonds/corner-sharing (CS) contributions. Reprinted figure with permission from [9]. Copyright (2013) by the American Physical Society.

The FPMD and HFMD sets of data differ by the the number of homopolar bonds Ge–Ge and S–S, more numerous in the FPMD case. This can be extracted from the reduced intensities of the first peaks in $g_{\mathrm{GeGe}}^{\mathrm{HFMD}}(r)$ and $g_{\mathrm{SS}}^{\mathrm{HFMD}}(r)$. This is a clear indication of different bonding characters, since HFMD promotes the ionic character that hampers the formation of homopolar bonds and important

Table 9.1. Atomic coordinations for glassy GeS_2 compared to the results from Harris functional calculation [7] and to results obtained for the $GeSe_2$ glassy system [15]. The results termed FPMD(1) and FPMD(2) have been obtained in [9]. The values for the RCN and CON models are also provided. Reprinted table with permission from [9]. Copyright (2013) by the American Physical Society.

	RCN	CON	HFMD	FPMD(1)	FPMD(2)	FPMD($GeSe_2$)
n_{Ge}^{Ge}	2	0	0.04	0.17	0.13	0.28
$n_{Ge}^{S(Se)}$	2	4	3.91	3.62	3.68	3.64
$n_{S(Se)}^{Ge}$	1	2	1.96	1.81	1.84	1.82
$n_{S(Se)}^{S(Se)}$	1	0	0.05	0.23	0.21	0.20

departures from chemical order. In a way similar to other chalcogenides already considered, one finds the three peak structure of Ge–Ge pair correlations standing for homopolar Ge–Ge bonds, Ge atoms involved in ES connections and Ge atoms involved in corner-sharing (CS) connections [12–14]. The coordination numbers n_α^β for the HFMD and the FPMD models of glassy GeS_2 are given in table 9.1, together with the values for glassy $GeSe_2$.

In both FPMD and HFMD results one observes the predominant contribution of tetrahedra. However, chemical disorder is higher within FPMD (more homopolar bonds), in agreement with the higher ionic character brought in by the HFMD approach. By resorting to the $\alpha(l)$ structural units (an atom α (Ge or Se) is l-fold coordinated to other atoms) we can further describe the short range order (see table 9.2).

In [16], it was observed that there is a profound difference between Ge and S in terms of short range order. There are sizeable proportions of S atoms that deviate from the twofold coordination, while most of the Ge atoms are fourfold coordinated (see table 9.2). This is due to unequal behaviors with respect to charge transfer when considering S and Ge atoms within the Harris functional framework [1]. Unlike S atoms, found in different coordination states, Ge atoms tend to form bonds having an exclusive fourfold character, as if the Ge atom behaved as a charged Ge^{4+} ion. The situation is different when using FPMD, since both the Ge and S subnetworks deviate from the standard tetrahedral configurations. For the S coordinations, FPMD gives more S atoms twofold coordinated than the HFMD case. Also, there is a vanishing number of S having one Ge neighbor, as opposed to 14.3% found within HFMD. Overall, the different topologies of the HFMD and FPMD networks rest on the large difference of Ge fourfold coordinated in the former case. When comparing glassy GeS_2 and glassy $GeSe_2$ as obtained within FPMD, one is unable to detect a clear pattern of structural differences, since in both systems the tetrahedra coexist with miscoordinations and homopolar bonds. This observation calls for a more refined tool of analysis in order to assess the nature of bonding in both systems and, in particular, to evaluate in a comparative fashion the degree of ionicity, as explained in section 9.2.3. One is also able to obtain the number of Ge atoms lying in

Table 9.2. Percentage of l-fold coordinated atoms, decomposed in term of each specific unit $n_{\alpha}(l)$, in glassy GeS_2 and glassy $GeSe_2$ from [15]. The results termed FPMD(1) and FPMD(2) have been obtained in [9]. Reprinted table with permission from [9]. Copyright (2013) by the American Physical Society.

		Proportion $n_{\alpha}(l)$ [%]			
			GeS_2		$GeSe_2$
		HFMD	FPMD(1)	FPMD(2)	FPMD
Ge atom					
$l = 2$					
	X_2	1.37	5.8	5.0	3.1
	Ge_1X_1	0.0	0.2	0.0	0.0
$l = 3$					
	Ge_1X_2	<0.1	1.6	1.0	1.4
	X_3	1.44	8.3	9.0	5.4
$l = 4$					
	Ge_1X_3	3.70	11.6	10.6	12.8
	Ge_2X_2	<0.1	1.4	0.6	2.3
	X_4	93.1	70.5	73.0	72.9
$l = 5$					
	Ge_1X_4	<0.1	0.4	0.2	0.0
	X_5	0.0	0.2	0.5	0.0
X = S, Se atom					
$l = 1$					
	Ge_1	14.3	0.8	0.6	0.2
$l = 2$					
	X_2	0.0	2.5	2.2	2.5
	X_1Ge_1	3.07	17.7	16.0	15.3
	Ge_2	67.9	73.9	76.2	82.8
$l = 3$					
	X_2Ge_1	<0.1	0.2	0.0	0.0
	X_1Ge_2	1.81	0.5	0.3	0.0
	Ge_3	12.8	4.5	4.7	1.4

zero, one, and two fourfold rings. According to the definition introduced in section 6.3.2 and recalled via figure 9.3, 52.3% of the Ge atoms do not belong to fourfold rings, 31.4% are found in a single fourfold ring, and 12.1% in two fourfold rings. This gives the total number of Ge atoms in ES configurations $N_{Ge}^{th}(ES) = 43.5\%$, in good agreement with the experimental estimate of $N_{Ge}^{exp}(ES) = 47\% \pm 5\%$ quoted in [5] by using *in situ* neutron and x-ray diffraction. The same number for glassy $GeSe_2$ was $N_{Ge}^{th}(ES) = 35\%$ [17].

By using the classification based on edge- and CS connections, it is possible to associate with a specific kind of Ge atom each one of the three first peaks appearing

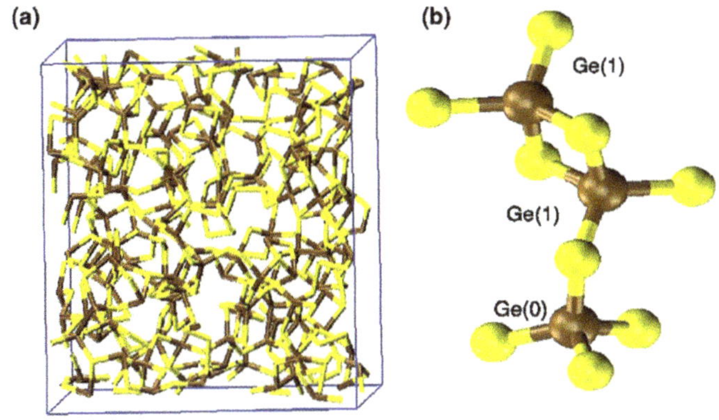

Figure 9.3. (a) Snapshot of glassy GeS_2 configuration. (b) Chain composed by two Ge(1) and one Ge(0). Reprinted figure with permission from [9]. Copyright (2013) by the American Physical Society.

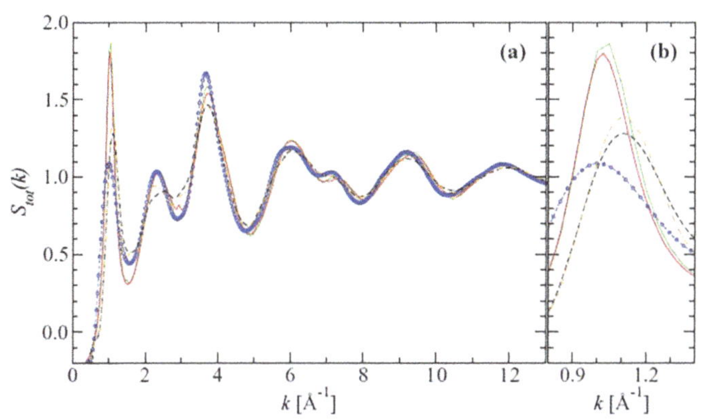

Figure 9.4. Total structure factor of glassy GeS_2. Comparison between the HFMD results from [7] obtained with the Harris functional (blue line), FPMD calculations (dashed black and orange lines for respectively FPMD(1) and FPMD(2), see [9]) and the experimental measurements from [5] (green line) and [10, 11] (red line). (a) The overall functions are presented on the left side, (b) a zoom on the FSDP region is presented on the right side. Reprinted figure with permission from [9]. Copyright (2013) by the American Physical Society.

in the Ge–Ge partial correlation function. The inset of figure 9.2 shows $g_{\mathrm{GeGe}}^{\mathrm{FPMD(1)}}(r)$ involving only Ge atoms in ES connections, with a peak at ~3 Å. By excluding these connections, we obtain (see the same figure) the peaks due to homopolar bonds and CS connections, the first and the third respectively in $g_{\mathrm{GeGe}}^{\mathrm{FPMD(1)}}(r)$ located at the left and right (smaller r and larger r) of the ES feature.

9.2.2 Reciprocal space properties

The calculated total neutron structure factors $S_{\mathrm{T}}^{\mathrm{FPMD(1)}}(k)$, $S_{\mathrm{T}}^{\mathrm{FPMD(2)}}(k)$ and $S_{\mathrm{T}}^{\mathrm{HFMD}}(k)$ are compared in figure 9.4 to experiments [5, 10, 11]. All calculated

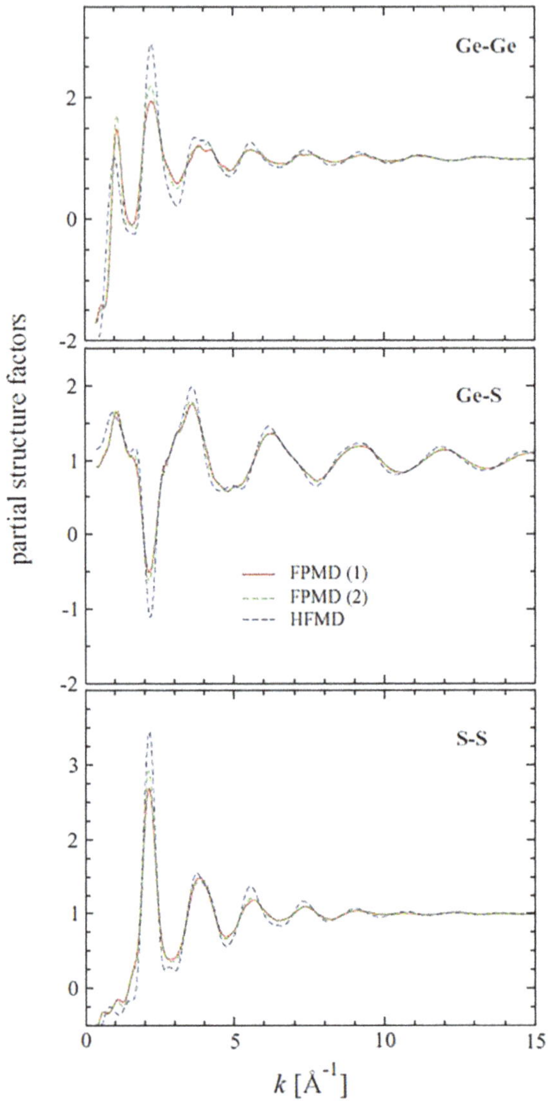

Figure 9.5. Partial structure factors of glassy GeS_2. Comparison between the results from [7] obtained with the Harris functional (dashed blue line), and FPMD calculations (red line for FPMD(1) and dashed green line for FPMD(2)). Reprinted figure with permission from [9]. Copyright (2013) by the American Physical Society.

structure factors considered hereafter (total and partials) have been obtained, from the pair correlation functions via a Fourier integration from real space (see section 2.3). They are shown in figures 9.4, 9.5 and 9.6.

Both $S_T^{FPMD(1)}(k)$ and $S_T^{FPMD(2)}(k)$ perform better than $S_T^{HFMD}(k)$ in reproducing the measured profiles over the entire range of wavevectors. The patterns observed for the Ge–S and S–S partial structure factors are quite close when comparing the FPMD and HFMD cases. There is no FSDP in $S_{SS}^{FPMD(1)}(k)$, $S_{SS}^{FPMD(2)}(k)$ and

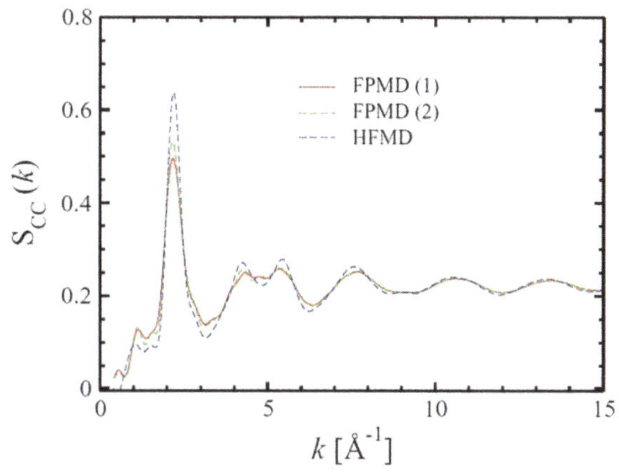

Figure 9.6. $S_{CC}(k)$ structure factor of glassy GeS_2. Comparison between the results from [7] obtained with the Harris functional (dashed blue line), and the FPMD calculations (red line for FPMD(1) and dashed green line for FPMD(2), see [9]). Reprinted figure with permission from [9]. Copyright (2013) by the American Physical Society.

$S_{SS}^{HFMD}(k)$. Also, similar intensities at the first-sharp diffraction peak (FSDP) location are found in $S_{GeS}^{FPMD(1)}(k)$, $S_{GeS}^{FPMD(2)}(k)$ and $S_{GeS}^{HFMD}(k)$. However, there is a more pronounced FSDP feature in $S_{GeGe}^{FPMD}(k)$ (FPMD(1) and FPMD(2)) compared to $S_{GeGe}^{HFMD}(k)$ (see figure 9.5). Therefore, we are facing here two different models for the same disordered network-forming material, sharing some similarities but also being substantially different in terms of manifestations of the intermediate range order. This allows establishing some correlations between the network topology and the extent of structural order. It appears that the FSPD is more noticeable when the signs of chemical disorder (homopolar bonds, coordinations differing from the tetrahedral one) involve both the Ge and S environments, as it occurs in the FPMD approach. The HFMD case does not fall into this category since as many as 93% of Ge atoms are fourfold coordinated to S atoms.

Finally, the Bhatia–Thornton concentration–concentration partial structure factor $S_{CC}(k)$ (figure 9.6) [2, 18] is characterized by a mark at the FSDP location for both models considered (FPMD and HFMD). As discussed in section 6.5 this is expected in view of the small departure from chemical order observed in these systems and the presence of the FSDP [13–15, 19–21].

9.2.3 Bonding properties

We mentioned earlier that the structural properties of glassy GeS_2 and glassy $GeSe_2$ look quite similar and do not allow to point out any stronger ionic character in either one of the two. By resorting to the maximally localized Wannier centers analysis [22, 23], we can identify two types of Wannier functions centers (WFCs) (see figure 9.7) and obtain information on the nature of bonding by looking at their positions.

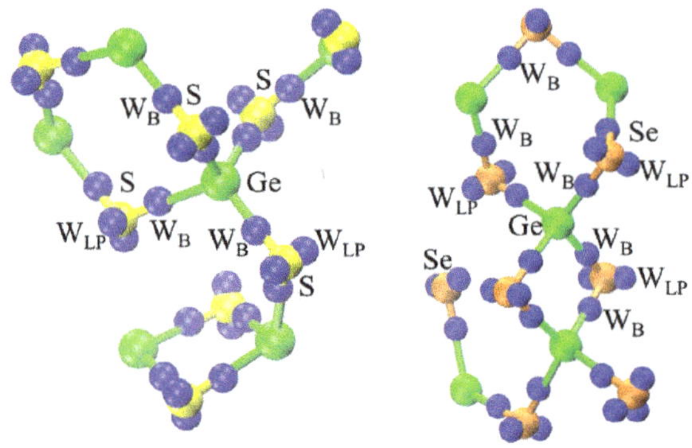

Figure 9.7. Details of the local bonding environment in the case of GeS_2 (left panel) and $GeSe_2$ (right panel). Ge atoms are shown in green, whereas S and Se atoms are in yellow and orange, respectively. The Wannier centers are blue spheres. Reprinted figure with permission from [9]. Copyright (2013) by the American Physical Society.

The first type (W_B) is indicative of electrons involved in Ge–S or Ge–Se bonds. The second type (W_{LP}) denotes a lone pair (LP) of valence electrons localized in the vicinity of the group VI tetra-valent atoms (S, Se). The distances between W_B centers and Ge atoms are very close when comparing GeS_2 and $GeSe_2$ (Ge–W_B = 1.331 ± 0.010 Å and Ge–W_B = 1.351 ± 0.008 Å, respectively) providing little information on any differences in bonding between the two systems. More instructive is the consideration of the relative position of the centers with respect to S and Se. In fact, the distances of W_B centers are longer in the case of Se, being S–W_B = 0.859 ± 0.012 Å and Se–W_B = 1.006 ± 0.005 Å. This indicates a lower ionic character of the $GeSe_2$ disordered material as opposed to GeS_2. Therefore, the higher proportion of ES connections found in glassy GeS_2 could be correlated to the higher ionic character of this system. To substantiate this assertion we observe that ES connections, within a qualitative description based on formal charges, are optimal to maximize the packing of anions and cations through the formation of rings. This concept was exploited in [24] to account for the polarizability of systems like glassy GeS_2 and glassy $GeSe_2$, confirming that a higher ionic behavior promotes the generation of ES units.

Working on glassy GeS_2 had the advantage of allowing for two kinds of investigations. First, we were able to compare the fully self-consistent FPMD schemes with a more qualitative approach proposed well before the publication of the FPMD results. Second, we could compare glassy GeS_2 to glassy $GeSe_2$ and take advantage of both structural and bonding properties. We found that FPMD does reproduce some of the HFMD results on the pair correlation functions, proving that a simplified, non self-consistent DFT model has also some legitimacy. However, there are important differences in the neighboring environment of the Ge atoms, departing from chemical order within FPMD and strongly chemically ordered (tetrahedra predominant) within HFMD. On the side of the comparison between glassy GeS_2 and glassy $GeSe_2$, we had to resort to an analysis of chemical bonding to unravel a more pronounced ionic character in GeS_2 and a correlation existing with the larger number of ES connections.

9.3 Glassy $GeSe_9$

When considering glassy $GeSe_9$ it is convenient to refer to glassy Se in which the Se atoms are found in Se_n chains ($n \geqslant 2$). Introducing Ge in this one-dimensional network causes the formation of linkages in between the Se_n chains, conferring to the Ge–Se network a larger dimensionality and the possibility to organize in a variety of structural motifs. We recall that Ge_xSe_{1-x} has a fairly large glass-forming region ($0 \leqslant x \leqslant 0.43$) [25]. FPMD simulations carried out in 2015–6 [26] were motivated by the analysis of the available literature on this system, pointing out interesting results based on methodological approaches that could be improved. To begin commenting on them, in [27], a periodic cell of $N_{at} = 400$ atoms was employed in conjunction with a quench rate bringing the system temperature down to $T = 300$ K from $T = 2200$ K in 5 ps. The glass thereby obtained contained both onefold coordinated ($\simeq$ 19.6%) and threefold coordinated ($\simeq$ 20.4%) Se atoms. Given our experience with glassy $GeSe_4$ produced via more extended relaxation times at room temperature and slower quench rates (see section 8.2 and [13, 28, 29]), we thought worth checking whether the large amount of mis-coordinated Se atoms could be due to a structure retaining a large memory of the liquid-state in which diffusion can induce such structural motifs. In [30] the trajectories were longer but the system definitely smaller ($N_{at} = 120$ i.e., 12 Ge atoms), with a number of ES Ge-centered tetrahedra leading to a ratio of ES to CS $Ge(Se_4)_{1/2}$ tetrahedra ES/CS = 0.86. The limited number of Ge atoms cannot be taken as satisfactory, calling for the application of the same FPMD protocol to a larger system.

In view of what mentioned above, new FPMD simulations were performed at constant volume on a system containing $N_{at} = 260$ atoms (26 Ge and 234 Se). The FPMD approach was implemented as described in this monograph, with all technical details being available in reference [26]. A configuration extracted from a fully equilibrated liquid at $T = 1000$ K was taken as initial point to make available three distinct models (referred to as II–IV) for the glass, differing by the thermal histories. Model I refers to the results of reference [30]. In model II the system spent $\simeq$ 5 ps at $T = 600$ K and $\simeq$ 5 ps at $T = 300$ K, while the intervals were $\simeq$ 8 ps at $T = 600$ K and $\simeq$ 8 ps at $T = 300$ K for model III. A three-step schedule was used in model IV with annealing times of $\simeq$ 12.5 ps at $T = 900$ K, $\simeq$ 25 ps at $T = 600$ K and $\simeq$ 30 ps at $T = 300$ K. The rates of reduction in the temperature are 7×10^{13} K s^{-1} (model II), 4.375×10^{13} K s^{-1} (model III) and 1.04×10^{13} K s^{-1} (model IV). After completion of these trajectories, models II and IV were extended in time for $\simeq$ 30 ps at $T = 300$ K. For model III relaxation took place on $\simeq$ 5 ps at $T = 300$ K.

These procedures were selected to investigate the impact on the atomic structure of (i) the quench rate and (ii) the length of the relaxation time at a given temperature, typically the final one (room temperature). In this context, model II differs by model III by the longer relaxation time at $T = 300$ K, while model II differs by model IV by a more extended trajectories for quenching from the melt. Thus, models II and III will provide information into the role of a longer relaxation time at room temperature, whereas models II and IV will allow to infer the effect of a longer procedure bringing the structure to the glassy phase from the liquid.

9.3.1 Comparing the structural models

The $S_T(k)$ function measured by neutron diffraction in the context of the work presented in [26] is given in figure 9.8 and compared with that obtained by Ramesh Rao *et al* [31] also by neutron diffraction. Both data sets exhibit a feature in the FSDP region at $k \simeq 1.3$ Å^{-1} [32] while there are differences in the low-k region, most likely due to a background scattering issue in [31]. All FPMD models considered (I to IV) are in good agreement with the experimental data on the whole k range.

The partial structure factors from models I–IV are shown in figure 9.9. The FSDP is noticeable in both $S_{GeGe}(k)$ and $S_{GeSe}(k)$ at $k_{FSDP} \simeq 1.02$–1.07 Å^{-1} and $k_{FSDP} \simeq 1.17$–1.29 Å^{-1}, respectively, more intense for $S_{GeGe}(k)$ than for $S_{GeSe}(k)$. In this respect changes between the models are minimal. Therefore, there is an order developing on an intermediate length scale that is due to Ge-β (β = Ge or Se) correlations exclusively. When looking at the neutron total pair correlation function $g_T(r)$ as defined in equations (2.2) and (2.4), it appears that this quantity will be insensitive to $g_{GeGe}(r)$ due to the limited concentration of this species (see figure 9.10). All of the models reproduce well the experimental $g_T(r)$ function since $g_{SeSe}(r)$ is largely predominant in $g_T(r)$ and the four realizations of $g_{SeSe}(r)$ are quite similar (see figure 9.11).

Table 9.3 contains the values for the average total coordination number $\bar{n}$ as defined in section 2.4.1, obtained from the experiments and the calculations detailed in [26]. These are in agreement with those obtained via the '8 − N' rule, which predicts that $n_{Ge} = 4$ and $n_{Se} = 2$ such that n('8 − N') = 2.2. The value of $\bar{n}$ issued from the experiments of [26] is smaller than $\bar{n} = 2.45(18)$ given in [31]. Also provided

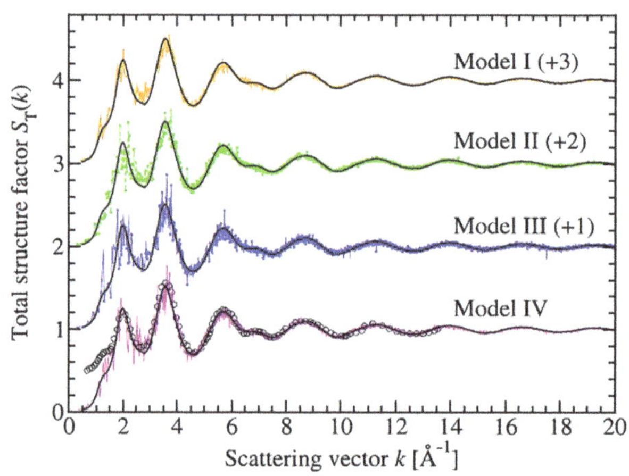

Figure 9.8. The neutron total structure factor $S_T(k)$ for glassy $GeSe_9$. Solid (black) curves with vertical error bars: [26], where the size of the error bars is smaller than the line thickness at most k values. Black circles: results from [31]. The experimental results are compared to those obtained from FPMD models I (solid orange curve), [30] II (broken green curve with squares), III (solid blue curve with squares) and IV (solid magenta curve) ([26] by a direct calculation in reciprocal space (see equation (2.11))). Reprinted figure with permission from [26]. Copyright (2016) by AIP Publishing.

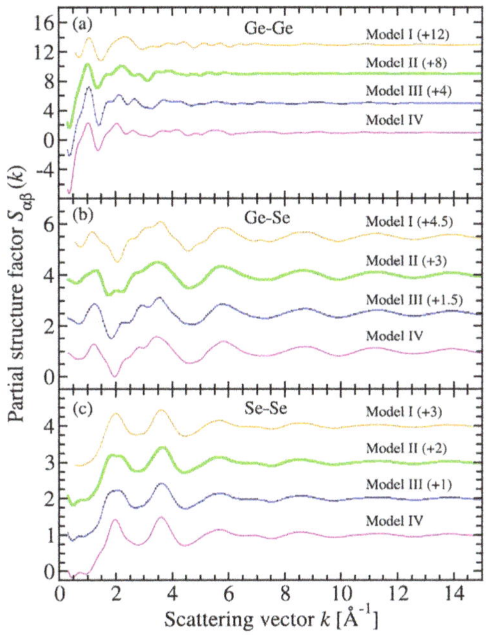

Figure 9.9. The (a) Ge–Ge, (b) Ge–Se and (c) Se–Se partial structure factors $S_{\alpha\beta}(k)$ for glassy $GeSe_9$ from FPMD models I (solid orange curve), [30] II (broken green curve with squares), III (solid blue curve with squares) and IV (solid magenta curve) [26]. The functions were obtained by Fourier transforming the $g_{\alpha\beta}(r)$ functions shown in figure 9.11. Reprinted figure with permission from [26]. Copyright (2016) by AIP Publishing.

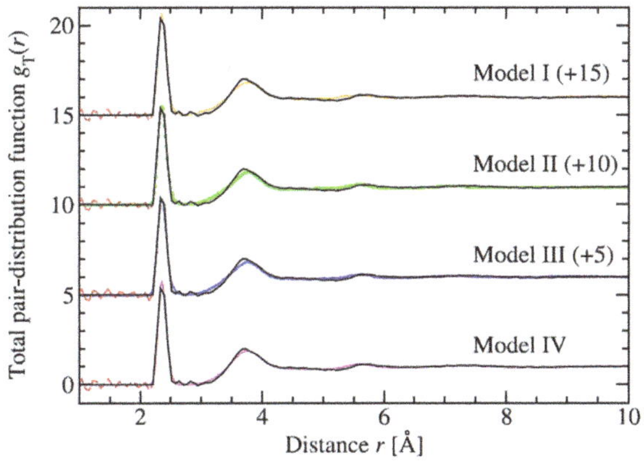

Figure 9.10. The neutron total pair correlation (named 'distribution' in this figure) function $g_T(r)$ for glassy $GeSe_9$. The measured $g_T(r)$ function given in solid (black) curves was obtained by Fourier transforming the spline-fitted measured $S_T(k)$ function shown in figure 9.8 with k_{max} = 30 Å^{-1}. The chained (red) curves shows the Fourier transform artifacts at unphysically small r values. The experimental results are compared to those obtained from FPMD models I (solid orange curve), [30] II (broken green curve with squares), III (solid blue curve with squares) and IV (solid magenta curve) [26]. Reprinted figure with permission from [26]. Copyright (2016) by AIP Publishing.

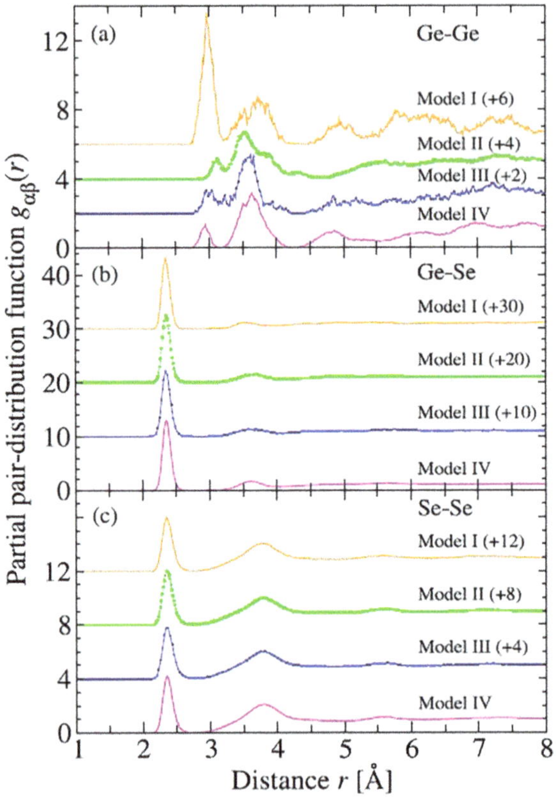

Figure 9.11. The (a) Ge–Ge, (b) Ge–Se and (c) Se–Se partial pair correlation functions $g_{\alpha\beta}(r)$ (named 'distribution functions') for glassy $GeSe_9$ from FPMD models I (solid orange curve), [30] II (broken green curve with squares), III (solid blue curve with squares) and IV (solid magenta curve) [26]. Reprinted figure with permission from [26]. Copyright (2016) by AIP Publishing.

Table 9.3. The mean Ge and Se coordination numbers n_{Ge} and n_{Se} as measured by using neutron diffraction (ND) or calculated by using FPMD with a cutoff distance $r_{\mathrm{cut}} = 2.75$ Å. The overall mean coordination number, $\bar{n}$, is also listed, and is compared to the value expected from the '8 − N' rule. Reprinted table with permission from [26]. Copyright (2016) by AIP Publishing.

Source	n_{Ge}	n_{Se}	$\bar{n}$	$\bar{n}$('8 − N')
ND [26]	–	–	2.20(1)	2.2
ND [31]	4.0	1.8(2)	2.45(18)	2.2
Models I–IV ([30](I)/[26](II–III–IV))	4.0	2.0	2.2	2.2

in the same table are n_{Ge} and n_{Se} (see section 2.4.1). The corresponding coordination numbers n_α^β are in table 9.4, where the integration range includes distances up to the first minimum in $g_{\mathrm{T}}(r)$. The partial correlation functions for models I–IV are displayed in figure 9.11. The first peak at $\simeq$ 2.35 Å in $g_{\mathrm{GeSe}}(r)$ can be ascribed to

Table 9.4. The coordination numbers n_{α}^{β} obtained from the FPMD models by using an integration range of 0–2.75 Å, where the upper limit corresponds to the minimum after the first peak in $g_T(r)$. The predictions of the CON and RCN models are also listed [32]. Reprinted table with permission from [26]. Copyright (2016) by AIP Publishing.

Model	n_{Ge}^{Ge}	n_{Ge}^{Se}	n_{Se}^{Ge}	n_{Se}^{Se}
I [30]	0.0	4.0	0.44	1.56
II–IV [26]	0.0	4.0	0.44	1.56
CON	0.0	4.0	0.44	1.56
RCN	0.73	3.27	0.36	1.63

Ge–Se heteropolar bonds and results in a coordination number n_{Ge}^{Se} = 4.0 not dependent on the model. The same holds for the first peak at ≃ 2.36 Å in $g_{SeSe}(r)$ that is indicative of Se–Se homopolar bonds. There is no feature at ≃ 2.4 Å in $g_{GeGe}(r)$, and hence Ge–Ge homopolar bonds are absent, in a way consistent with Se_n chains that are connected by $Ge(Se_4)_{1/2}$ tetrahedra. Therefore, all FPMD models are consistent with a chemically ordered network.

9.3.2 Sensitivity to size and production protocols

The partial pair correlation function $g_{GeGe}(r)$ is the one exhibiting the most visible changes when comparing the FPMD models I, II, III and IV (figure 9.11). This means that we are dealing with the most sensitive to the technical details of the different four simulations (quench rate and length of relaxation trajectories). All $g_{GeGe}(r)$ are characterized by a first peak at ~3 Å stemming from Ge–Ge distances in fourfold rings formed by ES $Ge(Se_4)_{1/2}$ tetrahedra, and the second peak at ~3.6 Å due to CS $Ge(Se_4)_{1/2}$ tetrahedra. The ES peak in $g_{GeGe}(r)$ has its largest intensity for model I, while it is strongly reduced for models II–IV. In the CS peak a double feature is visible in model II but it is absent for models III and IV. Model III has the largest statistical noise among those obtained with $N_{at} = 260$, this being attributed to the limited relaxation at room temperature. By considering that models II and IV have in common longer relaxation times at $T = 300$ K but differ by the length of the trajectories devoted to the quench, we can conclude that the double CS feature in $g_{GeGe}(r)$ for model II is due to an insufficient structural relaxation while bringing the system down to T = 300 K. As first introduced in section 6.3.2, Ge atoms in the network structure of a Ge_xSe_{1-x} disordered material can be classified as being in no fourfold rings Ge(0) (that is to say, involved in CS connections, definition valid when there are no homopolar Ge–Ge bonds), one fourfold ring Ge(1), or two fourfold rings Ge(2) [33], these two latter classes pertaining to Ge atoms involved in ES connections. Results for these Ge(ℓ) atoms (ℓ = 0, 1 or 2) are given in table 9.5, with cutoff distance r_{cut} = 2.75 Å. The ratio ES/CS = 1 obtained for model I is much larger than the values 0.08–0.18 found for models II–IV, experimental estimates

Table 9.5. The proportions of Ge atoms involved in 0, 1 or 2 fourfold rings, as obtained from the FPMD models of glassy $GeSe_9$ by using a cutoff distance r_{cut} = 2.75 Å. Note that there is a different value for model I in [30], but in that case the cutoff distance r_{cut} = 2.9Å was larger. Reprinted table with permission from [26]. Copyright (2016) by AIP Publishing.

Model	Ge(0)	Ge(1)	Ge(2)
I [30]	50	50	0.0
II [26]	92.3	7.7	0.0
III [26]	84.7	15.3	0.0
IV [26]	92.3	7.7	0.0

being 0.19(3) or 0.31(6) from Raman and ^{77}Se magic angle spinning (MAS) nuclear magnetic resonance (NMR) spectroscopy experiments, respectively [34, 35].

Model I was obtained with a smaller system (N_{at} = 120 versus N_{at} = 260) than models II, III, IV that also take advantage of independent trajectories for a better statistics. Therefore, the large amount of ES configurations predicted by model I [30] cannot be taken as a realistic feature of glassy $GeSe_9$. We recall that the N_{at} = 400 atom model of Tafen and Drabold [27], despite its exceedingly high content of defects, was also in favor of a much smaller number of ES connections.

In order to understand the reasons underlying the large number of ES tetrahedra in model I [30], one can invoke configurations of the liquid lasting during the quench process. This hypothesis is sustained by the consideration that model I was created by starting from four independent starting configurations of the liquid, well separated in time. One can argue that the number of Ge atoms in model I (N_{at} = 120 with 12 Ge atoms) was too limited to allow for an adequate sampling of all possible Ge coordination motifs.

The Ge–Se–Ge bond-angle distributions $B(\theta_{GeSeGe})$ for models I–IV are shown in figure 9.12(a). For the case of model I, ES motifs (table 9.5) are at the origin of a very sharp ES peak in $B(\theta_{GeSeGe})$ at ~80 °. As expected, the ES feature in $B(\theta_{GeSeGe})$ is less intense in models II–IV. The CS feature in $B(\theta_{GeSeGe})$ of model II is made of a peak at ~97 ° and a shoulder at ~109 °. These marks correspond in $g_{GeGe}(r)$ to a peak at ~3.5 Å and a second shoulder at ~3.8 Å, respectively (figure 9.11(a)). The type of CS structural motifs that are at the origin of this pattern are shown in figure 9.13, where one notices that there is a Se–Se homopolar bond in the left-hand panel. However, fivefold rings were not found in model IV, where a longer temporal trajectory during the quench was adopted. This is an indication that the fivefold rings of model II appear only for particularly rapid quench rates.

Figure 9.12(b) shows the Se–Ge–Se bond-angle distributions $B(\theta_{SeGeSe})$ with a peak centered at 109°, typical of CS $Ge(Se_4)_{1/2}$ tetrahedra. In model I one has also an additional feature at ~99°, due to smaller Se–Ge–Se angles with ES connections.

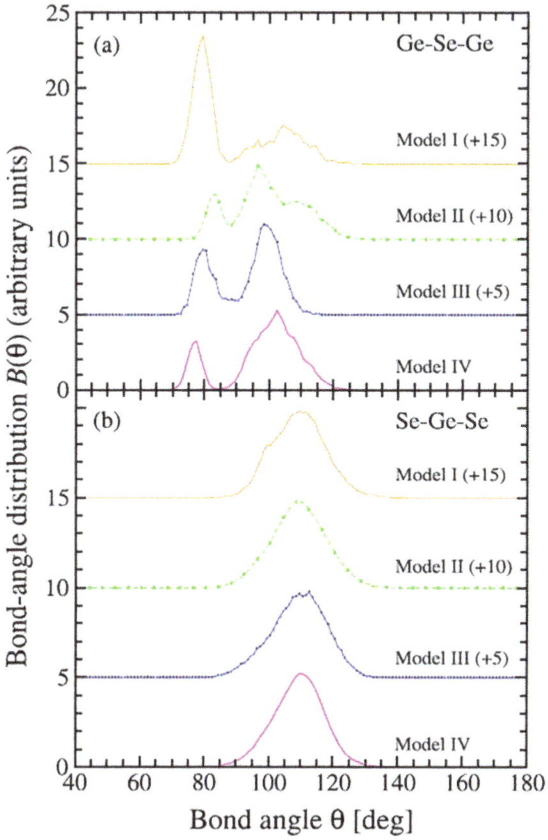

Figure 9.12. The bond-angle distributions (a) $B(\theta_{GeSeGe})$ and (b) $B(\theta_{SeGeSe})$ for glassy $GeSe_9$, where a cutoff distance $r_{cut} = 2.75$ Å was used in the analysis. The results correspond to FPMD models I (solid orange curve), (see [30]) II (broken green curve with squares), III (solid blue curve with squares) and IV (solid magenta curve). Reprinted figure with permission from [26]. Copyright (2016) by AIP Publishing.

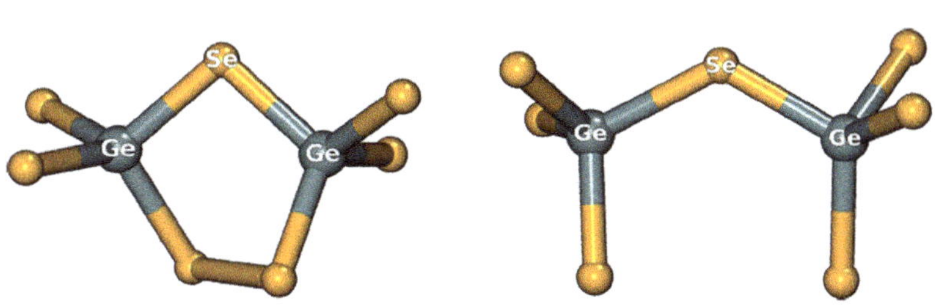

Figure 9.13. The CS motifs in model II that lead to bimodal CS features in both $g_{GeGe}(r)$ (figure 9.11(a)) and $B(\theta_{GeSeGe})$ (figure 9.12(a)). The Ge and Se atoms are represented by the dark (blue) and light (golden) balls, respectively. Reprinted figure with permission from [26]. Copyright (2016) by AIP Publishing.

By relying on the considerations introduced before (see for instance section 7.3.2 and figure 7.13) on the different kinds of connectivity for Se atoms, we can go back to the issue of phase separation invoked for glassy $GeSe_4$ on the basis of ^{77}Se NMR and Raman spectroscopy experiments [36, 37] but dismissed by further experimental and theoretical work [34, 38–41].

Based on table 9.6, it appears that all FPMD models are consistent with a coexistence of fourfold coordinated and twofold coordinated Ge and Se atoms respectively. There are 59%–65% of Se atoms in Se–Se–Se triads, in agreement with a value of 66(5)% from high-resolution isotropic ^{77}Se NMR spectra [40]. Se atoms are also in Se–Se–Ge triads (26%–35%) to indicate a non negligible fraction of them do not form chains and are not found in Ge–Se–Ge connections.

The results presented above provide a picture describing the role played by three specific factors in determining the structure of glassy $GeSe_9$: (i) the system size, (ii) the quench schedule, and (iii) the length of the relaxation time at $T = 300$ K.

Table 9.6. The proportions of the different structural units $n_\alpha(l)$ in the FPMD models I–IV for glassy $GeSe_9$, as obtained by using a cutoff distance $r_{\rm cut} = 2.75$ Å. The identity of the α atom (Ge or Se) at the center of a unit is given in bold font, and the identity of the l nearest neighbors is given in the second column. Reprinted table with permission from [26]. Copyright (2016) by AIP Publishing.

		Proportion of $n_\alpha(l)$ [%]			
		Model I	Model II	Model III	Model IV
Ge atom					
$l = 3$					
	Se_3	–	0.2	<0.1	–
$l = 4$					
	Se_4	>99.9	99.8	>99.9	>99.9
$l = 5$					
	$GeSe_4$	<0.1	–	–	<0.1
	Se_5	–	–	–	<0.1
Se atom					
$l = 1$					
	Ge	–	0.4	0.8	<0.1
	Se	0.2	1.1	0.8	<0.1
$l = 2$					
	Se_2	64.6	58.9	59.0	60.1
	SeGe	25.9	33.5	33.8	35.0
	Ge_2	9.3	4.7	4.4	4.7
$l = 3$					
	Se_2Ge	<0.1	1.2	0.3	<0.1
	Se_3	0.1	0.2	0.5	<0.1

As to the point (i) the proportion of ES units in model I (table 9.5) resulted from an inadequate statistical sampling, due to the small number $N_{Ge} = 12$ of Ge atoms. Focusing on the impact of the quench schedule and structural relaxation at room temperature for models II–IV, it is useful to label 'short' and 'extended' temporal trajectories by taking 10 ps as a threshold. Model III is the one with short quench steps and a short relaxation time, leading to the largest proportion of ES units (table 9.5). Model II has short quench steps and an extended relaxation time. Since this model has a smaller number of ES units, it appears that the relaxation at room temperature crucially affects the glass structure. Model IV corresponds to both extended quench steps and relaxation time. What are the structural dissimilarities between models II and IV? The main difference is the presence of a bimodal distribution of CS units for model II (figure 9.13) that is absent in model IV. This means that in model II there are configurations that are sensitive to the length of the quench steps, thereby emphasizing the importance of the quench schedule. Therefore, our findings highlight the roles played by two factors: the relaxation process itself and the quench schedule needed to bring the system to relaxation.

9.4 Glassy GeS_4 as compared to glassy $GeSe_4$

So far, we have seen that two specific compositions, $x = 0.33$ and $x = 0.2$ can be taken as prototypes of network structures for x ($x \leqslant 0.33$). Taking the Ge_xSe_{1-x} family as example, there are small deviations from a chemically ordered system at $x = 0.33$ [42], while for $x = 0.2$, one has Se_n chains and only a few Ge–Ge homopolar bonds [36, 43, 44] as it should be since most Ge atoms accommodate four Se neighbors. This basic description is supported by experimental evidence [32, 40, 42, 45–47], but it requires the quantitative input of an atomic-scale first-principles approach. Further information on the $x = 0.2$ networks can be obtained by comparing glassy $GeSe_4$, largely treated in the previous sections (see chapter 8 in particular) and glassy GeS_4, for which FPMD data were not available at the time of our study [29].

To implement the FPMD calculations on glassy GeS_4 and glassy $GeSe_4$ models made of $N_{at} = 480$ (96 Ge + 384 S/Se) and $N_{at} = 120$ (24 Ge and 96 S/Se) atoms were employed. In the case of liquid $GeSe_2$ we can refer to an investigation proving that the main structural features do not change appreciably when moving from $N_{at} = 120$ to $N_{at} = 480$ [17]. In section 8.3 when commenting on the transition from calculations performed with $N_{at} = 120$ to those carried out with a larger system, $N_{at} = 480$, we mentioned that this extension in size is more important for glasses than for liquids, due to the need of favoring structural relaxation in networks with reduced atomic mobility. As a further reason to avoid limiting our analysis of the glass structure on systems of minimal size, we can also add the improved statistics on the units found in the network. For these reasons the following comparative work of glasses $GeSe_4$ and GeS_4 is based on data for $N_{at} = 120$ and $N_{at} = 480$, providing useful hints on the sensitivity of glassy structures to the dimensions of the periodic simulation box.

In what follows, when using the NVT method (three out of the four cases considered here, see below), a further check at room temperature ensured that all densities correspond to pressure values as close as possible to 0 GPa. Concerning the methodological details of the calculations, for $N_{at} = 480$ and glassy GeS_4 the initial configuration was selected from the trajectories produced for glassy GeS_2 [9], by imposing the correct chemical composition and randomizing the system at 2000 K during $\simeq$ 25 ps. Then we took quenching steps of 500 K, lasting 10 ps each, the last one lowering the temperature from T = 500 K to T = 300 K. Similarly, in the $N_{at} = 120$ case of glassy GeS_4, we started by taking one configuration of liquid $GeSe_4$ followed by suitable changes of Se into S atoms in the simulation input. The thermal cycle encompassed 16 ps at 100 K, 10 ps at 300 K, 12 ps at 600 K, 33 ps at 900 K, 17 ps at 600 K and 22 ps at 300 K.

Turning to glassy $GeSe_4$, the system made of 120 atoms was studied in the isothermal-isobaric (NPT) ensemble, by using the fixed shape, variable size version of the Parrinello–Rahman technique [48–51]. In section 7.2.1 it was mentioned that changing the density in the NVT ensemble was less expensive and more practical than applying the NPT technique for the case of glassy $GeSe_2$. Among the reasons invoked, and in addition to those common to any system, we pointed out the slow convergence of the density. This problem was much less severe for the case of glassy $GeSe_4$ for which we could also take advantage of improved computational resources. For these reasons, we were in a position to produce a new set of NPT calculations for the specific case of glassy $GeSe_4$ in the context of the comparison with glassy GeS_4 that is the object of [29]. As initial configuration, we selected a disordered arrangement taken from the liquid [13] as a starting point at $T = 0$ K, with a new density found, after releasing the residual pressure, to be equal to $\rho_0^{new} = 0.031\,2$ Å^{-3} and box size = 15.66 Å. Trajectories were created by allowing for isotropic variation of the simulation box. This system is heated (16 ps at 100 K, 14 ps at 400 K, 10 ps at 600 K, 38 ps at 900 K) and then cooled at room temperature (26 ps at 600 K and 81 ps at 300 K). Final averages were taken over 60 ps. For $N_{at} = 480$, we came back to NVT simulations and we proceeded as in the case of glassy GeS_4, with identical thermal schedule and density as above for the $N_{at} = 120$ case.

9.4.1 Structure factors and pair correlation functions

The most interesting feature relative to the total neutron structure factors $S_T(k)$ in figure 9.14 is the difference in the height of the FSDP when comparing the two systems, to point out that the corresponding intermediate range order properties are far from being the same. The availability of two different sizes allows searching from improvements when adopting the larger $N_{at} = 480$ model. This is the case for glassy GeS_4 for which, in between the peaks located at $k\sim 1$ Å^{-1} and $k\sim 4$ Å^{-1}, a steady profile taking the form of a fairly wide peak (for $N_{at} = 480$) replaces an oscillating pattern (for $N_{at} = 120$). Also for glassy GeS_4, there is an unexpected feature at $k \simeq 0.5$ Å^{-1}, having no physical meaning and to be ascribed to some long living and yet statistically not significant extended range correlations.

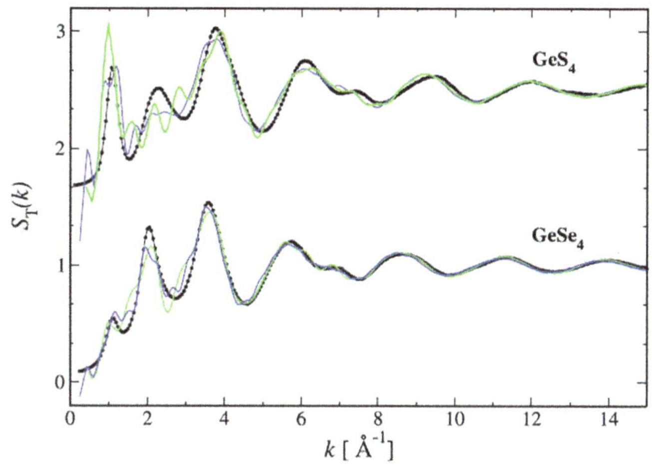

Figure 9.14. Total neutron structure factors at 300 K for glassy GeS_4 and glassy $GeSe_4$. For both systems, N_{at} = 120 atoms (green line), N_{at} = 480 atoms (black line). Experimental measurements (solid black line with circles) for glassy $GeSe_4$ from [32, 52] and glassy GeS_4 from [10]. Reprinted figure with permission from [29]. Copyright (2015) by AIP Publishing.

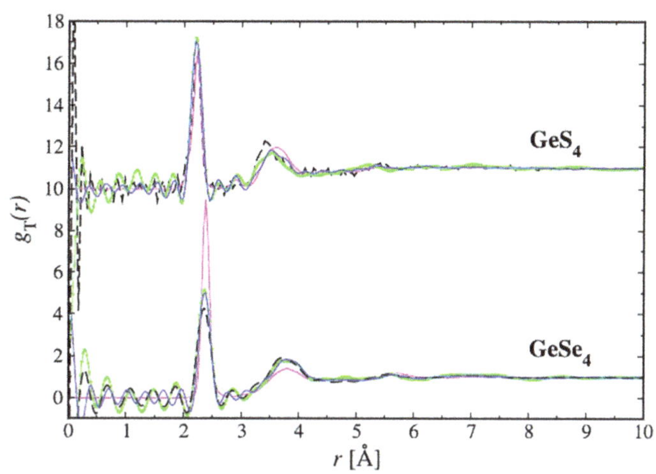

Figure 9.15. Total pair correlation functions at 300 K for glassy $GeSe_4$ and glassy GeS_4. These quantities have been obtained by Fourier integration from reciprocal space by taking the experimental upper limits. For both systems: N_{at} = 120 (solid green line with circles), N_{at} = 480 (solid blue line) compared to the experimental measurements for glassy GeS_4 from [10] (black broken line) and glassy $GeSe_4$ from [32, 52] (black broken line). In the case of N_{at} = 480 we also report the results obtained by direct calculation in real space (magenta lines). Reprinted figure with permission from [29]. Copyright (2015) by AIP Publishing.

Focusing on the data in direct space, the total pair correlation functions $g_T(r)$ of glassy GeS_4 and glassy $GeSe_4$ are given in figure 9.15 together with the experimental results, obtained by Fourier transforming the total neutron structure factors and using upper limits of integration l_{int}(glassy $GeSe_4$) = 19.95 Å^{-1} for glassy $GeSe_4$ and

l_{int}(glassy GeS_4) = 49.95 $Å^{-1}$ for glassy GeS_4 [10, 52]. For both glasses and for the N_{at} = 480 systems only, two methodologies were employed, i.e. direct calculation of the total pair correlation function and Fourier integration from the k-space data with the above experimental upper limits. Looking at glassy $GeSe_4$, it is clear that the best agreement corresponds to the use of l_{int}(glassy $GeSe_4$) = 19.95 $Å^{-1}$. In the case of glassy GeS_4, due to a much higher integration upper limit, (l_{int}(glassy GeS_4) = 49.95 $Å^{-1}$), the pair correlation function is not sensitive to the method of calculations. FPMD results are in very good agreement with experiments, confirming that the predominant units are the Ge–X_4 (X = S, Se) tetrahedra, the first peak in $g_T(r)$ being a mark of the intra-tetrahedra Ge–X (X = S, Se) connections and the second being associated to the inter-tetrahedra bonds.

The calculated partial structure factors of glassy GeS_4 and glassy $GeSe_4$ are given in figure 9.16. For glassy $GeSe_4$, there are limited changes when taking a larger size, the most noticeable being present in $S_{GeGe}(k)$. In the glassy GeS_4 case, the height of the FSDP intensity is smaller in the Ge–Ge partial structure factor, taking the form

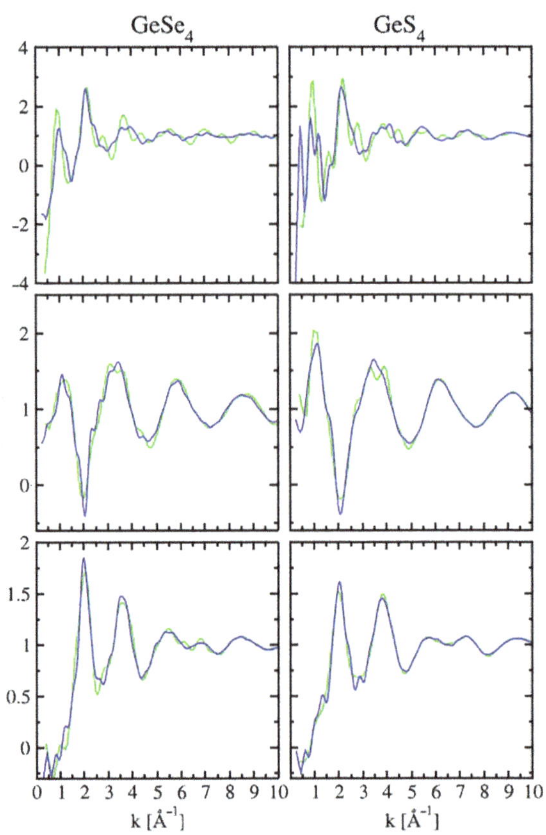

Figure 9.16. The partial structure factors $S_{GeGe}(k)$ (top panel), $S_{GeX}(k)$ (middle panel) and $S_{XX}(k)$ (bottom panel) (X = Se, S). From left to right: glassy $GeSe_4$ and glassy GeS_4 models. N_{at} = 120 atoms (solid green line), N_{at} = 480 (solid blue line). Reprinted figure with permission from [29]. Copyright (2015) by AIP Publishing.

of a double peak due to statistical noise. The same rationale accounts for the peak at $k \simeq 0.5$ Å^{-1} already observed in the total neutron structure factor, while no sign of such feature are present in either $S_{GeX}(k)$ or $S_{SeSe}(k)$ of glassy $GeSe_4$. In both glassy $GeSe_4$ and glassy GeS_4, the intermediate range order (which is closely related to the patterns recorded at around $k \simeq 1$ Å^{-1}) can be traced back to the Ge–Ge and Ge–Se (or Ge–S) correlations since no deviations from the baseline profiles of $S_{XX}(k)$ (X = Se, S) do exist in the vicinity of the FSDP region. The more intense FSDP in glassy GeS_4 when compared to glassy $GeSe_4$ observed in figure 9.14 corresponds to a higher peak in $S_{GeSe}(k)$ (figure 9.16). This is an indirect indication of the existence of Ge–Se correlations at intermediate range order distance, as it will be substantiated through an analysis of the rings patterns.

The partial pair correlation functions $g_{\alpha\beta}(r)$ for glassy GeS_4 and glassy $GeSe_4$ are shown in figure 9.17. For $N_{at} = 120$, the partial pair correlation functions $g_{GeGe}^{GeS_4}(r)$ and $g_{GeGe}^{GeSe_4}(r)$ do not exhibit any small peak at $r \simeq 2.45$ Å, indicating the absence of

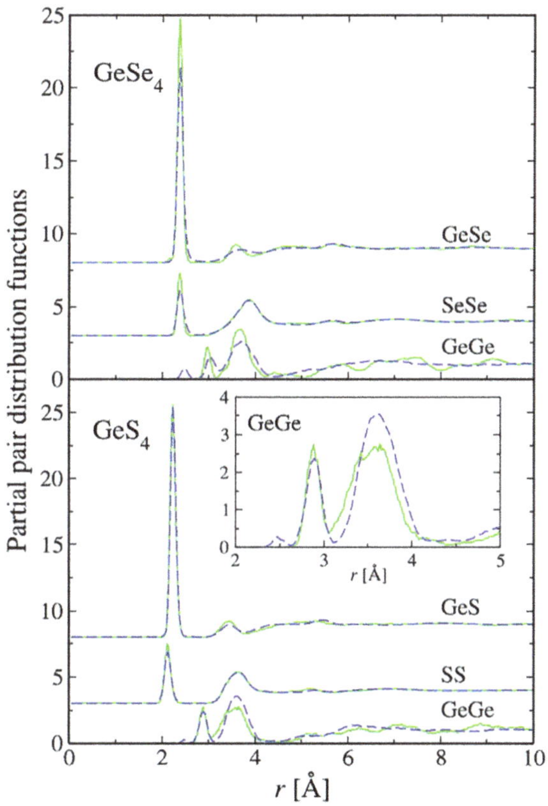

Figure 9.17. Partial pair correlation functions (termed 'distribution' in the figure) $g_{GeGe}(r)$, $g_{GeX}(r)$ and $g_{XX}(r)$ (X = S or Se) for glassy $GeSe_4$ (top panel) and glassy GeS_4 (bottom panel). We show both results for $N_{at} = 120$ (green line) and $N_{at} = 480$ (blue broken line). Reprinted figure with permission from [29]. Copyright (2015) by AIP Publishing.

homopolar Ge–Ge bonds. However, such peak is present for N_{at} = 480 and very much visible for glassy $GeSe_4$. The peaks denoting ES and CS tetrahedra are easily distinguishable at $r \simeq$ 2.9 Å, $r \simeq$ 3.6 Å ($g_{GeGe}^{GeS_4}(r)$) and $r \simeq$ 3.0 Å, $r \simeq$ 3.72 Å ($g_{GeGe}^{GeSe_4}(r)$), respectively. There are differences concerning the heights of the peak related to CS connections when comparing the cases N_{at} = 120 and N_{at} = 480 since the intensities either increase (glassy GeS_4) or decrease (glassy $GeSe_4$ case). The profile taken by $g_{GeGe}^{GeS_4}(r)$ and $g_{GeGe}^{GeSe_4}(r)$ changes with the larger size for $r > 5$ Å. In fact, especially for glassy $GeSe_4$, a clear maximum is visible around 6–6.5 Å for N_{at} = 480, contrasting with the oscillating sequence of maxima and minima found for N_{at} = 120. This can be taken as a size effect manifesting itself in the minority species (Ge).

The partial pair correlations functions g_{GeS} and g_{GeSe} differ in the position of the first peak (2.22 Å–2.23 Å and 2.36 Å–2.37 Å, respectively) but they are both consistent with the dominating Ge-centered tetrahedral motif. In a similar way, the presence of homopolar Se–Se and S–S bonds forming chains is underlined by the first peaks at $r \simeq$ 2.11 Å–2.12 Å in g_{SS} and $r \simeq$ 2.37 Å–2.38 Å in g_{SeSe}. For $r > 3$ Å, the peaks at $r \simeq$ 3.63 Å in glassy GeS_4, and at $r \simeq$ 3.85 Å–3.87 Å in the glassy $GeSe_4$ stand for the intra-tetrahedral X–X (X = S, Se) connections.

9.4.2 Coordination numbers, structural units and rings analysis: a rationale for intermediate range order

The coordination numbers, the total coordination numbers for Ge and X (Se or S) and the average coordination number (all defined in section 2.4.1) are listed in table 9.7. These values are in line with a predominant presence of tetrahedra, with the exception of n_{Ge}^{Ge} = 0.36 for glassy $GeSe_4$, N_{at} = 480, to be compared with n_{Ge}^{Ge} = 0.03 for glassy GeS_4, N_{at} = 480. The corresponding values for N = 120 are vanishingly small. As a first consideration, glassy GeS_4 is less chemically disordered than glassy $GeSe_4$. Second, there are no homopolar bonds in $GeSe_4$ for N_{at} = 120 as if the small number of Ge atoms prevented the occurrence of these configurations.

As a next step to describe the network we provide the average percentages of the individual $n_{\alpha}(l)$ structural units in table 9.8.

The tetrahedral motif $Ge(X_4)$ has the largest occurrences for both systems, even though in glassy $GeSe_4$ (N_{at} = 480) its impact lowers due to the presence of homopolar bonds. Chemical disorder in glassy $GeSe_4$ takes the form of threefold Ge-coordinated units for both N_{at} = 120 and N_{at} = 480. However, fourfold $Ge–GeX_3$ units are present in both systems. Considering the results for N_{at} = 120, one notices that the relative weight of AA, AB and BB linkages is similar in the two systems. These numbers change when moving to the N_{at} = 480 models and, in addition, the two glasses appear now quite different in terms of physical quantities, such as the AA, AB and BB values. While AB connections are largely the most frequent in glassy $GeSe_4$ (38% against 31% BB and 27% AA), this is not the case in glassy GeS_4 for which the percentages of AA, AB and BB connections fall within 2% around ~32% and ~34% (typical error bars being of the order of 3%). Our results demonstrate that the system size N_{at} = 120 is inadequate to account for correlations

Table 9.7. The first peak position (FPP), second peak position (SPP) in $g_{\alpha\beta}(r)$ and nearest neighbor coordination numbers n_α^β calculated for glassy GeS_4 and glassy $GeSe_4$. The coordination numbers n_α^β were obtained by integrating the corresponding partial pair correlation $g_{\alpha\beta}(r)$ functions for the GeX_4 models up to their the first minimum. Reprinted table with permission from [29]. Copyright (2015) by AIP Publishing.

$g_{\alpha\beta}(r)$	Model	FPP (Å)	SPP (Å)	n_α^β
$g_{GeGe}(r)$	$GeSe_4$-120	–	2.97–3.69	–
	$GeSe_4$-480	2.47	3.70	0.36
	GeS_4-120	–	2.77–3.64	–
	GeS_4-480	2.47	2.89–3.60	0.03
$g_{GeX}(r)$	$GeSe_4$-120	2.36	3.58	3.96
	$GeSe_4$-480	2.37	3.67	3.85
	GeS_4-120	2.22	3.43	4.00
	GeS_4-480	2.23	3.43	3.98
$g_{XGe}(r)$	$GeSe_4$-120	2.36	3.58	0.99
	$GeSe_4$-480	2.37	3.67	0.96
	GeS_4-120	2.22	3.43	1.00
	GeS_4-480	2.23	3.43	0.99
$g_{XX}(r)$	$GeSe_4$-120	2.38	3.87	0.99
	$GeSe_4$-480	2.37	3.85	1.04
	GeS_4-120	2.11	3.63	1.00
	GeS_4-480	2.12	3.63	0.99
		n_{Ge}	n_X	n_{ave}
	$GeSe_4$-120	3.96	1.99	2.386
	$GeSe_4$-480	4.21	2.00	2.44
	GeS_4-120	4.00	2.00	2.40
	GeS_4-480	3.99	1.98	2.385

involving next-nearest neighbors, at least when the concentration of one of two species is small, as it is the case for glassy GeS_4 and glassy $GeSe_4$.

Table 9.8 contains also the proportions of ES and CS connections, these latter by fare more numerous, as expected in view of the heights of the second and third peaks in the Ge–Ge pair correlation functions (see figure 9.17). There are more ES connection in glassy GeS_4. Therefore, based on the above results, both glasses can be described as a blend of Ge-centered tetrahedra and Se–Se (S–S) homopolar bonds.

Table 9.8. Different coordination units in glassy GeS_4 and glassy $GeSe_4$, Ge atoms involved in homopolar bonds N_{GeGe}, X atoms (S or Se) involved in homopolar bonds N_{XX} and Ge atoms making ES or CS connections. All values are given as percentages. The statistics of chains (number of chains of a given length) is also presented for the case N_{at} = 480. See also [29]. Reprinted table with permission from [29]. Copyright (2015) by AIP Publishing.

		$n_\alpha(l)$ [%]			
		$GeSe_4$		GeS_4	
		N_{at} = 120	N_{at} = 480	N_{at} = 120	N_{at} = 480
Ge atom					
$l = 2$					
	X_2	–	4.8	–	0.4
$l = 3$					
	GeX_2	–	–	–	–
	X_3	3.6	3.6	–	0.7
$l = 4$					
	GeX_3	–	6.2	–	2.0
	X_4	96.4	85.0	99.9	96.7
$l = 5$					
	Ge_2X_3	–	–	–	–
	GeX_4	–	–	0.1	0.1
	X_5	–	–	–	–
X atom					
$l = 1$					
	Ge	–	1.5	–	0.5
	X	1.0	1.7	–	1.3
$l = 2$					
	X_2 (BB)	28.1	30.7	28.1	32.0
	XGe (AB)	43.6	38.2	43.8	33.9
	Ge_2 (AA)	27.1	26.5	28.1	31.8
$l = 3$					
	X_2Ge	–	0.5	–	<0.1
	XGe_2	0.1	0.3	–	0.2
	Ge_3	–	0.3	–	0.2
	N_{Ge-Ge}	0.0	6.2	0.2	2.7
	N_{X-X}	71.9	70.1	71.9	67.4
	N_{Ge}(ES)	25	25.4	29.2	32.8
	N_{Ge}(CS)	75	68.3	70.6	64.6
Chains					
	Length		S–S	Se–Se	
	2		23	26.2	
	3		21	20	

4	8	15
5	4	5
6	5	4
7	2	2
8	1	2.3
9	2	1
12	1	<1%
16	1	<1%

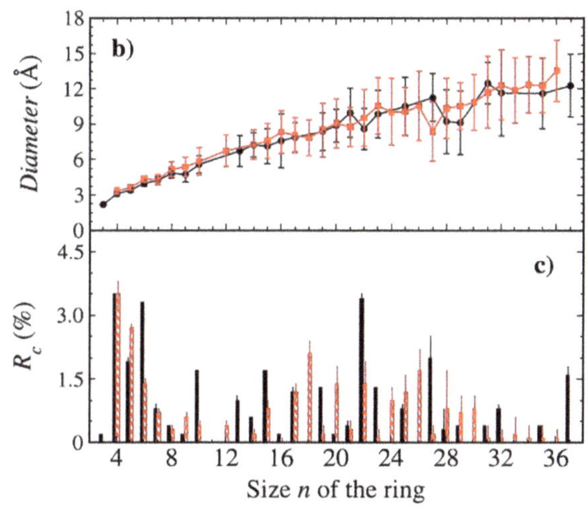

Figure 9.18. (b) Evolution of the ring diameter as a function of the size of the ring n. (c) $R_c(n)$, proportion of rings of size n, i.e. number of rings of size n normalized to the total number of atoms in the model. Reprinted figure with permission from [29]. Copyright (2015) by AIP Publishing.

In the search of an atomic-scale origin for the different behavior of our two systems with respect to intermediate range order properties, as shown in section 9.4.1, we have resorted to a further analysis tool, consisting of a statistics of rings, the main results of which being displayed in figure 9.18 [53–55]. The method has been implemented by imposing a shortest path criterion and allowing for homopolar bonds, with threshold distances being the first minima of the partial pair correlation functions. The distribution of rings exhibits higher rates of occurrence in both systems, for instance $n = 4, 5, 6$ for $n < 8$. In glassy GeS_4, $n = 6$ stands out as well as $n = 10, 13, 15$. These rings have diameters well beyond the first two shells of neighbors, 5 Å. This observation allows establishing a clear link between the presence of these rings and the enhancement of intermediate range order recorded in glassy GeS_4. In practical terms, one can envision Ge and S lying farther than bond distances in high n rings as being at the origin for the upsurge of the FSDP in $S_{GeS}(k)$. This is the main contribution to the FSDP in the total structure factor (see above in section 9.4.1).

9.4.3 Insight into electronic properties and correlation with structure

A first information on the bonding properties of glassy GeS_4 and glassy $GeSe_4$ is given in figure 9.19 for the two sizes. Even though the difference is by far not striking, it appears that both sets of results (N_{at} = 120 and N_{at} = 480) indicate a higher ionic character (larger gap) for glassy GeS_4.

In order to substantiate this first indication of a higher ionicity in glassy GeS_4, inspired by the Pauling bond ionicity definition [56] and more recent investigations [57, 58] one can resort to the analysis based on the maximally localized Wannier functions formalism (see section 2.5.2 and/or [22, 23]). 50 uncorrelated configurations at T = 300 K were considered to obtain the results shown in figure 9.20.

We recall that the inherent localized distribution in space of these quantities is instrumental to extract the relative impact of covalent vs ionic characters of bonding. To obtain this information we can rely on the correlation involving the distances between WF centers (labeled by W) and atomic positions, leading to the pair correlation functions $g_{GeW}(r)$ (in both systems), $g_{SeW}(r)$ (in $GeSe_4$) and $g_{SW}(r)$ (in GeS_4). As shown in figure 9.20, three types of WFCs can be distinguished (see also figure 9.21). The centers named W_B are those related to Ge–S or Ge–Se bonds. An unmistakable sign of them is the main peak in $g_{GeW}(r)$ at around 1.3 Å. Looking at $g_{Se(S)W}(r)$, the Se(S)–W correlations can be associated to to the peaks at ~0.9 Å (glassy GeS_4, see inset of figure 9.20) and ~1 Å (glassy $GeSe_4$, same inset). Let us focus now on the second type of WF center, labeled as W_H. This refers to homopolar S–S or Se–Se bonds and it gives the third peak in $g_{SeW}(r)$ (1.02 Å) or $g_{SW}(r)$ (1.18 Å) (see inset). Finally, there is a third type of WF center, W_{LP}, standing for the lone pair (LP) valence electrons not involved in any chemical bond and lying localized very

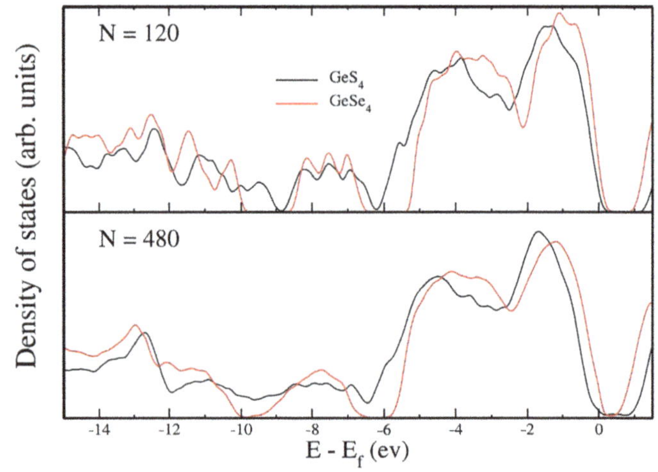

Figure 9.19. The electronic density of states extracted from the Kohn–Sham eigenvalues. Results for glassy GeS_4 (solid black curve) are compared to that obtained for glassy $GeSe_4$ (solid red curve). Results for N_{at} = 120 at the top and results for N_{at} = 480 at the bottom of the figure. Reprinted figure with permission from [29]. Copyright (2015) by AIP Publishing.

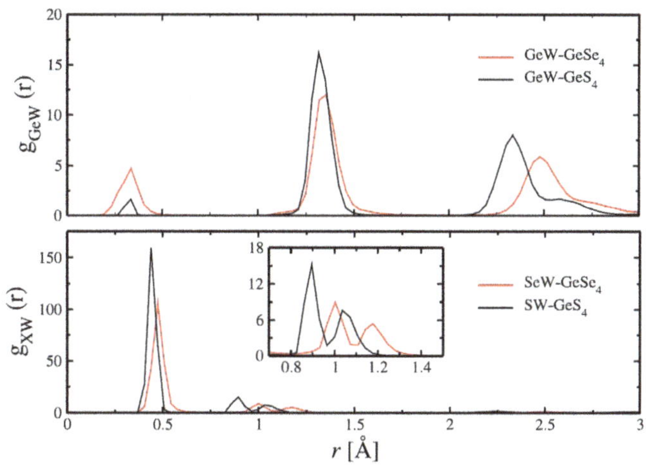

Figure 9.20. Partial pair correlation functions $g_{GeW}(r)$ (top panel) and $g_{XW}(r)$ (bottom panel). The results at N_{at} = 120 for the two glasses are compared. Reprinted figure with permission from [29]. Copyright (2015) by AIP Publishing.

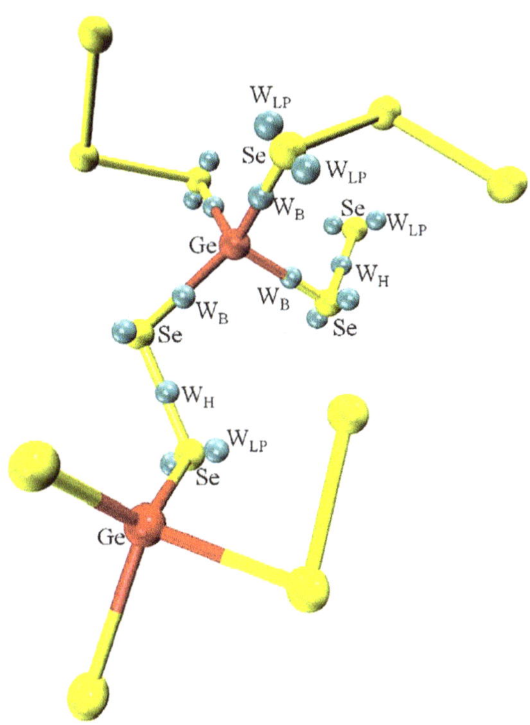

Figure 9.21. Details of the local bonding environment in the case of glassy $GeSe_4$. Ge atoms are shown in red, whereas Se atoms are in yellow. The Wannier centers are in blue. Reprinted figure with permission from [29]. Copyright (2015) by AIP Publishing.

close to the group VI atoms (S, Se). These lone pairs are responsible for the first peak at ~0.47 Å (in $g_{SeW}(r)$) and ~0.44 Å (in $g_{SW}(r)$) and the broad feature of $g_{GeW}(r)$ around ~2.33 Å (in glassy $GeSe_4$) and ~2.48 Å (in glassy GeS_4). In view of these definitions, some comment on the distance between the W_B centers and the Ge atoms is in order.

When looking at $g_{GeW}(r)$ for glassy GeS_4 or glassy $GeSe_4$ there are minimal differences between Ge–W_B = 1.32 ± 0.01 Å (GeS_4) and Ge–W_B = 1.36 ± 0.01 Å ($GeSe_4$). Therefore, no conclusions can be drawn on the ionic character based on these values. However, a better insight is provided by the distance separating the maxima at around 1.3 Å and those at ~2.3 Å (in glassy GeS_4) and ~2.5 Å (in glassy $GeSe_4$). Since a smaller value is found for the former, this is an indication of a Wannier center closer to the S atomic sites. In addition, the W_B centers, as highlighted in $g_{SeW}(r)$ and $g_{SW}(r)$, are found closer to the origin (the Se or the S atom) in the case of S. This is proved by the corresponding distances since S–W_B = 0.9 ± 0.01 Å and Se–W_B = 1.0 ± 0.1 Å. Therefore, two pieces of evidence exist showing that glassy GeS_4 has a higher ionic character than $GeSe_4$, in addition to those collected above on the basis of notions of chemical disorder or the observation of the electronic density of states. These amounts to (a) the relative position of the maxima (second and third for increasing distance) in $g_{GeW}(r)$ and (b) the shorter separation between the peaks in $g_{SW}(r)$. Overall, the WF centers of electronic localization (expressing heterogeneous bonding) are closer to the S sites than to the Se sites. This is a stringent fact allowing to complete our comparative analysis of the bonding character for these two glasses.

References

[1] Harris J 1985 *Phys. Rev.* B **31** 1770–9

[2] Salmon P S 1992 *Proc. R. Soc. Lond.* A **437** 591

[3] Petri I and Salmon P S 2001 *J. Non-Cryst. Solids* **293-295** 169–74

[4] Cai L and Boolchand P 2002 *Philos. Mag.* B **82** 1649–57

[5] Zeidler A, Drewitt J W E, Salmon P S, Barnes A C, Crichton W A, Klotz S, Fischer H E, Benmore C J, Ramos S and Hannon A C 2009 *J. Phys.: Condens. Matter* **21** 474217

[6] Blaineau S, Jund P and Drabold D A 2003 *Phys. Rev.* B **67** 094204

[7] Le Roux S and Jund P 2007 *J. Phys.: Condens. Matter* **19** 196102

[8] Sankey O F and Niklewski D J 1989 *Phys. Rev.* B **40** 3979–95

[9] Celino M, Le Roux S, Ori G, Coasne B, Bouzid A, Boero M and Massobrio C 2013 *Phys. Rev.* B **88** 174201

[10] Bychkov E, Miloshova M, Price D L, Benmore C J and Lorriaux A 2006 *J. Non-Cryst. Solids* **352** 63–70

[11] Bytchkov A, Cuello G J, Kohara S, Benmore C J, Price D L and Bychkov E 2013 *Phys. Chem. Chem. Phys.* **15** 8487–94

[12] Micoulaut M, Vuilleumier R and Massobrio C 2009 *Phys. Rev.* B **79** 214205

[13] Massobrio C, Celino M, Salmon P S, Martin R A, Micoulaut M and Pasquarello A 2009 *Phys. Rev.* B **79** 174201

[14] Celino M and Massobrio C 2003 *Phys. Rev. Lett.* **90** 125502

[15] Bouzid A and Massobrio C 2012 *J. Chem. Phys.* **137** 046101

[16] Blaineau S and Jund P 2004 *Phys. Rev.* B **69** 064201
[17] Micoulaut M, Le Roux S and Massobrio C 2012 *J. Chem. Phys.* **136** 224504
[18] Bhatia A B and Thornton D E 1970 *Phys. Rev.* B **2** 3004–12
[19] Massobrio C and Pasquarello A 2008 *Phys. Rev.* B **77** 144207
[20] Le Roux S, Zeidler A, Salmon P S, Boero M, Micoulaut M and Massobrio C 2011 *Phys. Rev.* B **84** 134203
[21] Le Roux S, Bouzid A, Boero M and Massobrio C 2012 *Phys. Rev.* B **86** 224201
[22] Marzari N and Vanderbilt D 1997 *Phys. Rev.* B **56** 12847–65
[23] Resta R and Sorella S 1999 *Phys. Rev. Lett.* **82** 370–3
[24] Mark Wilson and Salmon P S 2009 *Phys. Rev. Lett.* **103** 157801
[25] Azoulay R, Thibierge H and Brenac A 1975 *J. Non-Cryst. Solids* **18** 33–53
[26] Le Roux S, Bouzid A, Kim K Y, Han S, Zeidler A, Salmon P S and Massobrio C 2016 *J. Chem. Phys.* **145** 084502
[27] Tafen D N and Drabold D A 2005 *Phys. Rev.* B **71** 054206
[28] Sykina K, Furet E, Bureau B, Le Roux S and Massobrio C 2012 *Chem. Phys. Lett.* **547** 30–4
[29] Bouzid A, Le Roux S, Ori G, Boero M and Massobrio C 2015 *J. Chem. Phys.* **143** 034504
[30] Micoulaut M, Kachmar A, Bauchy M, Le Roux S, Massobrio C and Boero M 2013 *Phys. Rev.* B **88** 054203
[31] Ramesh Rao N, Krishna P S R, Basu S, Dasannacharya B A, Sangunni K S and Gopal E S R 1998 *J. Non-Cryst. Solids* **240** 221–31
[32] Salmon P S 2007 *J. Non-Cryst. Solids* **353** 2959–74
[33] Massobrio C and Pasquarello A 2007 *Phys. Rev.* B **75** 014206
[34] Gjersing E L, Sen S and Aitken B G 2010 *J. Phys. Chem.* C **114** 8601–8
[35] Bhosle S, Gunasekera K, Boolchand P and Micoulaut M 2012 *Int. J. Appl. Glass Sci.* **3** 205–20
[36] Bureau B, Troles J, Le Floch M, Guénot P, Smektala F and Lucas J 2003 *J. Non-Cryst. Solids* **319** 145–53
[37] Lucas P, King E A, Gulbiten O, Yarger J L, Soignard E and Bureau B 2009 *Phys. Rev.* B **80** 214114
[38] Kibalchenko M, Yates J R, Massobrio C and Pasquarello A 2010 *Phys. Rev.* B **82** 020202
[39] Kibalchenko M, Yates J R, Massobrio C and Pasquarello A 2011 *J. Phys. Chem.* C **115** 7755–9
[40] Kaseman D C, Hung I, Gan Z and Sen S 2013 *J. Phys. Chem.* B **117** 949–54
[41] Sykina K, Bureau B, Le Pollès L, Roiland C, Deschamps M, Pickard C J and Furet E 2014 *Phys. Chem. Chem. Phys.* **16** 17975–82
[42] Penfold I T and Salmon P S 1991 *Phys. Rev. Lett.* **67** 97–100
[43] Bychkov E, Benmore C J and Price D L 2005 *Phys. Rev.* B **72** 172107
[44] Tronc P, Bensoussan M, Brenac A and Sebenne C 1973 *Phys. Rev.* B **8** 5947–56
[45] Golovchak R, Shpotyuk O, Kozyukhin S, Kovalskiy A, Miller A C and Jain H 2009 *J. Appl. Phys.* **105** 103704
[46] Salmon P S and Petri I 2003 *J. Phys.: Condens. Matter* **15** S1509–15
[47] Salmon P S 2007 *J. Phys.: Condens. Matter* **19** 455208
[48] Parrinello M and Rahman A 1980 *Phys. Rev. Lett.* **45** 1196–9
[49] Parrinello M and Rahman A 1982 *J. Chem. Phys.* **76** 2662–6
[50] Parrinello M and Rahman A 1981 *J. Appl. Phys.* **52** 7182–90

[51] Focher P, Chiarotti G L, Bernasconi M, Tosatti E and Parrinello M 1994 *Europhys. Lett.* **26** 345–51
[52] Petri I and Salmon P S 2002 *Phys. Chem. Glasses* **43C** 185–90
[53] Le Roux S and Jund P 2011 *Comput. Mater. Sci.* **50** 1217
[54] Le Roux S and Jund P 2010 *Comput. Mater. Sci.* **49** 70–83
[55] Franzblau D S 1991 *Phys. Rev.* B **44** 4925–30
[56] Pauling L 1932 *J. Am. Chem. Soc.* **54** 3570–82
[57] Abu-Farsakh H and Qteish A 2007 *Phys. Rev.* B **75** 085201
[58] Qteish A 2015 *J. Phys. Chem. Solids* **82** 82–90

Chapter 10

Accounting for dispersion forces: glassy $GeTe_4$ and related examples

The treatment of dispersion (van der Waals) forces within electronic structure methods and, in particular, first-principles molecular dynamics (FPMD), has become increasingly important due to the development of schemes that can be straightforwardly integrated into the equations of motion. Here we go through a set of calculations (mostly on glassy $GeTe_4$) adopting dispersion forces under different proposals to check their impact on the structural properties. The main differences among these approaches amount to the account (or the neglect) of changes in the electronic structure when building the expression for the dispersion forces. In the first part, we also considered the impact of the exchange–correlation functional in conjunction with one specific choice for the dispersion interactions. In the second part (after digressing on the effect of dispersion forces on liquid $GeSe_2$), this study is extended by including calculations performed using the maximally localized Wannier functions scheme introduced along the lines sketched in section 2.5.2. Finally, five different dispersion schemes are compared by selecting different criteria for their performances. Overall, despite the limited impact of any of them on the topology of glassy $GeTe_4$, the indications collected concur to provide, for a given disordered structure, a quite unique set of reference calculations including van der Waals contributions.

10.1 Introduction

On the side of first-principles modeling of glasses by molecular dynamics (MD), this monograph has reported so far mostly on a single family of compounds (Ge_xSe_{1-x}), with consideration of a few additional systems employed mostly for comparative purposes. Ge_xSe_{1-x} networks possess the attractive feature of being highly instructive both as prototype of disordered network-forming materials and as ideal targets for testing and improving quantitative modeling, due to their peculiar iono/covalent

doi:10.1088/978-0-7503-2436-6ch10

bonding. In the previous chapters we have seen how the implementation of FPMD along the lines detailed in section 3.5, enriched by a rationale on the selection of exchange–correlation (XC) functional, has led to the choice of BLYP as a main tool, supplemented by alternative choices (PW, PBE) whenever useful to gather extra information on structure and bonding. In the area of disordered materials, application of the same strategy to systems other than Ge_xSe_{1-x} appeared worthwhile and substantiated by the observation that only a few cases had been considered with the same detail and completeness deployed to describe Ge_xSe_{1-x} networks. When we started, the choice of glassy $GeTe_4$ had a twofold interest, due to its technological relevance of ternary materials containing Ge and Te in various concentrations [1–6] and to its hypothesized sensitivity to dispersion forces, not included in the density functional theory (DFT) Kohn–Sham derivation of the total energy functional [7]. By leaving aside for a moment any consideration on the impact of Ge–Te systems as useful devices, it is important to devote some preliminary thoughts to the account of dispersion forces. These forces correspond to weak, long-range van der Waals (vdW) contributions to bonding, crucial to model interactions due to instantaneous dipole fluctuations of electrons in their ground state. Historically, DFT XC functional of the kind employed in most FPMD calculations were almost exclusively depending on the local electron density and its gradient within generalized gradient approximations (GGA, like PW, BLYP, PBE to mention those we have resorted to so far). Since dispersions forces act at distances for which there is essentially no overlap between electronic clouds, there is no hope to see them described within such frameworks. For further insight on these preliminary issues see, for instance, [8–10]. Having recognized that we felt it necessary to extend our range of FPMD-DFT applications to other chalcogenide glasses ($GeTe_4$), our interest for the inclusion of dispersion forces was also boosted by a quest of improved quantitative predictions since, for Ge_xSe_{1-x} network glasses, we were left with a highly reliable and yet not fully satisfactory description of the Ge–Ge nearest and next-nearest neighbor environment. Was there any chance that the inclusion of dispersion forces could improve the agreement between FPMD and the experimental Ge–Ge correlation functions? How about the absence of the first-sharp diffraction peak (FSDP) in the concentration–concentration structure factor? Could this possibly be related to the inclusion of this missing part? There was no better way to address these questions than planning some targeted calculations. These will be detailed in section 10.3.

In view of the amount of biological or condensed matter problem requiring the inclusion of the vdW interactions, theoretical efforts aiming at their treatment in the DFT framework have been numerous and oriented, from the beginning, along two main strategies. The first consisted in devising functionals that contain dispersion effects by construction ('from scratch'), this strategy being pioneered in seminal papers [11–18] and revisited later in more palatable forms by including seamlessly the dispersion part within a non-local construction built inside the energy functional [19, 20]. The second views the treatment of dispersion effects as a fully separable problem in which corrections taking the intuitive form $C_6{\cdot}R^{-6}$ are added to the Kohn–Sham original functionals. The most effective and widely employed prescriptions of this kind are those proposed by S Grimme [21] for which the dispersion energy to be added to the total energy expression takes the form

$$E_{\text{disp}} = -s_6 \sum_{i=1}^{N_{\text{at}}-1} \sum_{j=i+1}^{N_{\text{at}}} \frac{C_6^{ij}}{R_{ij}^6} f_{\text{dmp}}(R_{ij}) \tag{10.1}$$

with C^{ij} being the so-called dispersion coefficient for the atom pair ij (to be determined via highly accurate quantum chemical calculations on atomic orbitals performed upfront), s_6 accounts for the functional XC and is employed as a scaling factor, R_{ij} is the interatomic distance and f_{dmp} is a suitable damping factor to avoid singularities. In [21] (referred to as Grimme-D2 or simply D2 hereafter), C_6^{ij} is postulated to be equal to $\sqrt{(C_6^i C_6^j)}$ where each coefficient refers to a given atom i or j. In a further publication ([22], D3 hereafter) the D2 method has been improved to increase its 'first-principles' character and obtain 'less empiricism' while maintaining, as stated by Grimme himself, its 'non-electronic' nature, since the dispersion coefficients do not adapt to changes of the electronic structure during the time evolution of the system. An idea of the intense work of refinement inherent in [22] aimed at obtaining the best possible computational strategy for the vdW forces (being at the same time robust, tractable, affordable and compatible with a wealth of XC functionals) is given in the conclusive remarks of that paper. The results obtained for a large body of test calculations have prompted the authors to consider the D3 method very close to the limit of what it can be achieved via 'non-electronic' approaches.

We have taken advantage of the D2 and D3 methodology not only in our study of glassy $GeTe_4$ (sections 10.2, 10.5) but also (D2 only) to consider again the prototypical case of liquid $GeSe_2$ (see section 10.3). In addition to the Grimme approach (and keeping in mind the possibility of adopting schemes dealing with a treatment of vdW forces featuring no separation between the Kohn–Sham part and a dispersion part), we followed an alternative strategy proposed by P L Silvestrelli [23, 24]. This is rooted in the concept of the maximally localized Wannier functions (MLWF) $w_n(r)$, already introduced in this monograph as a tool to correlate atomic and electronic structure (see section 2.5.2 and several applications detailed in the previous chapters). Accordingly, the coefficients C_6^{ij} of equation (10.1) take the form describing the long-range interaction between two distinct and separated entities of matter

$$C_6^{ij} = \frac{3}{32\pi^{3/2}} \int_{r\leqslant r_{\text{cut}}} \int_{r'\leqslant r_{\text{cut}}} \frac{w_i(\mathbf{r}) \cdot w_j(\mathbf{r}')}{w_i(\mathbf{r}) + w_j(\mathbf{r}')} \mathrm{d}^3 r' \mathrm{d}^3 r. \tag{10.2}$$

Equation (10.2) has been obtained by exploiting the localized character of $w_n(r)$ since, in principle, the complete expression based on the local electronic density $\rho_n(\mathbf{r})$ reads as follows

$$C_6^{ij} = \frac{3}{32\pi^{3/2}} \int_{r\leqslant r_{\text{cut}}} \int_{r'\leqslant r_{\text{cut}}} \frac{\sqrt{\rho_i(\mathbf{r})} \cdot \sqrt{\rho_j(\mathbf{r}')}}{\sqrt{\rho_i(\mathbf{r})} + \sqrt{\rho_j(\mathbf{r}')}} \mathrm{d}^3 r' \mathrm{d}^3 r \tag{10.3}$$

but suffers from the fact that the total $\rho(\mathbf{r})$ cannot be rigorously defined locally.

Equation (10.2) allows bridging the gap between two opposite methodologies, a first one (*i*) based on the addition to the Kohn–Sham total energy of the dispersion parts having a form totally independent on changes in the electronic structure (dispersion coefficient not evolving with the electronic density in the DFT framework) and a second one (*ii*) including the dispersion contribution from the beginning in a theoretical framework where the dispersion part is neither added or separately considered *a posterori*. Equation (10.2) introduces the information on the electronic structure directly in the dispersion coefficients (making it prone to be used with any XC functional) by avoiding more elaborate and less tractable treatments of the long-range interactions. In particular, the construction based on localized orbitals is ideally suited to follow tiny electronic effects resulting from changes in the atomic configurations.

Part of the calculations on glassy $GeTe_4$ have also been based on a third approach (in addition to D2/D3 and MLWF): non-local (correlation) functionals able to account for the long-range character of the dispersion interactions. We have alluded to these methods above when mentioning functionals that contain dispersion effects by construction. The so-called VV10 method has been further refined and made accessible to the plane-waves basis set in a further implementation, to which we shall refer to as the rVV10 one [19, 20]. The idea of these approaches is to do away with what is left of the empirical character of the separation between total energy and dispersion part by achieving non locality built in the functional at a limited computational cost so as to afford extended FPMD trajectories. In what follows, we focus on glassy $GeTe_4$ in three different steps. In the first, PBE and BLYP functionals are employed with and without the D2 recipe by Grimme. In the second, we also consider the MLWF approach. Finally, the investigation is completed with the consideration of the D3 and rVV10 recipes. For all of them, FPMD trajectories have been produced independently, providing one of the few available cases of a system considered at finite temperature with as many as five different approaches for the dispersion forces within DFT. As an addition, after the first set of results on glassy $GeTe_4$ (section 10.2), we shall also apply the D2 and MLWF methodologies to the case of liquid $GeSe_2$, reaching conclusions on the limited effect of the dispersion corrections for this particular case.

10.2 Functional and dispersion forces: four models to understand their impact on glassy $GeTe_4$

As a further example of network organization for concentrations $x \sim 0.2$ in A_xX_{1-x} chalcogenides, the structure of glassy $GeTe_4$ in terms of a tetrahedral or an octahedral network (or a combination of the two) has stirred some controversy, motivating the use of predictive atomic-scale modeling [25, 26]. For instance, results by Akola and Jones on $Ge_{15}Te_{85}$ [25] point to Ge atoms fourfold coordinated and Te atoms threefold coordinated, compatible with the existence of voids and the absence of Te segregation. In the glassy state of $Ge_{15}Te_{85}$ there are both tetrahedral and defective octahedral configurations for Ge atoms [27]. The impact of dispersion forces in liquid $Ge_{15}Te_{85}$ and glassy $GeTe_4$ received some preliminary consideration as a worthwhile issue to be addressed in [7] and [26], prompting two considerations. First, it appears that dispersion forces do not have to be neglected at least in the case of some Ge_xTe_{1-x} disordered system. Second, the level of reliability of the FPMD approach based on DFT is lower when moving from disordered Ge_xSe_{1-x} systems to their Ge_xTe_{1-x} counterpart.

In [28], glassy $GeTe_4$ has been studied by resorting to *four* different options for the XC functional and the dispersion forces: PBE as XC functional with no vdW accounted for (named PBE hereafter) [29, 30], PBE as XC functional with vdW included (PBE–vdW), and two analogous approaches with BLYP [31, 32] instead of PBE, without and with dispersion forces. In this section, by inclusion of dispersion forces we mean the D2 Grimme recipe [21].

We started with a periodic cubic box of 185 atoms (37 Ge, 148 Te) of side equal to 18.30 Å for the two PBE schemes, with atomic positions generated at random. This will be called *model 1*. To randomize the system and obtain an amorphous structure two independent sets of trajectories (one for PBE and one for PBE–vdW) were produced at temperature $T = 1000$ K (15.2 ps), $T = 700$ K (15.2 ps), $T = 500$ K (15.2 ps), $T = 300$ K (35.6 ps/38.6 ps). A second 185 atom model (*model 2*) was obtained via those equilibrated in *model 1* with no vdW contributions. Then, 50 ps (46 ps) at $T = 300$ K were produced in the BLYP (BLYP + vdW) frameworks with the same starting configuration. Such *model 2* at room temperature suffers from the lack of important atomic rearrangements due to restricted mobility at that temperature. Therefore, the corresponding results can only be taken as preliminary indications of the impact of the different XC and vdW schemes employed. It appeared necessary to circumvent these limitations in order to obtain a comparison between the PBE and BLYP models totally significant and representative of uncorrelated equilibrium trajectories. For these reasons we constructed a third model *model 3* in the BLYP and the BLYP–vdW approaches, each one having its one independent trajectory. In this case we took 215 atoms (43 Ge, 172 Te) with edge of the cubic box equal to 19.24 Å, by choosing an initial configuration from a previous set of coordinates of glassy $GeSe_4$ [33] and replacing Se atoms with Te atoms. After getting rid of non-zero forces (resulting from the imposed substitutions) via annealing to 0 K, the BLYP and BLYP–vdW structures were brought to 100 K for about 4 ps. Then, the temperature was increased to 300 K with further equilibration lasting 12 ps. A liquid phase was produced with two runs at 600 K and 900 K of 22 and 20 ps, respectively, followed by quench to 600 K for 20 ps and to room temperature for 32 ps.

10.2.1 Total structure factors and pair correlation functions: a first insight

As a first observation, we can consider the results obtained for *model 1* and *model 2*. We recall that for *model 1*, the data give a reliable picture of the PBE case, since the trajectories bear no memory of the initial conditions or of the original liquid state, while this is not the case for *model 2* (BLYP scheme) since there was no significant evolution from the initial PBE configuration at $T = 300$ K. BLYP gives the best results when considering the heights of the maxima and minima, the positions being unaffected by changing from PBE to BLYP (figure 10.1). The vdW corrections

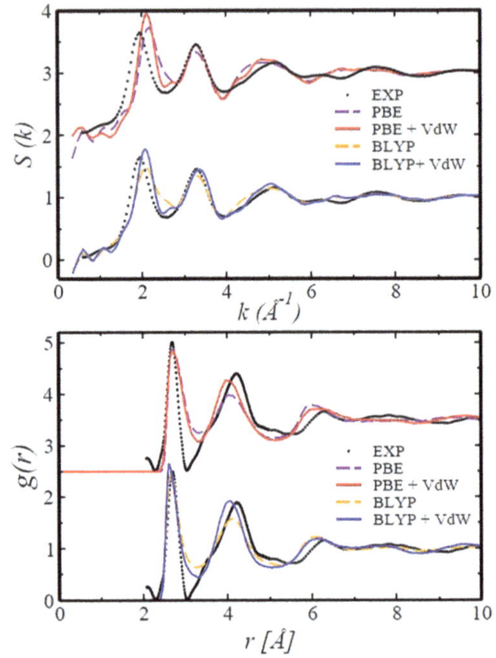

Figure 10.1. Total structure factor $S(k)$ (upper panel) and total pair correlation function $g(r)$ (lower panel) of glassy $GeTe_4$ for *model 1* (PBE) and *model 2* (BLYP). The experimental data [34] are represented by black dots. Reprinted figure with permission from [28]. Copyright (2015) by the American Physical Society.

improve the agreement with measurements for both functionals by consolidating the first minima, especially in the total pair correlation functions $g(r)$ (see figure 10.1). This quantity is in better agreement with experiments [34] for the BLYP case, the overall pattern being even closer to experiments when the vdW corrections are included. This is confirmed by the use of the goodness-of-fit parameter R_χ introduced by Wright [35],

$$R_\chi \equiv \left\{ \frac{\sum_i [g_{\exp}(r_i) - g_{\mathrm{FPMD}}(r_i)]^2}{\sum_i g_{\exp}^2(r_i)} \right\}^{1/2} \tag{10.4}$$

as made explicit by values of R_χ lower in the BLYP case than in PBE (16% (BLYP–vdW) against 22% (PBE–vdW)) when considering the total pair correlation functions. One can perform the same analysis by employing *model 3*, based on BLYP data at $T = 300$ K with a higher statistical accuracy. Our results are given in figure 10.2 for $S(k)$ and $g(r)$, where the PBE results displayed in figure 10.2 are the same as those of figure 10.1. In direct space (pair correlation function), the BLYP–vdW combination is the one performing best, as shown by the behavior of the first minimum. In contrast, in the PBE case the first minimum is much less intense and not substantially corrected by the vdW inclusion. The values of R_χ for this set of results are 17.4% (BLYP), 16.1% (BLYP–vdW), 24.5% (PBE), 21.6% (PBE–vdW).

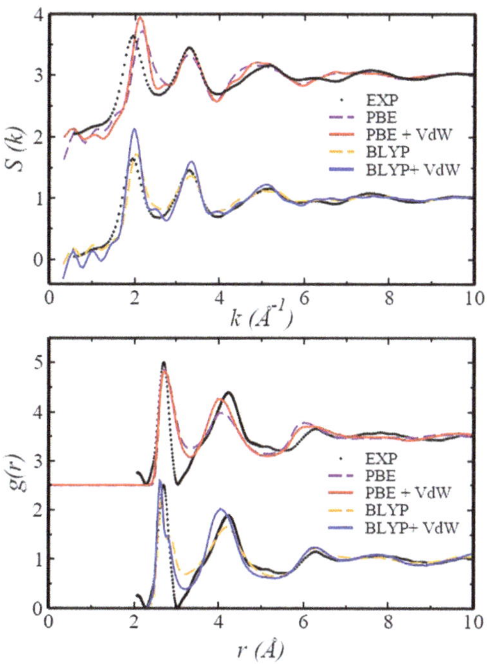

Figure 10.2. Total structure factor $S(k)$ (upper panel) and total pair correlation function $g(r)$ for glassy $GeTe_4$ calculated within *model 1* (PBE) and *model 3* (BLYP). The experimental data [34] are represented by black dots. Reprinted figure with permission from [28]. Copyright (2015) by the American Physical Society.

The behavior of *model 3* in reciprocal space deserves some further thoughts since the agreement with experiments on the main peak intensity is affected by inclusion of the vdW contributions, this peak increasing when compared to the BLYP only case (R_χ is equal to 7.3% (BLYP) and to 7.9% (BLYP–vdW)). This issue can be better understood by keeping in mind that the total structure factors shown in figures 10.1 and 10.2 are the result of Fourier integration from the total pair correlation functions.

As we have seen several times in this monograph and introduced in section 2.3, the most genuine and realistic definition of the structure factors (despite its intrinsically higher statistical noise, especially for results based on a single trajectory) is the one relying on a direct calculation in reciprocal space. Using this definition one obtains R_χ values equal to 10.3% (BLYP) and 9.8% (BLYP–vdW), showing that the vdW part is also a factor contributing to a better description of glassy $GeTe_4$ in reciprocal space (see figure 10.3).

10.2.2 Partial pair correlation functions, bond angles distributions and analysis of local environment

In what follows we shall analyze the atomic structure of glassy $GeTe_4$ as described through *model 1* (PBE, PBE–vdW) and *model 3* (BLYP, BLYP–vdW), *model 2* being statistically less meaningful. There is little to say (see figures 10.4, 10.5) on the partial

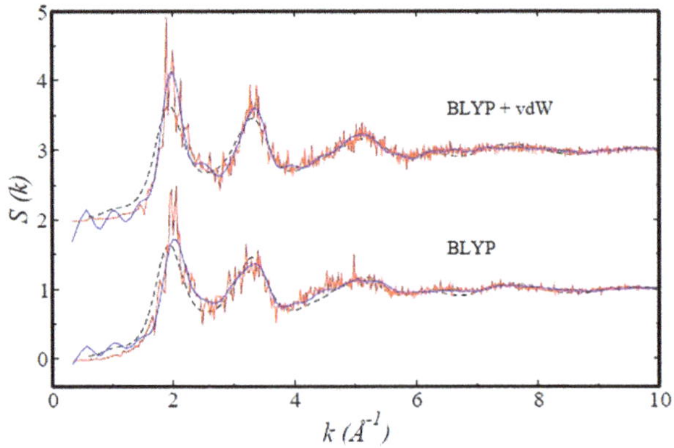

Figure 10.3. Total structure factor $S(k)$ of glassy $GeTe_4$ calculated within *model 3*. The black dots refer to the experiments [34], the solid blue curve refers to the total structure factor via integration from direct space of the total pair correlation functions and the solid red is the total structure factor calculated directly in the reciprocal space. Reprinted figure with permission from [28]. Copyright (2015) by the American Physical Society.

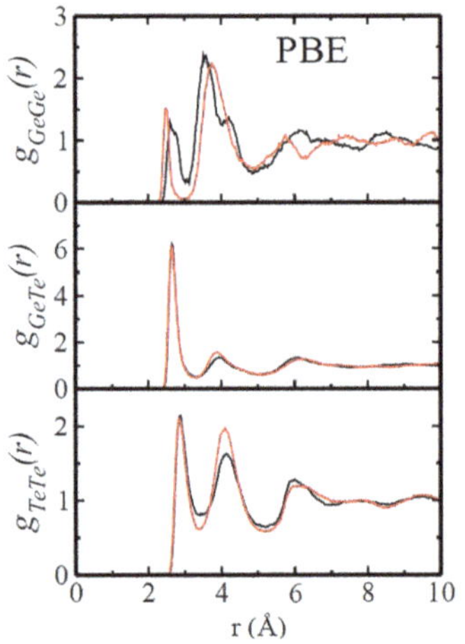

Figure 10.4. Partial pair correlation functions of glassy $GeTe_4$ for $g_{\mathrm{GeGe}}(r)$, $g_{\mathrm{GeTe}}(r)$ and $g_{\mathrm{TeTe}}(r)$ calculated with the XC PBE scheme. Results without VdW are the black lines and with vdW the red lines. Reprinted figure with permission from [28]. Copyright (2015) by the American Physical Society.

pair correlations function $g_{\mathrm{GeTe}}(r)$ that changes very little throughout the four different approaches for the XC and the dispersion contributions. The Ge–Ge correlations are much more affected by the vdW contributions that have a profound effect on the Ge environment. In the PBE case, there is a unique peak at about 4 Å instead of two, this marking an increase of corner-sharing connections (table 10.1).

Also, there are fewer Ge–Ge homopolar since the profile of the peak at 2.5 Å is more narrow. Turning to the BLYP and BLYP–vdW results for $g_{\mathrm{GeGe}}(r)$, vdW brings an increase in the intensities of the peaks while there is no trace of any double peak in between 3.5 Å and 4 Å. As a consequence, homopolar bonds are more

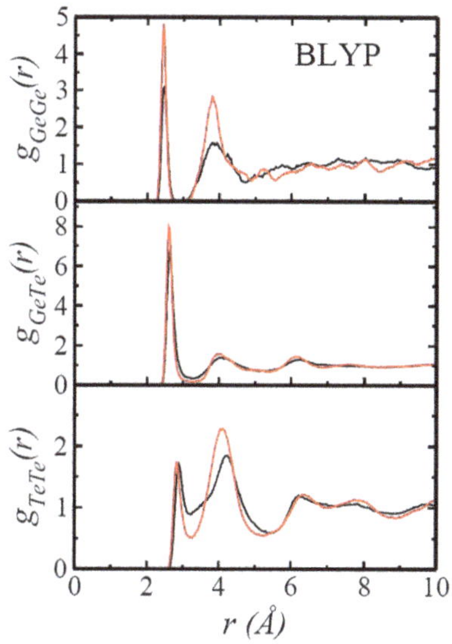

Figure 10.5. Partial pair correlation functions of glassy $GeTe_4$ for $g_{GeGe}(r)$, $g_{GeTe}(r)$ and $g_{TeTe}(r)$ calculated with the XC BLYP scheme. Results without vdW are the black lines and with vdW the red lines. Reprinted figure with permission from [28]. Copyright (2015) by the American Physical Society.

Table 10.1. Ge and Te atoms in homopolar bonds (N_{GeGe} and N_{TeTe}, given as percentages), together with Ge atoms involved in corner-sharing (CS) and edge-sharing (ES) connections and Ge and Te atoms fourfold and twofold coordinated, respectively (also as percentages). Reprinted table with permission from [28]. Copyright (2015) by the American Physical Society.

	N_{GeGe}	N_{TeTe}	N_{Ge}^{CS}	N_{Ge}^{ES}	Fourfold Ge	Twofold Te
PBE	26.3	85.6	34.7	39.0	68.6	57.5
PBE+VdW	16.0	87.9	65.6	18.4	67.3	55.8
BLYP	25.9	73.6	62.8	11.3	76.8	64.8
BLYP+VdW	30.3	72.5	53.0	16.6	85.9	73.9

numerous while the number of Ge atoms involved in corner-sharing connections decreases, affecting the network topology.

The pair correlation function $g_{TeTe}(r)$ is highly sensitive to the changes of the XC functional, especially when comparing BLYP and PBE. The number of Te–Te homopolar bonds is higher in the PBE case, but the overall shape is not altered significantly after inclusion of the vdW correction. The opposite occurs when using BLYP, for which the vdW contributions result in a deeper first minimum in between two fairly sharp peaks. On the basis of these observations, the PBE and BLYP Te sub-networks are more dissimilar than the corresponding Ge sub-networks, since there are more differences observed between PBE and BLYP for properties calculated based on the positions of the Te atoms. We have collected in table 10.2 information on coordination numbers (for each species and average) and nearest

Table 10.2. Bond distances r_{ij} (in Å) (position of the first maximum of the pair correlation functions), partial coordination numbers n_α^β, coordination numbers n_α and average coordination number $\bar{n}$. Note that for a tetrahedrally chemical ordered network $\bar{n}$ is equal to 2.4. The results of PBEsol+vdw are from [26]. Reprinted table with permission from [28]. Copyright (2015) by the American Physical Society.

	PBE	PBE+vdW	BLYP	BLYP+vdW	PBEsol+vdW
r_{GeGe}	2.67	2.49	2.46	2.43	2.48
r_{GeTe}	2.65	2.65	2.62	2.59	2.64
r_{TeTe}	2.88	2.83	2.89	2.84	2.90
n_{Ge}^{Ge}	0.27	0.16	0.28	0.37	0.33
n_{Te}^{Ge}	1.07	1.05	0.96	0.91	0.96
n_{Te}^{Te}	2.71	2.47	1.60	1.40	1.94
n_{Ge}	4.54	4.35	4.14	3.97	4.17
n_{Te}	3.78	3.51	2.57	2.31	2.90
$\bar{n}$	3.93	3.68	2.87	2.65	3.15

neighbor bond distances by following the definitions introduced in section 2.4.1 (note that $n_{GeTe} = 4n_{TeGe}$). Values of the bond distances are within 2% for the four schemes, with the exception of the largest r_{GeGe} distance found in the PBE case (differing by 7%). Focusing again on the Te subnetwork, the comparison between BLYP and PBE reflects different short range environments ($n_{Te}^{Te} = 2.71$, 2.47 with PBE vs $n_{Te}^{Te} = 1.60$, 1.40 with BLYP), with a reduction occurring when considering the dispersion contribution. The large values of n_{Te}^{Te} in the PBE case (2.71, 2.41) are a sign of strong departures from a tetrahedral network ($n_{Ge}^{Ge} = 0$, $n_{Ge}^{Te} = 4$, $n_{Te}^{Te} = 1$, $n_{Te}^{Ge} = 1$). On the other hand, the BLYP values for n_{Te}^{Te} (1.60, 1.40) are closer to 1, indicating a marked tendency toward a tetrahedral arrangement when using BLYP. The notion of BLYP enhancing the tetrahedral arrangement is reminiscent of what we discussed in section 6.7, where we showed that BLYP promotes this kind of short range order when compared to other XC functionals (as the Perdew–Wang one) based on the local density approximation as reference starting point [36].

In table 10.1 we notice that the number of Te atoms forming homopolar bonds is larger in the PBE case. This follows from the values of the n_{Te}^{Te} coordination numbers in between 2 and 3. As mentioned above, BLYP increases the tetrahedral order, leading to a number of Te atoms in homopolar bonds that decreases from ~85% to about 70%. This result is consistent with a chemically ordered network of the same concentration (glassy $GeSe_4$) for which we found a fraction of homopolarly bonded Se–Se atoms close to 70% [33, 37] (see also chapter 8).

The bond angle distributions (see section 2.4.2) accounting for Ge–Te–Ge angles (figure 10.6) contain information on the relative weight of corner- and edge-sharing tetrahedra. The consideration of vdW forces within PBE has the effect of increasing the number of corner-sharing connections observed in the BLYP approach (see table 10.1). Without vdW contributions, one peak is visible at about 80° together with a shoulder at ~110 °. Due to inclusion of vdW, the peak position is shifted to

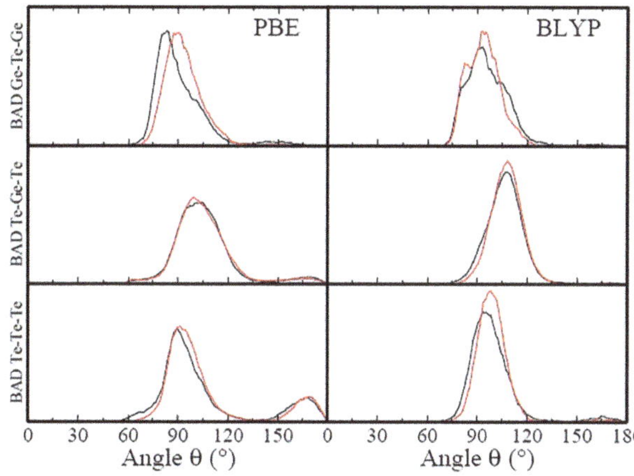

Figure 10.6. Bond angle distribution (BAD) Ge–Te–Ge, Te–Ge–Te and Te–Te–Te. Left panels refer to PBE calculations (*model 1*) and right panels to BLYP calculations (*model 3*). Black lines correspond to the neglect of vdW corrections and red lines to the inclusion of vdW corrections. Reprinted figure with permission from [28]. Copyright (2015) by the American Physical Society.

~90°, to signify more corner-sharing connections. When comparing the BLYP and the BLYP–vdW results one can also detect some modifications occurring in the profile of the Ge–Te–Ge bond angle distribution, as a peak arising from edge-sharing linkages, due to a higher overall ratio edge/corner. Also, the second peak is followed by a more rapid decrease, showing that connections other than tetrahedral have less impact in the BLYP–vdW case. Let's focus on the bond angle distribution Te–Ge–Te. This quantity is very much different when comparing PBE to BLYP, since in the latter there is a main peak located at a higher value (108°), standing for tetrahedral configurations. In the PBE–vdW case there is an additional peak quite small at 180° indicative of fivefold connected Ge atoms, somewhat precursor of octahedral-like structures. Finally, for the bond angle distribution Te–Te–Te one single peak appears in the BLYP–vdW case, while PBE has two mains peaks at around 90° and 180°. These peaks can be associated to Te chains with no less than three Te atoms. Again, there is a peak at 180° angle that is a signature of larger octahedral contributions.

In figure 10.7 we show the distribution of the different coordinations numbers for *n*-fold Ge and *n*-fold Te atoms in the BLYP and PBE cases. The number of miscoordinations is larger for PBE, marking the main difference between PBE and BLYP. Fourfold coordinated Ge sites are predominant in both BLYP–vdW and PBE–vdW, for which the percentages of threefold contributions are similar. However, PBE–vdW (as PBE) has more fivefold Ge atoms than BLYP. The most interesting information contained in figure 10.7 concerns the coordination numbers for Te. The twofold coordination is less present than the threefold one in both PBE and PBE–vdW. Overall, threefold, fourfold and even fivefold Te atoms have been found with twofold ones. In regard to the specific impact of the vdW forces, a higher structural organization is noticeable for both the XC functionals, as exemplified by a higher number of of fourfold Ge and twofold Te atoms.

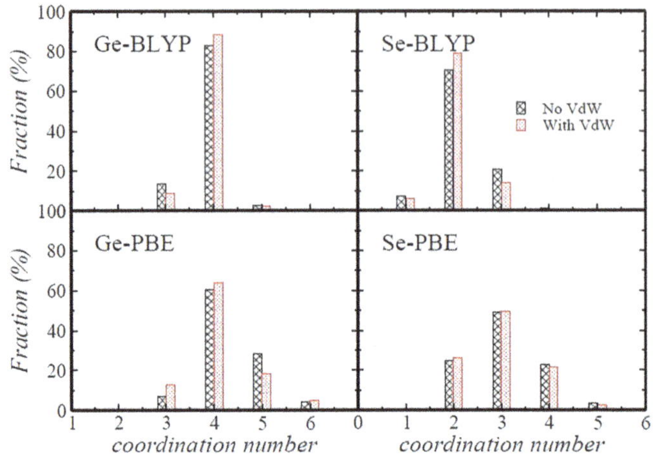

Figure 10.7. Fraction of *n*-fold Ge and *n*-fold Te atoms ($n = 1, 2, 3, 4, 5$ or 6) as calculated within BLYP *model 3* and PBE *model 1*. Reprinted figure with permission from [28]. Copyright (2015) by the American Physical Society.

The local order parameter **q** [38, 39] defined in section 2.4.2 (and taking a value of $\mathbf{q} = 1$ for an ideal tetrahedral network, while $\mathbf{q} = 0$ corresponds to sixfold coordinated octahedral sites) was also employed to shed some light into the structural organization of the network, in view of the fact that we have encountered some deviations from the typical tetrahedral network topologies. Values in between 0 and 1 are typical of a given defective octahedral configurations. The results are given in figure 10.8 for the schemes including vdW contributions, namely the PBE–vdW and BLYP–vdW models.

The PBE results are characterized by the occurrence of several possible structural motifs, as the perfect sixfold octahedral sites ($\mathbf{q} \sim 0$), while defective octahedra are responsible of broad peaks overlapping at $\mathbf{q} = 0.30$ and 0.45 and $\mathbf{q} = 0.75$. A main peak at $\mathbf{q} = 0.95$ is representative of tetrahedral configurations. In contrast, octahedral sites are not found in BLYP calculations as shown by the fact that the distribution goes to zero for $\mathbf{q} < 0.18$. Defective octahedra are responsible for a broad feature around $\mathbf{q} = 0.38$. The high intensity of the peak at $\mathbf{q} = 0.95$ demonstrates the strong stability of tetrahedra for this glassy material within the BLYP description, pointing out clear differences between PBE and BLYP structural organizations. In fact, octahedral coordinations are more favored within PBE.

10.2.3 Electronic properties and link with the structure

We follow the same procedure based on Wannier function and centers (WFCs) (see section 9.4.3 for instance), to elucidate the relative weights of the covalent and ionic contributions to bonding. To establish some useful analogy and interpret the presence of the observed features, we rely on the relationship between peculiar distances in the pair correlation functions and specific WFCs as we did for $g_{\mathrm{GeW}}(r)$ (glassy GeS_4 or glassy $GeSe_4$), $g_{\mathrm{SeW}}(r)$ (glassy $GeSe_4$) and $g_{\mathrm{SW}}(r)$ (glassy GeS_4) in section 9.4.3 when considering glassy $GeSe_4$ and glassy GeS_4.

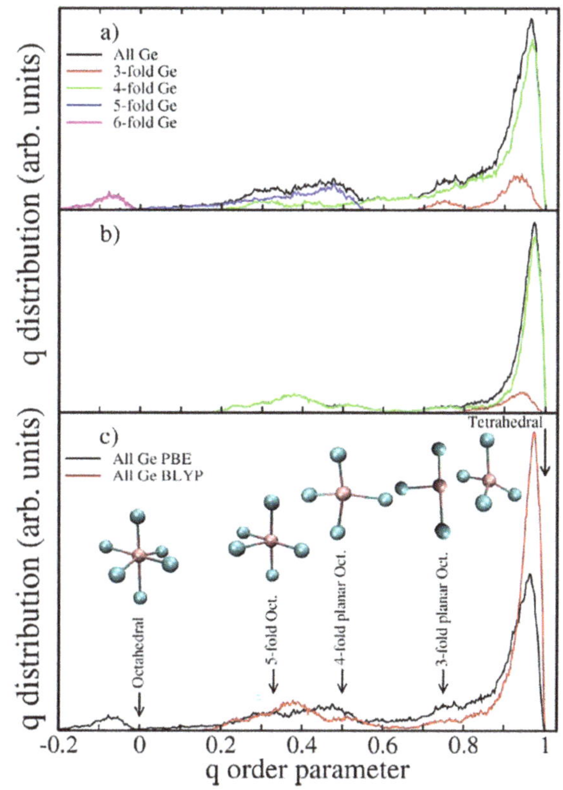

Figure 10.8. Distribution of the local atomic coordination motifs as a function of **q**. Panel (a) (PBE–vdW) *model 1*, Panel (b) *model 3* (BLYP–vdW). The colored lines show the contributions of *n*-fold Ge atoms ($n = 3$, 4, 5 and 6). In panel (c) we compare the results for *model 1* (PBE–vdW) and *model 3* (BLYP–vdW). The arrows shows the exact **q** value of tetrahedral, octahedral and some defective octahedral geometries. The distribution is normalized with respect to the number of Ge atoms. Reprinted figure with permission from [28]. Copyright (2015) by the American Physical Society.

We recall that three peaks were found in $g_{\mathrm{Se(S)W}}(r)$ when describing glassy $GeSe_4$ and glassy GeS_4, due to Ge–Se(S) bonds, homopolar Se–Se (S–S) bonds and lone pair valence electrons (see the inset of figure 10.9). Along the same lines, in figure 10.9 the $g_{\mathrm{TeW}}(r)$ pair correlation function issued from the BLYP–vdW and PBE–vdW frameworks is shown. The analogies with the shape of $g_{\mathrm{SeW}}(r)$ and $g_{\mathrm{SW}}(r)$ obtained in [40] are striking, indicating that glassy $GeTe_4$-BLYP is to a large extent a tetrahedral network. In the PBE case, $g_{\mathrm{WTe}}(r)$ has a shape with no double peak in between 1 Å and 1.5 Å, as if there were a larger variety of Ge–Te and Te–Te distances to be accounted for, as it occurs in the presence of a coexistence between tetrahedral and octohedral motifs.

10.2.4 What to learn about impact of dispersion forces and total energy schemes

We have determined in the case of glassy $GeTe_4$: (a) the role of vdW dispersion forces for each one of the two XC functionals (BLYP and PBE) and (b) the role of the XC functional with or without the use of the vdW dispersion part. We reiterate

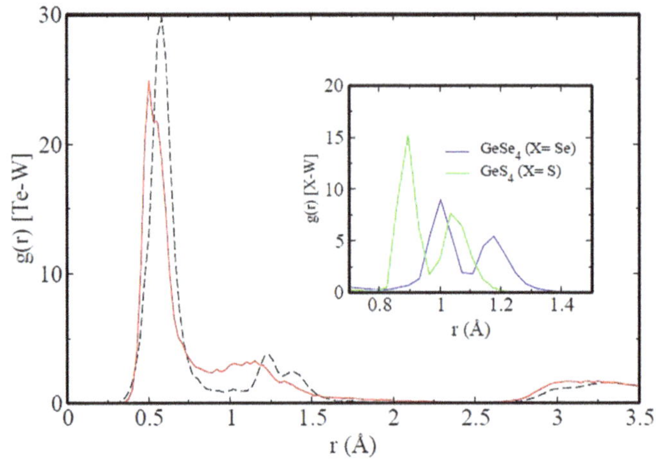

Figure 10.9. Partial pair correlation functions $g_{\mathrm{TeW}}(r)$ (W are the coordinates of the Wannier centers). Black dashed lines refers to BLYP+vdW and red solid lines to PBE+vdW calculations. The double peak in $g_{\mathrm{SeW}}(r)$ and $g_{\mathrm{SW}}(r)$ obtained in [40] is highlighted in the inset. Reprinted figure with permission from [28]. Copyright (2015) by the American Physical Society.

that by inclusion of dispersion forces we mean, in the specific section, the inclusion of the D2 Grimme recipe [21]. This strategy will be enlarged and amplified in the next sections of this chapter. Concerning (a), vdW contributions are useful to obtain better total pair correlations function and total structure factors (when compared to experiments). Looking at (b), the best performances is obtained with the BLYP XC functional, and even more when combined with the vdW forces. As a revealing property to capture the structural differences between the four models, one can take the total coordination number n_{ave} for PBE, PBE+vdW, BLYP, BLYP+vdW. n_{ave} is equal to 3.93, 3.68 (PBE) compared to 2.87, 2.65 (BLYP), approaching the reference value for a chemically ordered tetrahedral network, 2.4.

Globally, as a final comment to this first part on the effect of dispersion forces, the impact of these contributions cannot be neglected. However, adding the vdW forces does not ensure by itself the attainment of optimal results for a given DFT scheme. In other words, the impact of the XC functional turns out to be stronger. In the case of glassy $GeTe_4$, BLYP is once again the option of choice when comparing to experiments (pair correlation functions and structure factors). This choice corresponds to a defective tetrahedral network contrasting with the coexistence of octahedral and tetrahedral topologies.

10.3 Dispersion forces and disordered $GeSe_2$: can we make any progress?

We have seen in chapter 6 how liquid $GeSe_2$ and glassy $GeSe_2$, in addition to their intrinsic interest as disordered network-forming materials, can be quite useful to assess the validity of the whole FPMD computational scheme based on DFT and suitable choices of the XC functionals. We recall that we were able to achieve better performances in terms of comparison with experimental results for the structure by replacing LDA with the generalized gradient approximation (GGA) and, within GGA, by replacing PW with BLYP. This section is intended to pursue this kind of benchmark investigation by adding dispersion forces to the FPMD–BLYP scheme employed for liquid $GeSe_2$. The procedure followed is based on the comparison between four sets of results: experiments (extensively referred to in chapter 6), BLYP results (NvdW in what follows), BLYP with the dispersion contributions in the Grimme-D2 form [21] (vdW1) and BLYP with the MLWF contribution [23, 24] (vdW2). To this end, three new, independent trajectories extending at equilibrium over 30–40 ps have been created for a system of 240 atoms at T = 1050 K at experimental density [41] by following the principles of the FPMD theoretical scheme described and employed throughout this monograph.

We show in figure 10.10 the partial structure factors for liquid $GeSe_2$ as resulting from our three sets of calculations. Despite some small differences in $S_{\mathrm{SeSe}}(k)$ and $S_{\mathrm{GeGe}}(k)$, comparable to the statistical errors on the calculations (a few per cent), it is hard to detect any clearcut evidence on the possible occurrence of structural changes due to dispersion forces. As a consequence, changes in the behavior of the FSDP in $S_{\mathrm{CC}}(k)$ are also very minimal (see for details [41]), ruling out any substantial impact of dispersion forces on the intermediate range order in this system.

One can develop the same kind of considerations in direct space via the analysis of figure 10.11 that contains the pair correlation functions $g_{\mathrm{GeGe}}(r)$, $g_{\mathrm{GeSe}}(r)$ and $g_{\mathrm{SeSe}}(r)$. The corresponding coordination numbers (Ge–Se case) are ($n_{\mathrm{Ge}}^{\mathrm{Se}}(\mathrm{exp}) = 3.5$, $n_{\mathrm{Ge}}^{\mathrm{Se}}(\mathrm{NvdW}) = 3.7$, $n_{\mathrm{Ge}}^{\mathrm{Se}}(\mathrm{vdW1}) = 3.7$, $n_{\mathrm{Ge}}^{\mathrm{Se}}(\mathrm{vdW2}) = 3.6$), demonstrating the little sensitivity sensitivity of Ge–Se correlations to dispersion forces. This result is not totally unexpected since in liquid $GeSe_2$ the predominant bonding contributions are the ionic and covalent ones, thereby reducing the impact of long-range residual interactions.

For Se–Se one has $n_{\mathrm{Se}}^{\mathrm{Se}}(\mathrm{exp}) = 0.23$. Our calculations give $n_{\mathrm{Se}}^{\mathrm{Se}}(\mathrm{NvdW}) = 0.20$, $n_{\mathrm{Se}}^{\mathrm{Se}}(\mathrm{vdW1}) = 0.19$, $n_{\mathrm{Se}}^{\mathrm{Se}}(\mathrm{vdW2}) = 0.25$. These numbers together with the profiles in figure 10.11 concur to point a large agreement among NovdW, vdW1 and vdW2. However, the Grimme recipe (vdW1) exhibits the largest disagreement with experimental prediction in terms of the second main peak at about 4 Å. Turning to $g_{\mathrm{GeGe}}(r)$ we have learned that this case is highly intriguing, since theory is somewhat unable to approach experiments as it does for the other correlations (see section 6.7), despite a broad agreement on the coordination number relative to homopolar bonds. We have $n_{\mathrm{Ge}}^{\mathrm{Ge}}(\mathrm{NvdW}) = 0.18$, $n_{\mathrm{Ge}}^{\mathrm{Ge}}(\mathrm{vdW1}) = 0.19$, $n_{\mathrm{Ge}}^{\mathrm{Ge}}(\mathrm{vdW2}) = 0.23$ against $n_{\mathrm{Ge}}^{\mathrm{Ge}}(\mathrm{exp}) = 0.25$. Let us have a look at the intensities of the second and of the third peak of $g_{\mathrm{GeGe}}(r)$. They are clearly different in the vdW1 case, a third peak exceeding any hypothesized error bar, meaning an excess of corner-sharing tetrahedra within the vdW1 description. In contrast, MLWF does not differ from the NvdW result, showing better performances than vdW1 for these particular correlations.

To complement this information we can rely on the number of Ge atoms in corner-sharing and edge-sharing connections, $N_{\mathrm{Ge}}(\mathrm{CS})$ and $N_{\mathrm{Ge}}(\mathrm{ES})$ respectively and the number of Ge atoms forming Ge–Ge contacts $N_{\mathrm{Ge-Ge}}$ (see [43] where it was proposed that $N_{\mathrm{Ge}}(\mathrm{CS}) = 1 - N_{\mathrm{Ge}}(\mathrm{ES}) - N_{\mathrm{Ge-Ge}}$).

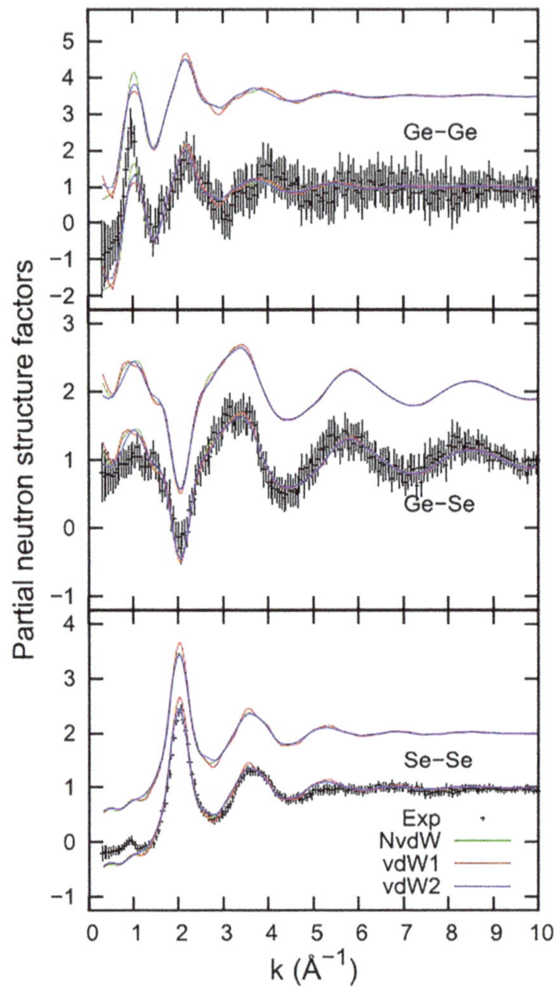

Figure 10.10. The partial structure factors of liquid $GeSe_2$ at $T = 1050$ K. Experimental data from [42] (crosses with error bars) are compared to results for the models NvdW (green line), vdW1 (red line) and vdW2 (blue line). For sake of clarity, we made the choice of including (lower curve) or not including (higher curve) the experimental results in the comparison with the calculations. Reprinted figure with permission from [41]. Copyright (2017) by AIP Publishing.

The values of N_{Ge}(CS) and N_{Ge}(ES) (see table 10.3) allow obtaining the ratios $R^{NvdW}_{CS/ES} = N_{Ge}(CS, NvdW)/N_{Ge}(ES, NvdW) = 0.8$ and in an analogous way, $R^{vdW1}_{CS/ES} = 1.1$, $R^{vdW2}_{CS/ES} = 0.87$. Close amounts of ES and CS connections are predicted by experiments [42], while a tendency to values larger than 1 is found in vdW1. Therefore, $R^{vdW1}_{CS/ES} = 1.1$ expresses an artificial contribution induced in the FPMD model by Grimme-D2.

To summarize the pieces of evidence collected when studying the influence of dispersion forces (Grimme-D2 and MLWF) on liquid $GeSe_2$, we can conclude that the

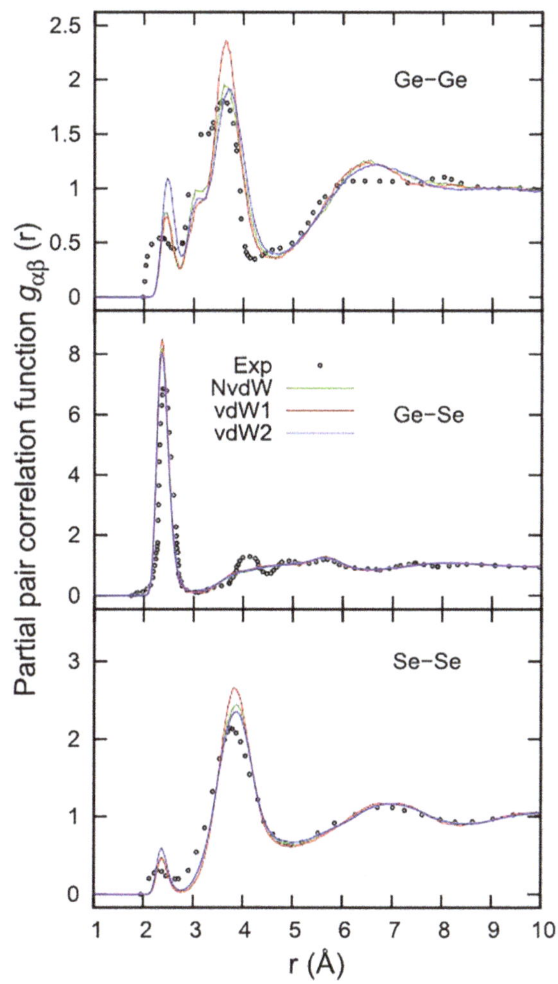

Figure 10.11. The partial pair correlation functions $g_{GeGe}(r)$ (top panel), $g_{GeSe}(r)$ (middle panel) and $g_{SeSe}(r)$ (bottom panel) at $T = 1050$ K. Open small circles are the experimental data from [42]. They are compared with results for the models NvdW (green line), vdW1 (red line) and vdW2 (blue line). Reprinted figure with permission from [41]. Copyright (2017) by AIP Publishing.

Table 10.3. Percentages of Ge atoms in edge-sharing, corner-sharing and homopolar bonding configurations for liquid $GeSe_2$ as reported in [41]. Reprinted table with permission from [41]. Copyright (2017) by AIP Publishing.

	N_{Ge}(ES)	N_{Ge}(CS)	N_{Ge-Ge}
NvdW	45	36	18
vdW1	39	43	18
vdW2	41	36	23

set of standard structural properties considered here are not altered significantly by either one of these contributions. However, specific indications have been collected. On the one hand, the Grimme-D2 recipe approach manifests itself through some spurious effects in the partial pair correlation functions and an overestimate of corner-sharing connections. On the other hand, the maximally localized Wannier scheme does not modify the results obtained in the absence of dispersion forces. The question arises on whether such results are of general applicability, for instance to other chalcogenide systems such as $GeTe_4$. This issue will be addressed in the next section.

10.4 How to select the best dispersion prescription for glassy $GeTe_4$? Part I

We have seen in the previous section that the MLWF approach reproduces existing first-principles atomic structures of liquid $GeSe_2$ while the Grimme-D2 scheme is not equally performing in the case of the Ge–Ge pair correlation function [41]. Given these findings, two considerations are in order. On a general side, different schemes for the dispersion forces might not act in the same manner on the structure of disordered chalcogenides. More specifically, a specific behavior arises when the vdW coefficients changes in time with the electronic structure, as in the MLWF approach. In view of these thoughts, it is worthwhile to pursue the study on the role of dispersion forces by applying to the case of glassy $GeTe_4$ the calculation strategy implemented above for liquid $GeSe_2$. This has been achieved in [44]. To this purpose we have produced a new trajectory (same system size and statistics as in [28]) with the BLYP exchange–correlation functional but using instead the MLWF methodology. These new results [44] have been compared to experiments and to the Grimme-D2 results of section 10.2 and [28]. Figure 10.12 contains the experimental data [34] for the total pair correlation function as well as the data presented in

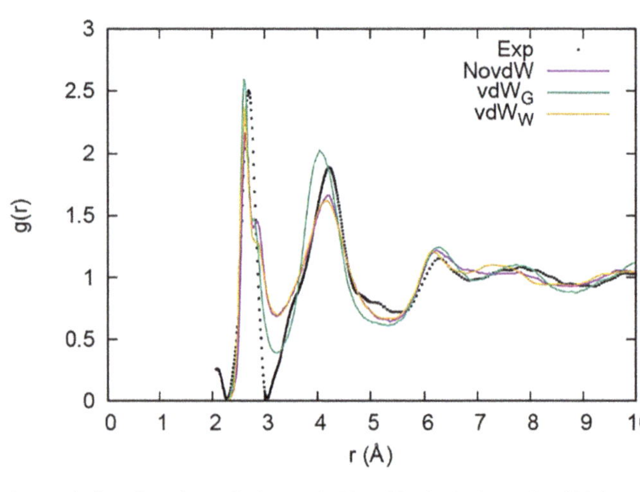

Figure 10.12. Total correlation function of glassy $GeTe_4$. No inclusion of vdW interactions: purple line, inclusion of vdW interactions following the Grimme-D2 scheme: green line, vdW interactions in the MLWF formalism: orange line. Experiments are from reference [34]. Reprinted figure with permission from [44].

section 10.2, enriched by calculations performed in the MLWF framework (we termed them NovdW, no dispersion included, vdW_G, Grimme-D2 [28] and vdW_W (MLWF recipe), respectively). It turns out that vdW_W and NovdW data are very close, pointing out a sort of artificial contribution affecting the use of the vdW_G recipe. However, in terms of comparison with experimental data, vdW_G is the closest among the three, especially if one looks at the first minimum of the total pair correlation function.

To complete this analysis in terms of distances between pairs of atoms, we can consider the partial pair correlation functions (figure 10.13).

In the Ge–Te partial pair correlation function there are essentially no differences between the three methodologies, this meaning that short range interactions between atoms playing (in a qualitative description) the role of cations and anions are not affected by the dispersion contributions. In the Ge–Ge case, there are small variations relative to homopolar bonds when comparing NovdW, vdW_G and

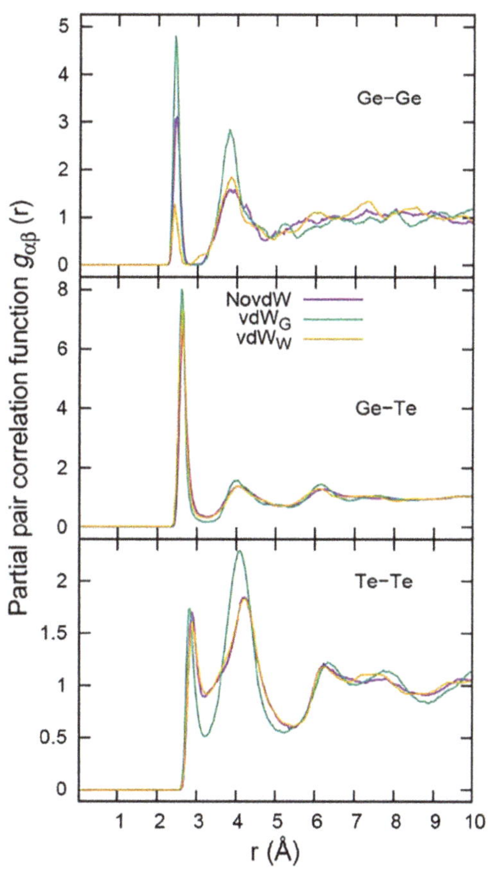

Figure 10.13. Partial pair correlation functions of glassy $GeTe_4$. Results with no inclusion of vdW interactions (purple line) and with inclusion of the Grimme-D2 scheme (green line) are compared to those including the MLWF formalism (orange line). Reprinted figure with permission from [44].

vdW_W. However, the most intense peak is found in vdW_G. Also, around $r \simeq 4$ Å, vdW_G features the highest second peak. The g_{TeTe} pair correlation function exhibits a deep first minima in the vdW_G case, at the opposite of vdW_W that is superposed to the NovdW result. Overall, vdW_W and NovdW are very close, thereby differing from vdW_G and confirming the indications provided by the work on liquid $GeSe_2$. However, and in a quite intriguing way, this latter set of results were found in better agreement with experimental data (see figure 10.12).

In order to fully appreciate the limits and the impact of these findings, one should keep in mind that they are based on only one equilibrium trajectory for glassy $GeTe_4$ for each of the different schemes employed for the dispersion forces. Also, while they are representative of different conceptions for the vdW interaction (Grimme-D2: coefficients not depending on the changes of the electronic structure resulting from the dynamical evolution, unlike MLWF) they cannot be taken as conclusive, since alternative approaches as those having a full non-local character [19, 20], mentioned in the introduction, could also be considered. It remains true that we have encountered here a second case of glassy chalcogenide for which vdW_W reproduces the results with no dispersion forces, as if they were somehow unnecessary. This could be a drawback leading to artifacts when dispersion recipes are used to complete the total energy Hamiltonian. A more exhaustive set of calculations on these issues were planned in this context and are presented below [45].

10.5 How to select the best dispersion prescription for glassy $GeTe_4$? Part II

The purpose of this section is to achieve a more complete assessment of the impact of the dispersion forces on glassy $GeTe_4$. This can be achieved by considering more extensively the behavior of schemes adapting to changes in the electronic structure as well as more empirical ones. Also, it is important to obtain some indications on whether dispersion schemes contribute to the atomic structure even when they are not deemed to do it, since other predominant interactions make their presence unimportant. For instance, in the case of glassy $GeTe_4$, the weight of iono-covalent interactions is expected to be largely superior. In [45] two additional vdW schemes have been considered for a total of 11 independent FPMD trajectories of glassy $GeTe_4$: the Grimme D3 expression (vdW_{D3} hereafter), known for ensuring better accuracy and a wider range of applications than D2 [22] and the rVV10 functional that is not characterized by any separation between the DFT Kohn–Sham functional and the dispersion part (vdW_{VV}) [20, 46, 47]. This allows dealing with dispersion schemes belonging to either one of there three categories: those based on coefficients (a) accounting (MLWF) or (b) not accounting (BLYP-D2 and BLYP-D3) for changes in the electronic structure along the time trajectories and (c) the rVV10 method including dispersion forces via a non-local construction built in the energy functional.

By concentrating on the total pair correlation function $g_T(r)$ one can establish, in a comparative fashion, the degree of accuracy and the predictive power of the different vdW schemes. To this purpose, our choice was not to include short range contributions for distances lower than the first minimum of the pair correlation function (typically 3 Å). Therefore, we shall not comment on the shape of $g_T(r)$ in the indicated range, not meaningful to appreciate the role of the dispersion forces. For $g_T(r)$, by focusing on a comparison among all cases, we introduce the function Δ (exp,tot) equal to the difference between $g_T(r)$ and the experimental counterpart, and this for each choice of the dispersion forces. Similarly, Δ(NovdW, tot) is the difference between all calculated $g_T(r)$ and the NovdW result. The Δ functions are used to obtain the integrals of their absolute values within [3–9] Å knowing that such quantities (their inverse) allow to infer the level of agreement between the sets of data to be compared.

Focusing on $g_T(r)$ (figure 10.14), one notes that vdW_{D3} performs better than vdW_{D2}, both of them improving upon the NovdW scheme. vdW_W is also closer to the experimental data than NovdW while vdW_{VV} is the farthest. These conclusions are exemplified in table 10.4 and in figure 10.15 (Δ(exp, tot) function).

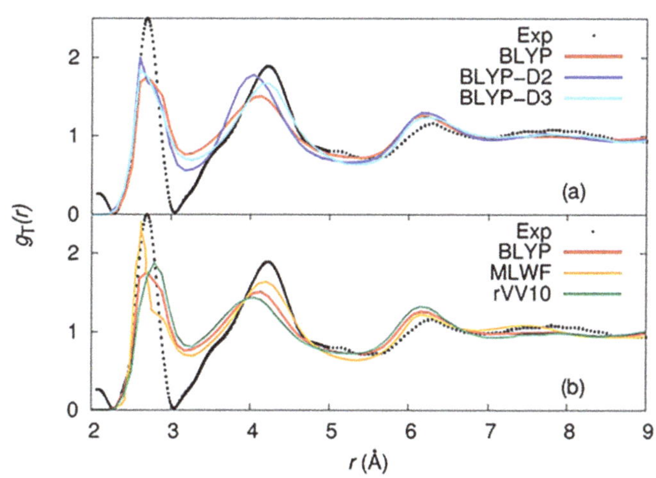

Figure 10.14. Comparing the five different dispersion schemes with experiments (from [34]) for the total pair correlation function. Reprinted figure with permission from [45]. Copyright (2020) by the American Chemical Society.

Table 10.4. Integral of the absolute value of the differences between a given $g_T(r)$ and the experimental data on the range [3–9] Å. See also [45]. Reprinted table with permission from [45]. Copyright (2020) by the American Chemical Society.

Functional	$I_{\|\Delta(exp,tot)\|}$ (a.u.)
BLYP-D3 (vdW_{D3})	1.044
BLYP-MLWF (vdW_W)	1.125
BLYP-D2 (vdW_{D2})	1.397
BLYP (NowdW)	1.439
rVV10 (vdW_{VV})	1.864

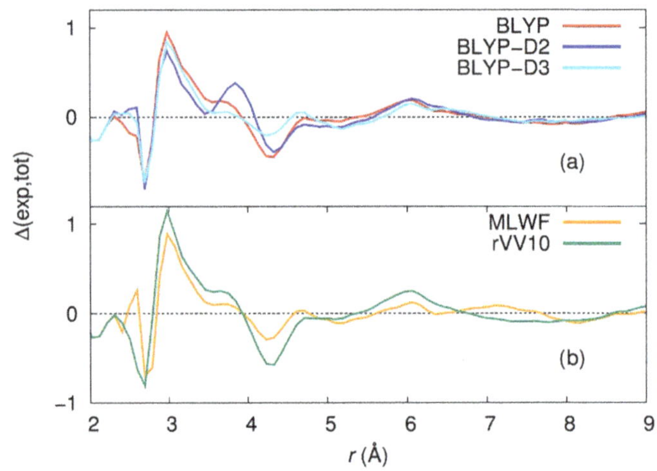

Figure 10.15. $g_T(r)$. Difference functions obtained by subtracting the experimental data from each calculated total pair correlation function. Experiments from [34]. Reprinted figure with permission from [45]. Copyright (2020) by the American Chemical Society.

Table 10.5. Integral of the absolute value of the differences between each one of the calculated $g_T(r)$ and the BLYP (NovdW) data on the range [3–9] Å. Reprinted table with permission from [45]. Copyright (2020) by the American Chemical Society.

Functional	$I_{\lvert\Delta(\text{NovdW,tot})\rvert}$ (a.u.)
rVV10 (vdW_{VV})	0.175
BLYP-D2 (vdW_{D2})	0.515
BLYP-MLWF (vdW_W)	0.684
BLYP-D3 (vdW_{D3})	0.942

So far, we have assumed that we take the experimental data as the reference ones to evaluate the reliability of the different dispersion schemes. However, we can also take as reference data the BLYP ones with no dispersion forces included. Our results are collected in table 10.5, again by exploiting the integrals of the absolute value of Δ (NovdW, tot) (figure 10.16).

The vdW_{VV} results are the closest to NovdW, followed by vdW_{D2}. Also, vdW_W is not too far from these two. This demonstrates that this scheme not only improves upon BLYP approaching the experimental results but it does it without departing drastically from the same BLYP data. Therefore, the dispersion schemes considered here can be divided in two classes, those reducing the differences between the experimental data and the NovdW solution (vdW_{D3}, vdW_W) and those differing less from NovdW than from the experimental data (vdW_{VV} and to a smaller extent vdW_{D2}). Given these results, vdW_W appears as an optimal compromise between these two tendencies, knowing that ideally one would like to do better with respect to the mere BLYP NovdW data and approach experiments as much as possible.

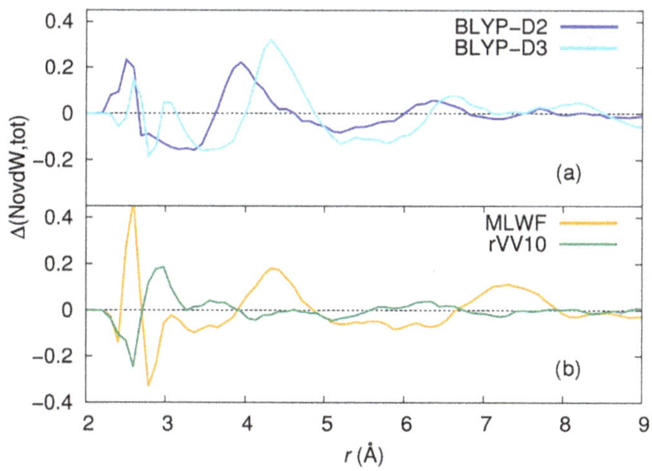

Figure 10.16. Difference functions obtained by subtracting the BLYP (NovdW) data from each one of the other calculated total pair correlation functions. Reprinted figure with permission from [45]. Copyright (2020) by the American Chemical Society.

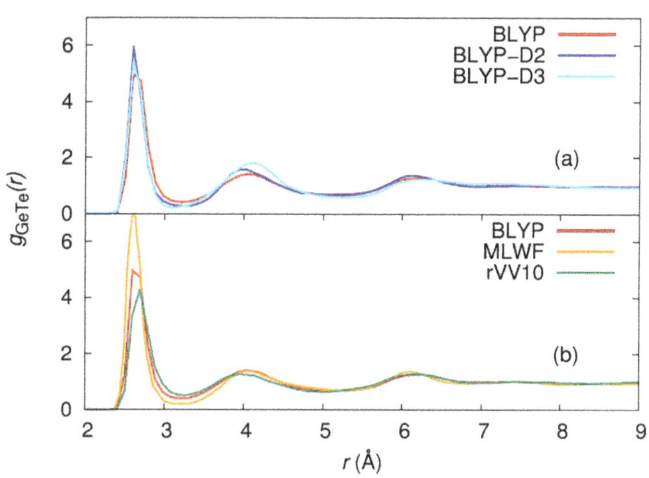

Figure 10.17. GeTe partial pair correlation functions. Reprinted figure with permission from [45]. Copyright (2020) by the American Chemical Society.

We have also employed the partial pair correlation functions to compare the different schemes to BLYP NovdW (figures 10.17, 10.18 and 10.19).

By skipping the case of $g_{GeTe}(r)$ (Figure 10.17) for which the five results are quite similar, $g_{TeTe}(r)$ (figure 10.19) is responsible for the close behavior of NovdW and vdW_{VV} in $g_T(r)$, since this partial pair correlation function has the largest contribution in $g_T(r)$. The largest discrepancies between BLYP NovdW and the other schemes (see figure 10.18) is recorded in $g_{GeGe}(r)$ but this partial pair correlation function is the one with the smallest impact on the total one. Based on

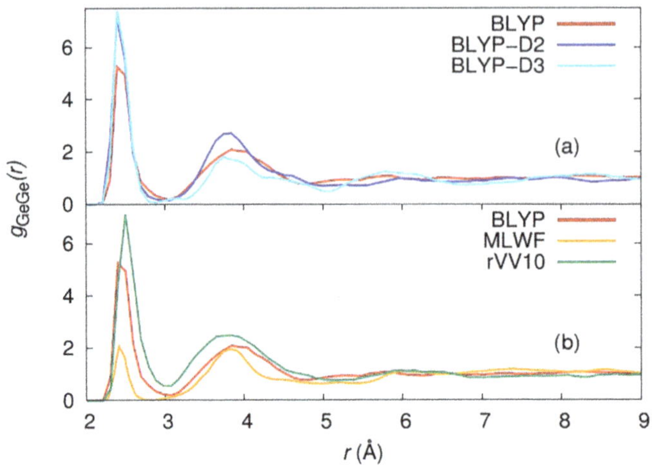

Figure 10.18. GeGe partial pair correlation functions. Reprinted figure with permission from [45]. Copyright (2020) by the American Chemical Society.

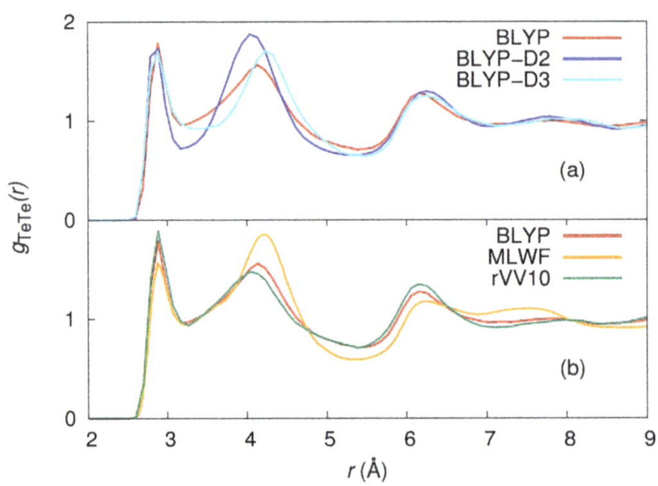

Figure 10.19. TeTe partial pair correlation functions. Reprinted figure with permission from [45]. Copyright (2020) by the American Chemical Society.

these results it appears clear that there are non negligible differences among the four schemes employed above. Worthy of mention is the similarity between NovdW and vdW_{VV} for Te–Te correlations, knowing that in vdW_{VV} the dispersion contribution is treated within the energy functional by not adding a separate term.

To summarize our findings on the effect of dispersion schemes on the structure of glassy $GeTe_4$, we can recall that in terms of agreement with experiments (total pair correlation function), the optimal choices are D3 or MLWF schemes. D3 has a reduced empirical character when compared to D2 since a molecular environment is taken into account via atom-pair specific C_6 coefficients by including some local information on the coordination numbers. It is also interesting to see what happens when the reference results are those obtained without considering dispersion forces (BLYP results). In this case, rVV10 scheme has to be preferred. When searching for a compromise by putting together both criteria (agreement with experiments and with the BLYP results) one can come to the conclusion that the MLWF scheme stands out as the best option. This holds at least for systems in which the modification due to dispersion forces are not expected to be determinant to achieve a realistic description of bonding. The behavior of the partial pair correlation functions confirms this statement.

References

[1] Maurugeon S, Bureau B, Boussard-Plédel C, Faber A J, Lucas P, Zhang X H and Lucas J 2011 *Opt. Mater.* **33** 660–3

[2] Lucas P, Boussard-Pledel C, Wilhelm A, Danto S, Zhang X-H, Houizot P, Maurugeon S, Conseil C and Bureau B 2013 *Fibers* **1** 110–8

[3] Li R, Tang S-Y, Bai G, Yin Q-N, Lan X-X, Xia Y-D, Yin J and Liu Z-G 2013 *Chin. Phys. Lett.* **30** 058101

[4] Jiang H, Guo K, Xu H, Xia Y, Jiang K, Tang F, Yin J and Liu Z 2011 *J. Appl. Phys.* **109** 066104

[5] Chen M, Rubin K A and Barton R W 1986 *Appl. Phys. Lett.* **49** 502–4

[6] Wuttig M and Yamada N 2007 *Nature Mater* **6** 824–32

[7] Micoulaut M 2013 *J. Chem. Phys.* **138** 061103

[8] Anatole von Lilienfeld O, Tavernelli I, Rothlisberger U and Sebastiani D 2004 *Phys. Rev. Lett.* **93** 153004

[9] Zimmerli U, Parrinello M and Koumoutsakos P 2004 *J. Chem. Phys.* **120** 2693–9

[10] Wu X, Vargas M C, Nayak S, Lotrich V and Scoles G 2001 *J. Chem. Phys.* **115** 8748–57

[11] Kristyán S and Pulay P 1994 *Chem. Phys. Lett.* **229** 175–80

[12] Pérez-Jordá J M and Becke A D 1995 *Chem. Phys. Lett.* **233** 134–7

[13] Meijer E J and Sprik M 1996 *J. Chem. Phys.* **105** 8684–9

[14] Kohn W, Meir Y and Makarov D E 1998 *Phys. Rev. Lett.* **80** 4153–6

[15] Rydberg H, Lundqvist B I, Langreth D C and Dion M 2000 *Phys. Rev.* B **62** 6997–7006

[16] Rydberg H, Dion M, Jacobson N, Schröder E, Hyldgaard P, Simak S I, Langreth D C and Lundqvist B I 2003 *Phys. Rev. Lett.* **91** 126402

[17] Hult E, Andersson Y, Lundqvist B I and Langreth D C 1996 *Phys. Rev. Lett.* **77** 2029–32

[18] Andersson Y, Langreth D C and Lundqvist B I 1996 *Phys. Rev. Lett.* **76** 102–5

[19] Vydrov O A and Van Voorhis T 2009 *Phys. Rev. Lett.* **103** 063004

[20] Sabatini R, Gorni T and de Gironcoli S 2013 *Phys. Rev.* B **87** 041108

[21] Grimme S 2006 *J. Comput. Chem.* **27** 1787–99

[22] Grimme S, Antony J, Ehrlich S and Krieg H 2010 *J. Chem. Phys.* **132** 154104

[23] Silvestrelli P L 2008 *Phys. Rev. Lett.* **100** 053002

[24] Silvestrelli P L and Ambrosetti A 2019 *J. Chem. Phys.* **150** 164109

[25] Akola J and Jones R O 2008 *Phys. Rev. Lett.* **100** 205502

[26] Gunasekera K, Boolchand P and Micoulaut M 2014 *J. Appl. Phys.* **115** 164905

[27] Kalikka J, Akola J, Jones R O, Kohara S and T Usuki 2011 *J. Phys.: Condens. Matter* **24** 015802

[28] Bouzid A, Massobrio C, Boero M, Ori G, Sykina K and Furet E 2015 *Phys. Rev.* B **92** 134208

[29] Perdew J P, Burke K and Ernzerhof M 1996 *Phys. Rev. Lett.* **77** 3865–8

[30] Perdew J P, Burke K and Ernzerhof M 1997 *Phys. Rev. Lett.* **78** 1396

[31] Becke A D 1988 *Phys. Rev.* A **38** 3098–100

[32] Lee C, Yang W and Parr R G 1988 *Phys. Rev.* B **37** 785–9

[33] Sykina K, Furet E, Bureau B, Le Roux S and Massobrio C 2012 *Chem. Phys. Lett.* **547** 30–4

[34] Kaban I, Halm T, Hoyer W and Jóvári P 2003 *J. Non-Cryst. Solids* **326-327** 120–4

[35] Wright A C 1993 *J. Non-Cryst. Solids* **159** 264–8

[36] Micoulaut M, Vuilleumier R and Massobrio C 2009 *Phys. Rev.* B **79** 214205

[37] Micoulaut M, Kachmar A, Bauchy M, Le Roux S, Massobrio C and Boero M 2013 *Phys. Rev.* B **88** 054203

[38] Errington J R and Debenedetti P G 2001 *Nature* **409** 318–21

[39] Chau P-L and Hardwick A J 1998 *Mol. Phys.* **93** 511–8

[40] Bouzid A, Le Roux S, Ori G, Boero M and Massobrio C 2015 *J. Chem. Phys.* **143** 034504

[41] Lampin E, Bouzid A, Ori G, Boero M and Massobrio C 2017 *J. Chem. Phys.* **147** 044504

[42] Penfold I T and Salmon P S 1991 *Phys. Rev. Lett.* **67** 97–100

[43] Salmon P S and Petri I 2003 *J. Phys.: Condens. Matter* **15** S1509–15

[44] Massobrio C, Martin E, Chaker Z, Boero M, Bouzid A, Le Roux S and Ori G 2018 *Front. Mater* **5** 78

[45] Silvestrelli P L, Martin E, Boero M, Bouzid A, Ori G and Massobrio C 2020 *J. Phys. Chem.* B **124** 11273–79

[46] Stöhr M, Van Voorhis T and Tkatchenko A 2019 *Chem. Soc. Rev.* **48** 4118–54

[47] Vydrov O A and Van Voorhis T 2010 *J. Chem. Phys.* **133** 244103

IOP Publishing

Chapter 11

Ternary systems for applications: meeting the challenge

In this chapter we consider two ternary systems having a high technological interest and exhibiting, despite their close chemical nature and composition, different physical behaviors. $Ge_2Sb_2Te_5$ is a prototype of phase change material allowing for rapid structural changes between the crystalline and the amorphous phases. $Ga_{10}Ge_{15}Te_{75}$ is very much stable as a disordered system and has remarkable infrared properties. On the methodological point of view, the study of glassy $Ge_2Sb_2Te_5$ reveals the impact of the pseudopotential construction on the structural properties that were found quite sensitive to a precise choice of the electronic structure description, even within the same selection for the exchange–correlation (XC) functional. The structure of glassy $Ge_2Sb_2Te_5$ is mostly tetrahedrally coordinated. In the case of $Ga_{10}Ge_{15}Te_{75}$, for which the tetrahedral arrangement is also predominant, the influence of the addition of the dispersion forces is analyzed in detail and proved not to be negligible.

11.1 Introduction

All disordered systems considered so far within first-principles molecular dynamics (FPMD) were binary and, by looking at the totality of the glasses (or liquids) presented in this monograph, none of them is made of more (or less) than two species, knowing that the example provided in section 5.1 on monoatomic Lennard-Jones has mostly an academic character. However, there is no point in limiting the application of quantitative methods to binary systems, since most useful disordered materials feature three (or more) components, as those belonging to two distinct families of chalcogenides, very much similar with respect to their chemical identities while being drastically different in terms of physical properties. We are referring here to two classes of materials, having in common the presence of Ge and

doi:10.1088/978-0-7503-2436-6ch11

(in predominant concentrations) Te but differing by their nature of 'bad' (very much unstable on short timescales) and 'good' (stable as disordered structures and exploitable as such) glasses, respectively. Phase change materials (PCM, [1]) are based on the occurrence of rapid structural switches from an amorphous to one or more crystalline phases conferring them exceptional storage capabilities. From the standpoint of glass science, and despite their huge importance in so many aspects related to energy saving, PCM are by far not the dreamland, since there is no way of manipulating them as glasses on accessible laboratory time scales. In this chapter, we shall first report results on the structure of amorphous $Ge_2Sb_2Te_5$ (GST hereafter) [2], most likely the most popular and well known phase change material due to its switching speed and contrast (optical and electrical) between the crystalline and amorphous phase [3, 4]. In a further section, our focus will be on a ternary system with drastically different physical properties, a stable glassy materials with remarkably larger infrared (IR) transparency windows, $Ga_{10}Ge_{15}Te_{75}$ (GGT hereafter) [5–7]. The case of GST has a twofold interest. We proposed a structural model with an unprecedented level of agreement with experimental probes like the total neutron structure factor and, in addition, we were able to demonstrate that this system is quite sensitive to specific details of the electronic structure description, namely the construction of the pseudopotential. Being able to achieve quantitative agreement while addressing a conceptual issue relevant to the foundations of density functional theory makes these results very valuable, granting them the character of a milestone toward full knowledge of glassy (chalcogenide) materials. A further observation is in order concerning the size of the system. While we were able to obtain meaningful results for $Ge_2Sb_2Te_5$ with a minimal number of atoms (32 Ge, 32 Sb and 80 Te), we had to avoid falling into inconsistencies arising when these numbers become too small, as encountered and commented on in the case of $GeSe_9$ (see chapter 9 and section 9.3) for $N_{at} = 120$. We studied $Ga_{10}Ge_{15}Te_{75}$ and we took $N_{at} = 480$ (48 Ga, 72 Ge, 360 Te), by deferring to a later work [8] the consideration of larger $Ge_2Sb_2Te_5$ glassy systems, with structural properties very close to those presented in what follows and in [2].

11.2 $Ge_2Sb_2Te_5$

The potential value for applications of a phase change material rests on the optimization of those properties enhancing the contrasting behavior between the amorphous and the crystalline phase while helping to improve the efficiency of the related devices [3, 4, 9, 10]. Despite the existence of information on the impact of processing on the sample topology, the atomic structure and the bonding properties of glassy GST are still contentious, due to a longstanding debate fostered by diverging experimental predictions on the structure (mostly octahedral, tetrahedral or containing both) [11–20]. At the time of our first involvement in the atomic-scale study of glassy GST via FPMD modeling, density functional theory (DFT) had already been employed to rationalize the amorphous state on the basis of a tetrahedral geometry, or of the coexistence between tetrahedral and octahedral arrangements for the Ge sites, at odds with some early EXAFS (x-ray absorption fine structure) measurements [21–24]. By reviewing the available FPMD literature, we concluded in [2] that some important advances had been made to understand GST and its phase transformation but the agreement with experiments in terms of bond distances, neutron structure factors and pair correlations functions could be improved [22–25]. This prompted us to carry out an investigation of glassy GST in the XC/FPMD framework described in this monograph as the best suited to tackle the atomic structure of chalcogenides (BLYP) complemented by the use of the PBE recipe. The choice of suitable (norm conserving) pseudopotentials (PPs), that proved highly crucial in this case, was carried out by selecting first two different constructions, namely the one due to Troullier and Martins (TM) [26] and that proposed by Goedecker, Teter and Hutter (GTH) [27, 28].

In section 4.2.2 we mentioned that the calculation of $U_{\mathrm{pp}}(\mathbf{r})$ (equation (3.46)) requires special care when handling the separation into local and non-local parts. Typically, within planewaves pseudopotential calculation, one adopts the Kleinman–Bylander (KB) construction [29] in which there is a local component $U^{\mathrm{loc}}(\mathbf{r})$ and an expansion in pseudo-atomic eigenstates $|\phi_{\mathrm{lm}}\rangle$ for the non-local term

$$U_{\mathrm{KB}} = U_{\mathrm{loc}}(r) + \sum_{\mathrm{lm}}^{l^{\max}} \frac{\left|U_l^{'}(r)\phi_{lm} > < U_l^{'}(r)\phi_{\mathrm{lm}}\right|}{<\phi_{\mathrm{lm}}\left|U_l^{'}(r)\right|\phi_{\mathrm{lm}}>} \tag{11.1}$$

where $U_{\mathrm{loc}}(r)$ is the local part of the potential, $U_l^{'}(r) = U_{\mathrm{PS}}(r) - U_{\mathrm{loc}}(r)$ is the non-local part of the potential and $U_{\mathrm{PS}}(r)$ is the atomic potential of the reference state. To implement this construction, one has to choose the local part $V_{\mathrm{loc}}(r)$ (quite arbitrary), and up to which value of the angular momentum $l^{\max}$ to run the sum in equation (11.1). The GTH construction of the pseudopotentials was found independent on specific values for the local part and $l^{\max}$. In the case of Sb and Te, for the TM PPs, we selected loc $= p$ and $l^{\max} = d$, in line with the most common choice found in the literature. For Ge we took loc $= p$, $l^{\max} = p$, as we did in all the FPMD calculations presented in this monograph. Other relevant details specific to these simulations can be sketched as follows. The thermal cycle was made of NVT trajectories at $T = 300$ K (40 ps), 600 K (20 ps), 900 K (70 ps), 600 K (50ps), and 300 K (42 ps, averages taken on the last 30 ps) with a small expansion of volume to release the residual pressure at the experimental density. This schedule was identically applied to three different systems, namely BLYP–TM, PBE–TM and BLYP–GTH (two different XC functionals, BLYP or PBE and two different pseudopotentials schemes, TM or GTH).

11.2.1 Total structure factors and pair correlation functions

In figure 11.1 the total neutron structure factors as obtained within the BLYP–TM and PBE–TM schemes are compared to experimental data. BLYP–TM and PBE–TM are equally well performing for $k < 4$ Å^{-1}, while in the range 4–9 Å^{-1} PBE–TM is slightly diverging, unlike BLYP–TM that behaves remarkably well. BLYP–GTH is the worst over the entire range of k values. By relying on the Wright parameter R_χ [30, 31] (equation (10.4)) in the form

$$R_\chi \equiv \left\{ \frac{\sum_i \left[S_{\exp}(k_i) - S_{\mathrm{FPMD}}(k_i) \right]^2}{\sum_i S_{\exp}^2(k_i)} \right\}^{1/2} \tag{11.2}$$

we obtain R_χ equal to 6.84% (BLYP–TM case) and 7.31% (PBE–TM case).

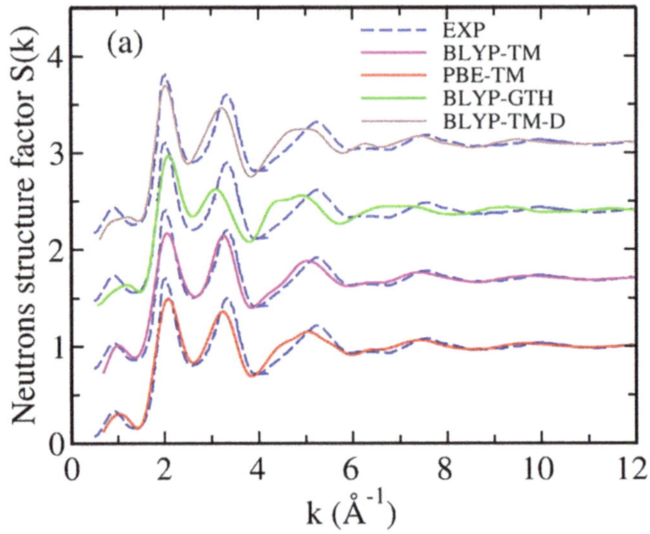

Figure 11.1. Calculated and experimental neutrons total structure factor $S(k)$ of glassy $Ge_2Sb_2Te_5$. Experimental data are taken from [32]. BLYP–TM: magenta line, PBE–TM: red line, BLYP–GTH: green line, BLYP–TM-D: brown line (see text). All data have been obtained in the framework of the FPMD calculations described in [2]. Reprinted figure with permission from [2]. Copyright (2017) by the American Physical Society.

Figure 11.2 provides information on the results obtained via the different DFT recipes used to produce FPMD trajectories, via the comparison with the neutron diffraction measurements from [32]. This figure includes the FPMD results of [2] as well as those obtained in [23] and [24] R_χ is equal to 12.93% for BLYP–GTH [2], 12.68% for PBETM [23] and to 12.09% for PB–GTH [24].

The question arises on why R_χ (6.84%) is much lower in the BLYP–TM case (figure 11.2(b)). This motivates a detailed analysis focused on the construction of the pseudopotential since, in principle, all the results we quoted above are very close and it is unlikely that any crucial role can be played by the exchange–correlation functionals. There are differences between our pseudopotential for Ge-TM and the one used in [23, 24] at the level of the Kleinman–Baylander (KB) [29] construction implemented in each realization. As shown in table 11.1, our choice was $l^{\max} = p$ for the maximum channel in equation (11.1), while in [23, 24] one has $l^{\max} = d$.

To enrich this analysis and see what happens if one replaces $l^{\max} = p$ by $l^{\max} = d$ we implemented the corresponding model, termed BLYP–TM-D. This allows a thorough comparison to be carried out with all schemes employed in [2] and, in particular, with the BLYP–TM case ($l^{\max} = p$) as given in figure 11.1.

The Wright parameter R_χ is equal to 10.95% in the BLYP–TM-D case, larger than the value recorded in [2] for BLYP–TM, 6.84% as defined when using $l^{\max} = p$. Therefore, it appears that the quantitative character of the FPMD when employed to model glassy GST improves when adopting $l^{\max} = p$ in the Ge PPs.

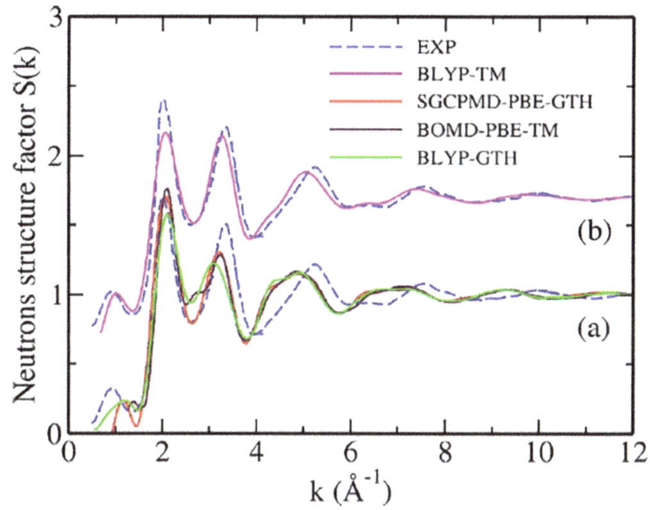

Figure 11.2. Calculated and measured total neutrons structure factor $S(k)$ of glassy $Ge_2Sb_2Te_5$. (a): The BLYP–GTH results (green line [2]), PBE–GTH simulations from [24] (red line) and PBE–TM simulations from [23] (maroon line) are compared to experimental data from [32] (dashed blue line). (b): The BLYP–TM results (magenta line [2]) is compared to experimental data (dashed blue line). SGCPMD stands for 'second generation Car–Parrinello molecular dynamics' (see [33]), while BOMD stands for 'Born–Oppenheimer molecular dynamics'. Reprinted figure with permission from [2]. Copyright (2017) by the American Physical Society.

Table 11.1. Maximum angular momentum $l^{\max}$ used in the KB construction of pseudopotentials. Reprinted table with permission from [2]. Copyright (2017) by the American Physical Society.

FPMD scheme	Kleinman–Baylander construction		
	Ge	Sb	Te
TM [2]	$l^{\max} = p$	$l^{\max} = d$	$l^{\max} = d$
TM [23]	$l^{\max} = d$	$l^{\max} = d$	$l^{\max} = d$
GTH [2]	$l^{\max} = d$	$l^{\max} = d$	$l^{\max} = d$
GTH [24]	$l^{\max} = d$	$l^{\max} = d$	$l^{\max} = d$

11.2.2 Partial pair correlation functions and analysis of local environment

Partial pair correlation functions are shown in figure 11.3, while the coordination numbers are given in table 11.2. The first peak in the Ge–Ge and Ge–Sb pair correlation functions is due to the Ge–Ge and Ge–Sb 'wrong bonds', namely bonds not occurring in the crystal, as defined in [34]. The results of [2] are consistent with a conspicuous number of such bonds, in agreement with experimental high energy x-ray and K-edge EXAFS measurements [35]. The BLYP–TM results are those best reproducing the r_{GeGe}, r_{GeTe} and r_{SeTe} distances, confirming their superior quantitative character.

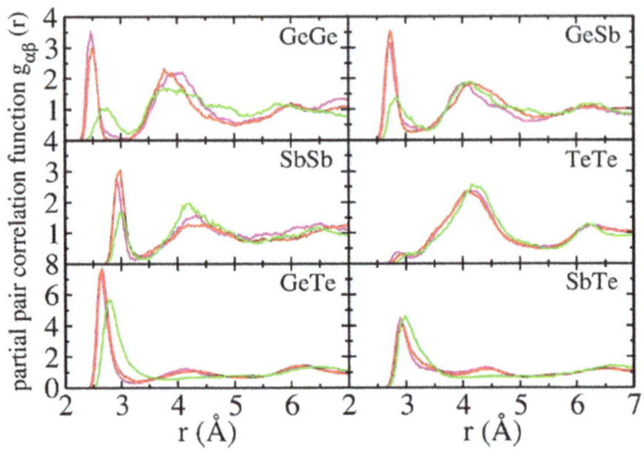

Figure 11.3. Partial pair correlation functions of glassy $Ge_2Sb_2Te_5$ for the BLYP–TM (magenta line), PBE–TM (red line) and BLYP–GTH (green line) models. All data have been obtained in the framework of the FPMD calculations described in [2]. Reprinted figure with permission from [2]. Copyright (2017) by the American Physical Society.

Table 11.2. Bond lengths ($r_{\alpha\beta}$) and partial coordination numbers (n_α^β) extracted from the partial pair correlation functions ($g_{\alpha\beta}(r)$) of models generated with BLYP and PBE XC functionals. α and β denote the two atomic species. *T*Ge is the fraction Ge atoms in tetrahedral geometry. SGCPMD stands for 'second generation Car–Parrinello molecular dynamics', while BOMD stands for 'Born–Oppenheimer molecular dynamics'. Reprinted table with permission from [2]. Copyright (2017) by the American Physical Society.

	Exp	Reference [2]			SGCPMD		BOMD
	a: [15], *b*: [16]	BLYP–TM	PBE–TM	BLYP–GTH	BLYP–GTH [36]	PBE–GTH [24]	PBE–TM [23]
r_{GeGe}	2.47[b]	2.47	2.49	2.75	–	–	–
r_{GeSb}	–	2.73	2.73	2.82	–	–	–
r_{GeTe}	2.61[a], 2.63[b]	2.63	2.66	2.8	–	–	2.78
r_{SbSb}	–	2.92	2.99	3.0	–	–	–
r_{SbTe}	2.85[a], 2.83[b]	2.89	2.92	3.0	–	–	2.93
n_{Ge}^{Ge}	0.6 ± 0.2[b]	0.42	0.32	0.29	0.29	0.29	–
n_{Ge}^{Sb}	–	0.59	0.51	0.43	0.29	0.36	–
n_{Ge}^{Te}	3.3 ± 0.5[b]	3.02	3.30	4.27	3.16	3.31	–
n_{Sb}^{Sb}	–	0.41	0.56	0.36	0.41	0.43	–
n_{Sb}^{Te}	2.8 ± 0.5[b]	2.99	3.09	3.85	3.30	3.36	–
n_{Te}^{Te}	–	0.18	0.16	0.20	0.26	0.30	–
*T*Ge (%)	–	68.7	65.7	22.06	33	27	34

The different weights of the coordination numbers for Ge, Sb and Te are shown in figure 11.4. Similar data are obtained for the BLYP–TM and PBE–TM schemes, whereas BLYP–GTH data are radically different. There is a clear amount of predominant (~80%) fourfold Ge sites. Among those, Ge–Ge_nTe_{4-n} ($n = 0, 1$) are the most frequent, i.e. 38.6% for BLYP–TM and 38.9% for PBE–TM. Te is mostly 2-fold (~60%) with ~30% of 3-fold atoms and Sb is mostly 3-fold (~60%) with ~30% of 4-fold configurations.

We can also employ the order parameter [37] **q** introduced in section 2.4.2 to identify tetrahedral vs octahedral arrangements or combinations of both, knowing that, for instance, defective octahedra correspond to **q** = 5/8 for fourfold coordination and **q** = 7/8 for threefold pyramidal ones [22].

For both the BLYP–TM and PBE–TM cases, the predominant fourfold Ge atoms give rise to the corresponding peak in figure 11.5, meaning that most fourfold Ge sites are found in tetrahedra, while the broad feature up to **q** = 0.5 is due to

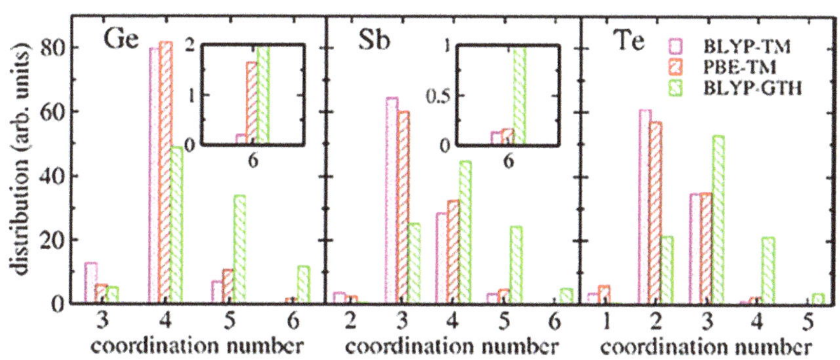

Figure 11.4. Fractions of n-fold Ge, Sb and Te atoms ($n = 1, 2, 3, 4, 5$ or 6) for the BLYP–TM (magenta bar), PBE–TM (red bar) and BLYP–GTH (green bar) models. All data have been obtained in the framework of the FPMD calculations described in reference [2]. Reprinted figure with permission from [2]. Copyright (2017) by the American Physical Society.

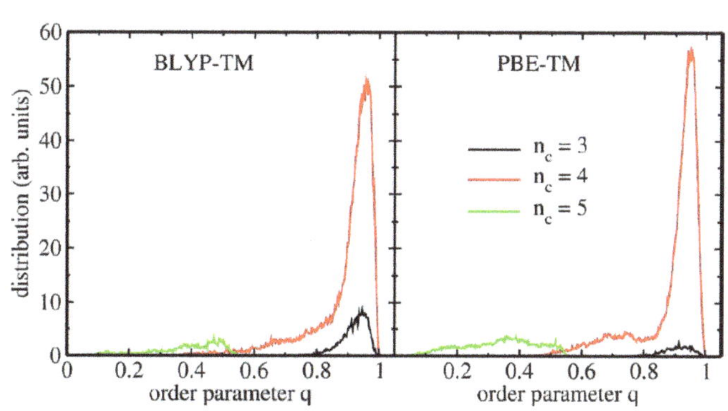

Figure 11.5. Distribution of the local order parameter around Ge atom for the BLYP–TM and PBE–TM models produced in [2]. Results are given for different coordination numbers (n_c). Reprinted figure with permission from [2]. Copyright (2017) by the American Physical Society.

defective octahedral fourfold Ge. The percentage of tetrahedral Ge can be evaluated via integration of the **q** ($n_c = 4$) distribution in between 0.8 and 1. One obtains 68.7% for BLYP–TM and 65.7% for PBE–TM, the rest being defective octahedral configurations. An additional signature of defective octahedral configuration can be found in the range **q** = 0–0.6. This can be attributed to fivefold Ge atoms.

Overall, the BLYP–TM model (and, to a minor extent, the PBE–TM model) developed for glassy GST in reference [2], based on a specific choice for the pseudopotential construction, is very much consistent with experimental data on the atomic structure and leads to a combined tetrahedral/octahedral network favoring the tetrahedral arrangement. This prediction for the network is in agreement with some experiments but it is in contrast with some others [11, 12, 15–17, 19, 20, 25]. In this context we have to quote the observation made in [25] where it was pointed out that Raman features due to Ge atoms tetrahedrally coordinated are lacking in experiments. Therefore, we can achieve the conclusion that the debate on the structure of glassy $Ge_2Sb_2Te_5$ requires additional calculations involving additional physico/chemical properties, as vibrational ones. In this context, the results and the rationale presented in section 12.1 are quite instructive.

11.3 $Ga_{10}Ge_{15}Te_{75}$

In this section the focus will be on a ternary system with drastically different physical properties, a stable glassy materials with remarkably larger infrared (IR) transparency windows, $Ga_{10}Ge_{15}Te_{75}$ (GGT hereafter) [5–7].

When looking for glassy materials of enhanced stability and potentially high technological impact, Te-rich ternary compounds are very good candidates, especially in combination with Ge [6]. This stems from the existence of well known Ge–Te binary materials, as $GeTe_4$, for which the inclusion of Se, Ga or I in small concentrations contributes to increase glass formation and stability against crystallization [6, 7, 38]. For this reason, in [5] we considered glassy $Ga_{10}Ge_{15}Te_{75}$ by applying our FPMD methodology and comparing with experiments [39] and other DFT results by Voleská and coworkers [40] for a close GaGeTe system ($Ga_{11}Ge_{11}Te_{78}$). These last results were obtained by using a mixed DFT–FPMD/reverse Monte Carlo approach (RMC–FPMD–PBE) bringing into the calculation the information of an external structural refinement. By sticking to a methodology not including any information from available measurements (in line with the strategy of all FPMD applications presented in this monograph), we have considered a periodic GGT system with N_{at} = 480 atoms (48 Ga, 72 Ge, 360 Te) by constructing two models. For the first, the initial coordinates were taken from a previous sample of glassy $GeSe_2$. The trajectory is made of 5 ps at T = 300 K, 4.5 ps at T = 600 K, 10 ps at T = 900 K, 6.6 ps at T = 600 K and 18.5 ps at 300 K. For the second (initial coordinates taken from glassy $GeSe_4$), we had 6 ps at T = 300 K, 4.2 ps at T = 600 K, 12.3 ps at T = 900 K, 9.5 ps at T = 600 K and 16.4 ps at T = 300 K. Published results [5] are mean values of these two temporal trajectories at T = 300 K. The FPMD protocol was implemented by adding dispersion forces according to the Grimme-D2 expression [41] extensively invoked in chapter 10. Also relevant to [5] is the observation that the quantities in reciprocal space were obtained as in equation (2.12).

11.3.1 Structural properties

In figure 11.6, the total neutron structure factor is compared to the experimental data for $Ga_{11}Ge_{11}Te_{78}$ [39]. The agreement is excellent along the entire k range and, in addition, calculations reproduce well the feature in the FSDP region around 1 $Å^{-1}$, to indicate some existing degree of intermediate range order in the network.

The partial pair correlation functions have been compared to those given in [40] (see figure 11.7) where FPMD techniques were employed after the selection of an initial structure obtained via Reverse Monte Carlo at the concentration ($Ga_{11}Ge_{11}Te_{78}$, RMC–FPMD–PBE results). Focusing on those pair correlation functions $g_{\alpha\beta}(r)$ associated with the homopolar bonds Ga–Ga and Ge–Ge, there are fewer Ga–Ga homopolar bonds than Ge–Ge ones. The situation changes within RMC–FPMD–PBE, for which the second peaks in $g_{GaGa}(r)$ and $g_{GeGe}(r)$ are weaker to indicate a shell of second neighbors less organized. The most important differences between FPMD–BLYP and RMC–FPMD–PBE concern Ge–Ga correlations. FPMD–BLYP gives a regular second large peak, much less intense for distances larger than 4 Å. In RMC–FPMD–PBE instead, the Ge–Ga second neighbors are found at values higher than 4 Å. Figure 11.7 can also shed light on Ge–Ge, Ge–Te, Te–Te correlations via a comparison with glassy $GeTe_4$ (section 10.2). The pair correlation functions g_{GeTe} are very similar as the two networks have in common predominant heteropolar connections forming Ge centered units. The same occurs for homopolar Ge–Ge bonding when comparing g_{GeGe} ($GeTe_4$) and g_{GeGe} (GST). Equally close are the pair correlation functions g_{TeTe}, with well distinguishable maxima and minima, even though the depth of the first minimum is more pronounced in the GGT case.

Since we are including dispersion forces following the D2 expression of Grimme [41] one can add some thoughts to the rationale presented in chapter 10 by considering their impact on the structure of glassy GGT. For this reason we can also include in the comparison the pair correlation functions of $GeTe_4$ at the BLYP level with no van der Waals (vdW) corrections, as given in [38] (see section 10.2).

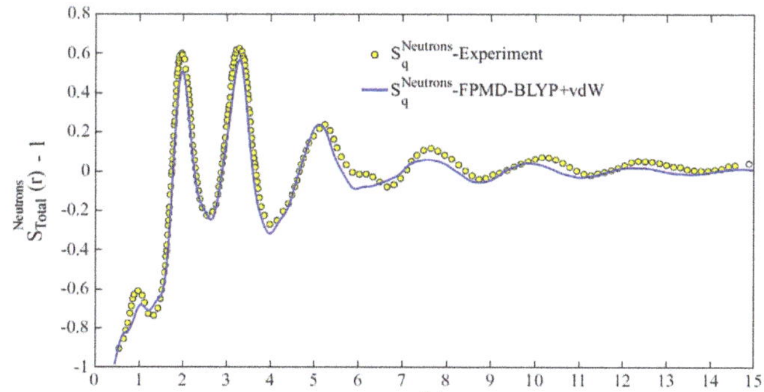

Figure 11.6. Calculated total neutron structure factor of glassy $Ga_{10}Ge_{15}Te_{75}$ (blue solid line). The experimental data (circle symbols) are obtained from [39]. Reprinted with permission from [5]. Copyright (2018), with permission from Elsevier.

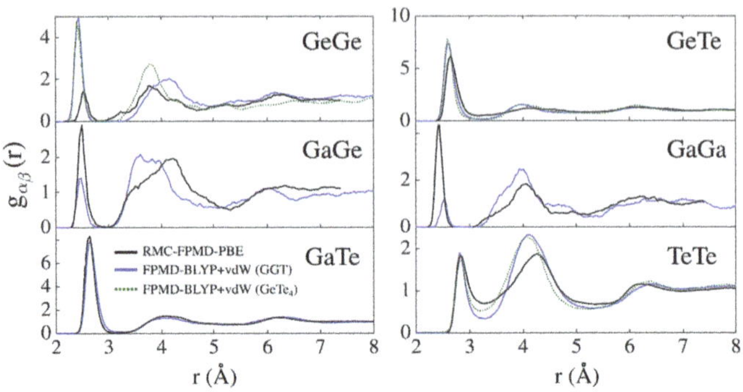

Figure 11.7. Partial pair correlation functions $g_{\alpha\beta}$ (α, β = Ge, Ga, Te) for the GGT system. Calculations of [5]: blue lines, RMC–FPMD–PBE results of Voleska *et al* [40]: black lines. FPMD–BLYP calculations on $GeTe_4$ presented in section 10.2: green dotted lines. Reprinted with permission from [5]. Copyright (2018), with permission from Elsevier.

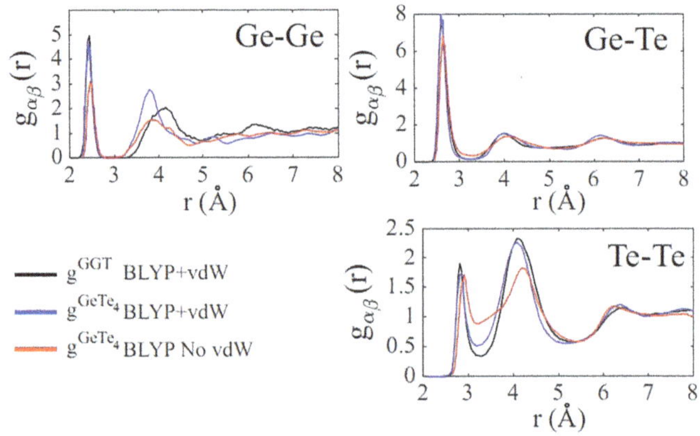

Figure 11.8. Partial pair correlation functions for GeGe, GeTe and TeTe interactions. The results of [5] (including vdW correction) are highlighted in black solid lines while the results for $GeTe_4$ presented in section 10.2 are reported in two situations: including (blue lines) or not (red solid lines) vdW interactions. Reprinted with permission from [5]. Copyright (2018), with permission from Elsevier.

As it appears clearly in figure 11.8 and focusing on distances larger than the nearest neighbors, the vdW part enhances the depth of the first minimum and the height of the second maximum in the Te–Te case. To learn more about these trends, we have reported in figure 11.9 three results for the Ge–Ge, Ge–Te and Te–Te pair correlation functions. These refer to glassy $GeTe_4$ and glassy GGT via the RMC–FPMD–PBE approach (both without inclusion of vdW corrections, for GGT see [40]) as well as to the results of section 10.2 that account for vdW corrections. One notes that in g_{TeTe} and g_{GeGe}, the patterns found for glassy $GeTe_4$ and glassy GGT (RMC–FPMD–PBE, no vdW) are very close while they are quite different from the GGT data with vdW correction. Therefore, the origin of the dissimilar profiles in g_{GeGe} and g_{TeTe} found in our GGT model with respect to [40] can be ascribed to the account of the dispersion forces. It should be added that at the time of publication of

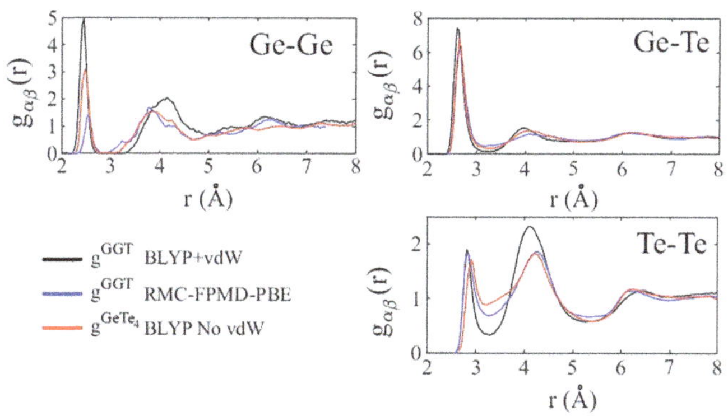

Figure 11.9. Partial pair correlation functions for GeGe, GeTe and TeTe interactions. The results of [5] (including vdW correction) are highlighted in black solid lines while the results obtained by Voleska *et al* [40] for glassy $Ga_{11}Ge_{11}Te_{78}$ correspond to the blue solid lines. Results presented in section 10.2 for the $GeTe_4$ glassy chalcogenide with no vdW interactions are given in red solid lines. Reprinted with permission from [5]. Copyright (2018), with permission from Elsevier.

[5] we were considering only one expression for the vdW interactions, this meaning that all conclusions drawn above reflect the use of Grimme-D2. In view of what established in chapter 10 on the effect of the different recipes, it would be worthwhile to apply some of them, and especially the MLWF one, to the case of glassy GGT.

11.3.2 Network topology

First we consider the local order via the corresponding coordination numbers (table 11.3). As customary, with c_α we indicated the concentration of each species and n_{ave} is equal to $c_{Ge}n_{Ge} + c_{Ga}n_{Ga} + c_{Te}n_{Te}$, where the coordination numbers for each species are obtained by summing up the corresponding partial ones.

The three sets of results agree well on the bond distances, showing the good predictive level of FPMD calculations. The topology of the network as described via the coordination numbers is based on Ga and Ge atoms in a defective fourfold environment. Te atoms are found around Ge or Ga or in homopolar connections. However, experiments indicate an average number of 3 Te atoms coordinating Ga atoms. BLYP–vdW favors a higher number of Ge–Ge homopolar bonds, with a few Ga–Ga ones. In a way consistent with the pair correlation functions, RMC–FPMD–PBE does the opposite (higher number of Ga–Ga homopolar bonds). The local coordination analysis, shown in table 11.4 is indicative of fourfold and threefold Ge coordinations, threefold connected Ge atoms corresponding to deviations from bonding of heterogeneous kind. Both BLYP–vdW and RMC–FPMD–PBE point toward the predominant fourfold nature of the Ga connections.

However, the two FPMD approaches differ by the distribution of the units, since RMC–FPMD–PBE features Ga–Ga bonding and, in particular, the Ga–$GaTe_3$ coordination (13.3%). This is responsible for the Ga–Ga high peak in the Ga–Ga pair correlation function. BLYP–vdW results are characterized by the impact of

Table 11.3. Bonds properties (interatomic distances and the corresponding coordination numbers) for the GGT system calculated with the FPMD–BLYP+vdW method (second column) compared to the experimental results of [39] and the RMC–FPMD–PBE of [40]. Reprinted with permission from [5]. Copyright (2018), with permission from Elsevier.

	Experiment	BLYP+vdW	RMC–FPMD–PBE
r_{Ge-Ge}	—	2.45 Å	2.53 Å
r_{Ge-Ga}	—	2.49 Å	2.49 Å
r_{Ge-Te}	2.60 Å	2.59 Å	2.65 Å
r_{Ga-Ga}		2.52 Å	2.43 Å
r_{Ga-Te}	2.60 Å	2.67 Å	2.63 Å
r_{Te-Te}	2.80 Å	2.81 Å	2.83 Å
n_{Ge-Ge}		0.24	0.07
n_{Ge}^{Ga}		0.06	0.14
n_{Ge}^{Te}	3.97	3.44	3.55
n_{Ga}^{Ga}		0.02	0.17
n_{Ga}^{Te}	3.00	3.78	3.77
n_{Te}^{Te}	1.36	1.00	1.52
n_{Ge}		3.73	
n_{Ga}		3.89	
n_{Te}	2.36	2.19	
n_{ave}	2.61	2.59	

$GaTe_4$ units, together with a few homopolar bonds. In the case of Te (not shown here but made available with the corresponding data in [5]) there are a variety of different twofold coordination subsets, such as Ge–Te–Ge, Ge–Te–Te, Te–Te–Te, Ga–Te–Ga, Ga–Te–Te and Ga–Te–Ge. Sizeable amounts of these units are typical of networks in which tetrahedral (or defective octahedral) units are cross-linked through Te atoms arranged in chains of various lengths [42, 43].

The nature of the coordination for Ge and Ga atoms is identified by using a local order parameter **q** introduced in section 2.4.2 and, for instance, employed by Bouzid *et al* [38] (see section 10.2) to describe the tetrahedral/octahedral local atomic environments in glassy $GeTe_4$.

The resulting distributions are the object of figure 11.10(a) for both the Ge- and Ga-coordination. The environment of the Ge and Ga species is largely of tetrahedral nature, leading to sharp peaks close to $\mathbf{q} = 1$. The main contributions are due to the threefold (especially for Ge atoms) and fourfold coordinations in agreement with the coordinations given in table 11.4. These observations are confirmed by visual inspection of the configurations in figure 11.10(c), where one can also detect the large number of homopolar Te–Te bonds (figure 11.10(b)).

To summarize the content of this section, comparison with experimental structural data (neutron diffraction) for glassy $Ga_{10}Ge_{15}Te_{75}$ studied via FPMD–BLYP–vdW is very satisfactory, showing the tetrahedral network character of this system. Glassy $Ga_{10}Ge_{15}Te_{75}$ is a predominantly tetrahedral network (Ga and Ge

Table 11.4. $n_\alpha(l)$ structural units expressing the different coordination motifs in glassy GGT for the species (Ga, Ge or Te). The atomic species are indicated in bold and the coordination numbers are classified with respect to the number (l) of neighbors and their type. Values smaller than 0.1 are not reported. In the bottom part, the quantification of homopolar bonds for each species is extracted from FPMD–BLYP calculations. We provide the percentage of Ge, Ga and Te involved in at least one homopolar bond. Reprinted with permission from [5]. Copyright (2018), with permission from Elsevier.

		$n_\alpha(l)$ [%]	
		BLYP–vdW	RMC–FPMD–PBE [40]
Ge atom			
$l = 2$			
	Te_2	1.07	
$l = 3$			
	Te_3	22.0	23.0
	$GeTe_2$	2.5	
$l = 4$			
	Te_4	51.6	54.5
	$GaTe_3$	5.2	10.5
	$GeTe_3$	13.5	5.0
	Ge_2Te_2	3.5	3.5
	$GaGeTe_2$	0.7	1.7
Ga atom			
$l = 0$			
		0.6	
$l = 1$			
	Te	1.0	
$l = 2$			
	Te_2	0.6	
$l = 3$			
	Te_3	4.7	<1
	$GaTe_2$	0.1	
$l = 4$			
	Te_4	82.3	69.7
	$GaTe_3$	8.6	13.3
	$GeTe_3$	1.7	16.1
$l = 5$			
	$GaTe_4$	0.2	
Homopolar bonds			
	Ge	Ga	Te
	20.3	2.1	73.5

fourfold coordinated to Te atoms as most frequent structural unit). Not only do Te atoms form Te–Te homopolar bonds but also Ge atoms are found in homopolar configurations, while Ga–Ga contacts are less numerous. If one limits the consideration of dispersion forces to the Grimme-D2 recipe, as we saw in section 10.2, this

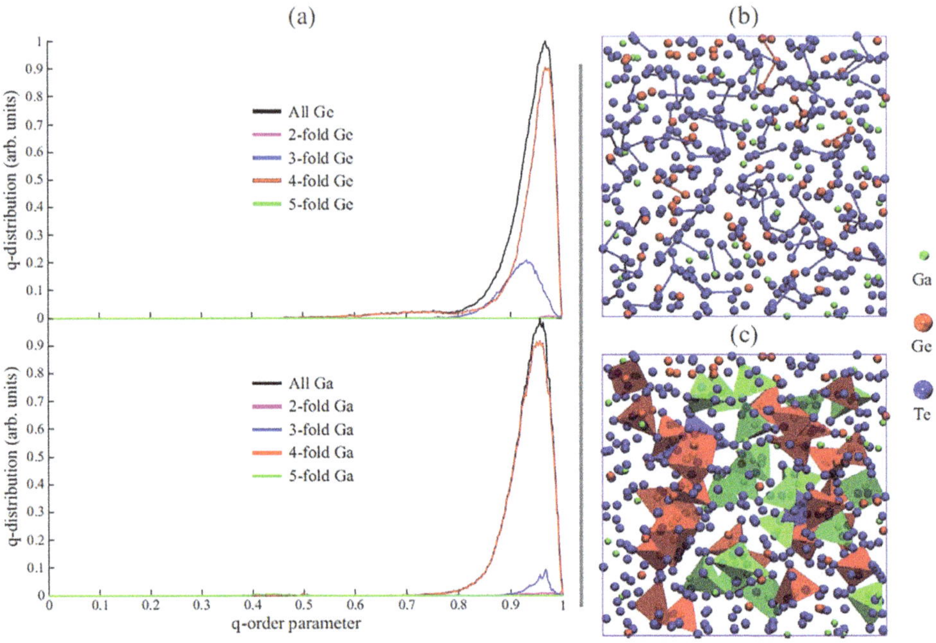

Figure 11.10. (a) Distribution of the local atomic coordination motifs as a function of the order parameter **q**. (b) Snapshot of the GGT system studied via BLYP–vdW calculations. We highlight the predominance of the Te–Te homopolar bonds (blue) and the lower amount of Ge–Ge (green) and Ga–Ga (red) bonds. (c) Visualization of the tetrahedral arrangements characteristic of the GGT structure for both Ge and Ga atoms. Reprinted with permission from [5]. Copyright (2018), with permission from Elsevier.

contribution plays a role in determining the atomic structure, in agreement with the indications collected for glassy $GeTe_4$.

References

[1] Raoux S 2009 *Annu. Rev. Mater. Res.* **39** 25–48
[2] Bouzid A, Ori G, Boero M, Lampin E and Massobrio C 2017 *Phys. Rev.* B **96** 224204
[3] Wuttig M and Yamada N 2007 *Nature Mater* **6** 824–32
[4] Zalba B, Marín J, Cabeza L F and Mehling H 2003 *Appl. Therm. Eng.* **23** 251–83
[5] Chaker Z, Ori G, Boero M, Massobrio C, Furet E and Bouzid A 2018 *J. Non-Cryst. Solids* **498** 338–44
[6] Danto S, Houizot P, Boussard-Pledel C, Zhang X-H, Smektala F and Lucas J 2006 *Adv. Funct. Mater.* **16** 1847–52
[7] Lucas P, Boussard-Pledel C, Wilhelm A, Danto S, Zhang X-H, Houizot P, Maurugeon S, Conseil C and Bureau B 2013 *Fibers* **1** 110–8
[8] Martin E, Ori G, Duong T-Q, Boero M and Massobrio C 2022 *J. Non-Cryst. Solids* **581** 121434
[9] Mitrofanov K V, Kolobov A V, Fons P, Wang X, Tominaga J, Tamenori Y, Uruga T, Ciocchini N and Ielmini D 2014 *J. Appl. Phys.* **115** 173501

[10] Philip Wong H-S, Raoux S, Kim S, Liang J, Reifenberg J P, Rajendran B, Asheghi M and Goodson K E 2010 *Proc. IEEE* **98** 2201–27
[11] Carria E, Mio A M, Gibilisco S, Miritello M, d'Acapito F, Grimaldi M G and Rimini E 2011 *Electrochem. Solid-State Lett.* **14** H480
[12] Krbal M, Kolobov A V, Fons P, Tominaga J, Elliott S R, Hegedus J and Uruga T 2011 *Phys. Rev.* B **83** 054203
[13] Mukhopadhyay S, Sun J, Subedi A, Siegrist T and Singh D J 2016 *Sci. Rep.* **6** 25981
[14] Xu M, Cheng Y Q, Sheng H W and Ma E 2009 *Phys. Rev. Lett.* **103** 195502
[15] Kolobov A V, Fons P, Frenkel A I, Ankudinov A L, Tominaga J and Uruga T 2004 *Nat. Mater.* **3** 703
[16] Baker D A, Paesler M A, Lucovsky G, Agarwal S C and Taylor P C 2006 *Phys. Rev. Lett.* **96** 255501
[17] Paesler M A, Baker D A, Lucovsky G, Edwards A E and Taylor P C 2007 *J. Phys. Chem. Solids* **68** 873–7
[18] Sen S, Edwards T G, Cho J-Y and Joo Y-C 2012 *J. Phys. Chem. Solids* **108** 195506
[19] Hosokawa S, Ozaki T, Hayashi K, Happo N, Fujiwara M, Horii K, Fons P, Kolobov A V and Tominaga J 2007 *Appl. Phys. Lett.* **90** 131913
[20] Hada M, Oba W, Kuwahara M, Katayama I, Saiki T, Takeda J and Nakamura K G 2015 *Sci. Rep.* **5** 13530
[21] Wełnic W, Pamungkas A, Detemple R, Steimer C, Bluegel S and Wuttig M 2006 *Nat. Mater.* **5** 56–62
[22] Caravati S, Bernasconi M, Kühne T D, Krack M and Parrinello M 2007 *Appl. Phys. Lett.* **91** 171906
[23] Akola J and Jones R O 2007 *Phys. Rev.* B **76** 235201
[24] Caravati S, Bernasconi M, Kühne T D, Krack M and Parrinello M 2009 *J. Phys.: Condens. Matter* **21** 2555
[25] Sosso G C, Caravati S, Mazzarello R and Bernasconi M 2011 *Phys. Rev.* B **83** 134201
[26] Troullier N and Martins J L 1991 *Phys. Rev.* B **43** 1993–2006
[27] Goedecker S, Teter M and Hutter J 1996 *Phys. Rev.* B **54** 1703–10
[28] Krack M 2005 *Theor. Chem. Acc.* **114** 145–52
[29] Kleinman L and Bylander D M 1982 *Phys. Rev. Lett.* **48** 1425–8
[30] Wright A C 1993 *J. Non-Cryst. Solids* **159** 264–8
[31] Wright A C, Sinclair R N and Leadbetter A J 1985 *J. Non-Cryst. Solids* **71** 295–302
[32] Jóvári P, Kaban I, Steiner J, Beuneu B, Schöps A and Webb M A 2008 *Phys. Rev.* B **77** 035202
[33] Kühne T D 2014 *WIREs Computational Molecular Sci.* **4** 391–406
[34] Akola J and R O Jones 2008 *J. Phys.: Condens. Matter* **20** 465103
[35] Jóvári P, Kaban I, Steiner J, Beuneu B, Schöps A and Webb A 2007 *J. Phys.: Condens. Matter* **19** 335212
[36] Caravati S and Bernasconi M 2015 *Phys. Status Solidi* B **252** 260–6
[37] Errington J R and Debenedetti P G 2001 *Nature* **409** 318–21
[38] Bouzid A, Massobrio C, Boero M, Ori G, Sykina K and Furet E 2015 *Phys. Rev.* B **92** 134208
[39] Jóvári P, Kaban I, Bureau B, Wilhelm A, Lucas P, Beuneu B and Zajac D A 2010 *J. Phys.: Condens. Matter* **22** 404207

[40] Voleská I, Akola J, Jóvári P, Gutwirth J, Wágner T, Vasileiadis T, Yannopoulos S N and Jones R O 2012 *Phys. Rev.* B **86** 094108
[41] Grimme S 2006 *J. Comput. Chem.* **27** 1787–99
[42] Bouzid A, Le Roux S, Ori G, Boero M and Massobrio C 2015 *J. Chem. Phys.* **143** 034504
[43] Sykina K, Furet E, Bureau B, Le Roux S and Massobrio C 2012 *Chem. Phys. Lett.* **547** 30–4

Chapter 12

Past, present and future

This final chapter is structured in three sections. In the first (section 12.1) we look critically at the global strategy employed in this book by considering the impact of properties other than those stemming from a direct knowledge of the atomic coordinates. In particular, the focus is on vibrations as a tool that could be more selective since allowing, better than diffraction probes and their first-principles molecular dynamics (FPMD) counterpart, to discriminate among different structural models of a given system (glassy $GeSe_2$ in the case discussed here). In the second section (section 12.2) the emphasis is on the calculation of thermal properties as obtained by coupling FPMD to a technique explicitly devised to make possible calculations of heat transport on relaxation times accessible to FPMD schemes. We show how to establish a correlation between thermal conductivities and the existence of propagative modes in disordered structures. The final section (section 12.3) contains some considerations on the future of atomic-scale modeling of materials and of glasses in particular, by providing some recommendations on the guidelines to be followed together with examples of current research geared toward quantitative understanding of technologically useful materials.

12.1 Past: what else beyond structure?

Working on the final chapter of a monograph can be either celebrative or self-critical. Since most of this book (for the achievements parts) has been constructed on my own results, there is no point in further extolling what obtained to build a nice ending. I learned from science that nothing is perfect or, better, that anything can be improved, even the nicest realization. So I prefer to go through the exercise of finding the weak points of the whole construction, being aware of what could have been done better does not mean at all that what was done was wrong. Looking back at the strategy followed throughout this book, namely a strong emphasis on atomic

doi:10.1088/978-0-7503-2436-6ch12

structure of glassy and amorphous systems via the calculation of their structural properties as carried out over more than three decades of scientific work, a legitimate question is in order. Are we neglecting some important piece of information on a glassy structure by focusing almost exclusively on atomic-scale structural properties and on the comparison with diffraction measurements? A paper reporting on structural and *vibrational* properties of glassy $GeSe_2$ provides important clues for addressing this issue [1].

Three models of glassy $GeSe_2$ were created to reproduce different conditions of chemical order, having in mind to obtain, indications on their performances in terms of vibrational properties. Models I and model II are essentially identical in terms of structural organization, differing only by the number of atoms (model I, $N_{at} = 180$, model II, $N_{at} = 120$). They have been obtained by taking advantage of classical molecular dynamics (CMD) potentials proposed by Vashishta and coworkers [2–4] (see also section 3.3) to create an initial trajectory in the diffusive regime. FPMD is applied only at the end of the quench procedure, by leading to the formation of Se–Se homopolar bonds and other minor modifications, as a small increase in the bond lengths. This model of glassy $GeSe_2$ is highly chemically ordered and, in principle, is far less realistic than those presented in section 6.2.3 since it does not contain specific structural motifs (miscoordinations, homopolar bonds for both species) compatible with experimental evidence. However, as shall see, this statement has to be softened in light of the pieces of evidence collected via the study of vibration properties. Model III is very close to the obtained reported in section 6.2.3. It results from a rapid quench of the corresponding liquid, with a final structure obtained from a quench to $T = 0$ K. The analysis of pair correlations functions and partial structure factors for the three model ends with a strong statement on the inability of diffraction problems (despite the absence of homopolar bonds for both species in model I and model II when compared to model III) to distinguish between different degrees of chemical order. This is exemplified in figure 12.1 via the comparison between the concentration–concentration structure factors $S_{CC}(k)$ obtained for the three different cases. There is no point in arguing with the observation that chemical order cannot be associated to a specific pattern observed in reciprocal space for a highly elusive quantity as $S_{CC}(k)$. However, it remains true

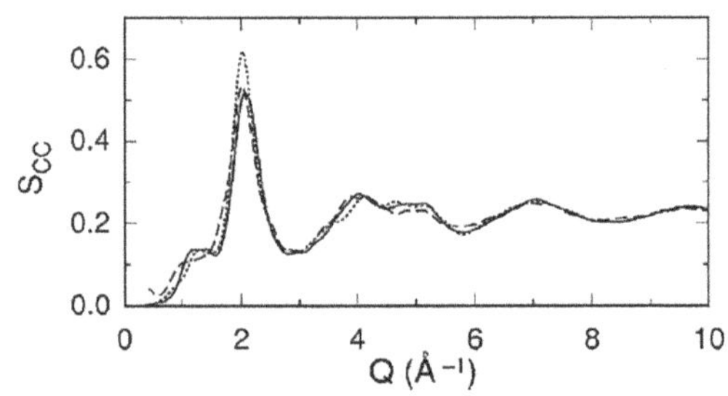

Figure 12.1. Concentration–concentration structure factors $S_{CC}(k)$ of glassy $GeSe_2$ as calculated in [1] by relying on three different models. Model I (solid line) and model II (dotted line): high degree of chemical order. Model III (dashed line): low degree of chemical order. Reprinted figure with permission from [1]. Copyright (2007) by the American Physical Society.

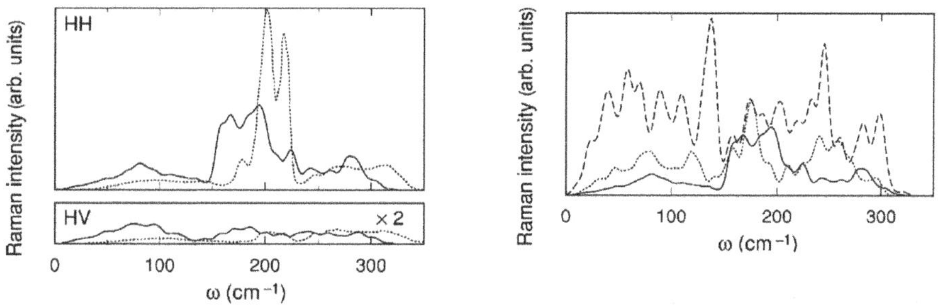

Figure 12.2. Left part: reduced HH and HV Raman spectra of glassy $GeSe_2$ for model I (solid line), compared to experimental data (dotted line). Right part: comparison between the calculated HH Raman spectra of glassy $GeSe_2$ for model I (solid line), model II (dotted line) and model III (dashed line). (HH stands for the proportion of horizontally transmitted waves coming back horizontally to the sensor, HV those coming back vertically.) Reprinted figure with permission from [1]. Copyright (2007) by the American Physical Society.

that models not featuring homopolar bonds remain highly questionable and, perhaps, not worthy of any further use. By leaving aside any further consideration on this debate, it is of interest to move to vibrational properties and see how and to what extent they are sensitive to chemical order.

One revealing example is provided by the Raman spectra of glassy $GeSe_2$ calculated as detailed in references [5–7] and shown in figure 12.2. We are referring here to incoming and outgoing photons with parallel HH or perpendicular polarizations HV. Experimentally, the HH Raman spectrum of glassy $GeSe_2$ exhibits a marked doublet with a principal peak located at 201 cm^{-1} and another peak at 218 cm^{-1} (figure 12.2, left). Model I (figure 12.2, left) is able to reproduce the doublet in the HH spectrum, with bands having intensities not too dissimilar to experiments. Concerning the the HV spectrum the intensities are much weaker, featuring three bands, only two of them being visible in the results obtained for model I (the second and the third merging). In figure 12.2 (right) a comparison is carried out for the HH results of models I, II, and III. The worst performing is model III showing a very specific pattern highly fluctuating differing strongly from the experimental one. Therefore, we are able to collect strong evidence demonstrating that a system highly chemically disordered does not describe glassy $GeSe_2$ in a way compatible with Raman vibrational properties[1]. This assertion questions the validity of conclusions on the structural organization of this network based on the mere analysis of atomic-scale probes like the pair correlation functions and the partial structure factors. How can we reconcile the results of chapter 6, favoring enhanced chemical disorder (essentially based on structural properties and the presence or absence of homopolar bonds and miscoordinations) with the findings of this section, pointing out a much better agreement in terms of Raman features when adopting a highly chemically ordered network? It looks like the amount of chemical disorder found in glassy $GeSe_2$ by quenching from the liquid is (moderately) overestimated, as it can be

[1] It is important to point out that, as far as other vibrational properties are concerned, a comparison made available in [1] among the vibrational densities of states of model I, model II and model III does not show any particular sensitivity to their different structural organizations.

expected in view of what we have discussed in terms of incomplete structural relaxation. However, a model structure for this network devoid of homopolar bonds for both species cannot be taken as realistic, despite its better performances when the focus in on vibrations. The truth lies in between, I would conclude, with no need to undermine either one of the two tracks taken to envision the optimal structure to be targeted in future, improved calculations.

12.2 From past to present, from structural to thermal properties: thermal conductivity

Working on thermal properties has become an unavoidable task in materials science, due to the request of efficient devices for which optimal thermal management is a target. In this book, a few words have already been spent on phase change materials (PCMs, see section 11.2) by pointing out their impact on memory devices. By achieving a better understanding of thermal properties from the theoretical side one can contribute to limit the amount of energy involved through heat transport, thereby limiting resources waste when realizing the switch from the crystalline to the amorphous phases. The connection between FPMD and thermal properties has been ensured by the 'Approach to equilibrium molecular dynamics' technique (AEMD) [8, 9] that is briefly recalled in what follows. AEMD is based on the notion of a thermal transient and its associated decay time. To this purpose, a given system is divided in two parts (see figure 12.3, each one at a specific temperature (hot and cold) during a first step of the methodology (phase 1)).

Due to periodicity, hot and cold blocks are infinitely propagated in space. At the end of phase 1 the temperature difference between the blocks is lifted, inducing the so-called phase 2 (transient regime) during which the temperature of the hot block decreases and the temperature of the cold block increases (see figure 12.4). Under the assumptions detailed, for instance, in [10], the decay time τ of the thermal transient leads to to the thermal conductivity κ via:

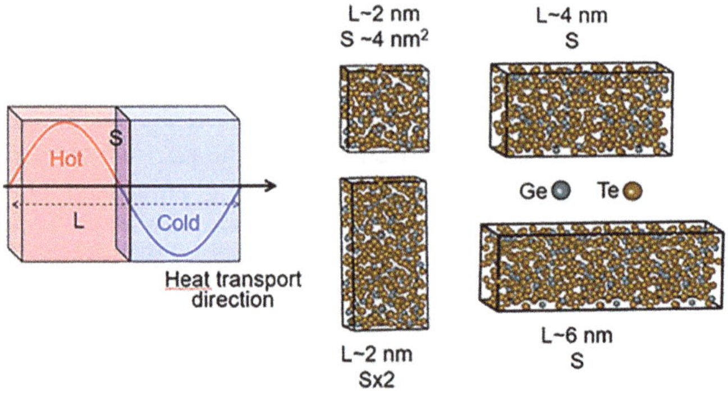

Figure 12.3. The four atomic models studied in [10] for the case of glassy $GeTe_4$. L is the cell length in the heat transport direction. S is the cross section of the system in the direction of the heat flux. Courtesy of E Martin (ICube, Strasbourg).

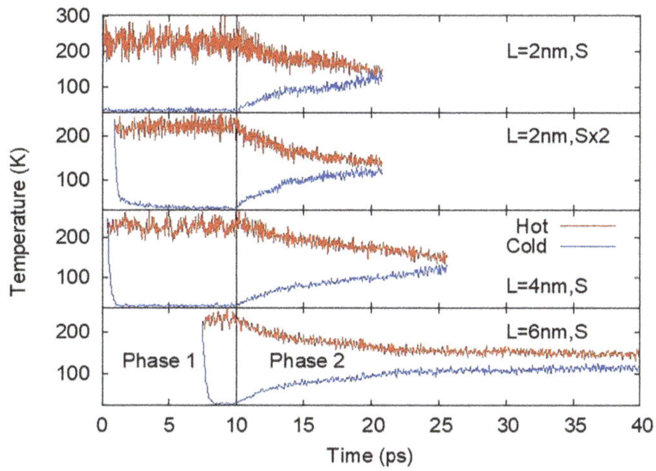

Figure 12.4. Temperatures in the hot (red line) and cold (blue line) block during the two phases of AEMD for the four different models of glassy $GeTe_4$ considered in [10] and shown in figure 12.3. Courtesy of E Martin (ICube, Strasbourg).

$$\kappa = \frac{L^2 C\rho}{4\pi^2\tau} \tag{12.1}$$

where L is the length of the system in the direction of thermal transport, C is the specific heat and ρ the number density. The thermal conductivity is characterized by a size dependence over the heat transport dimension L, confirming previous results obtained on a smaller range of values of L [11]. The fact that one is able to access values of κ depending on L entails a wealth of interesting consequences in terms of nanoscale heat properties, since the thermal conductivity of glassy $GeTe_4$ increases even when the box length in the direction of heat transport is 6 nm. The notion of propagons can be invoked to identify heat carriers in amorphous materials, the results on $GeTe_4$ being the first ever reported for chalcogenide glasses [10].

As a second example of application of the FPMD/AEMD strategy to disordered network, and in view of its technological importance and our past experience with its structural properties (section 11.2), we turned our attention to amorphous $Ge_2Sb_2Te_5$ [15]. By following the same methodological route implemented for $GeTe_4$, we were able to highlight the persistence of heat transport via propagative modes as well as of size effects (figure 12.5). The most interesting feature is the extent of these phenomena beyond medium range order distances inherent in disordered network-forming materials. In addition, the calculated thermal conductivities are in agreement with the experimental data, showing a marked size dependence into the 1.7–10 nm range. This range is compatible with the dimensions of PCMs-based devices.

The above features of disordered materials are not limited to chalcogenide systems but are quite general. An analogous calculation carried on amorphous SiO_2 [16], based on heat propagation for lengths comprised in between 2 and 8

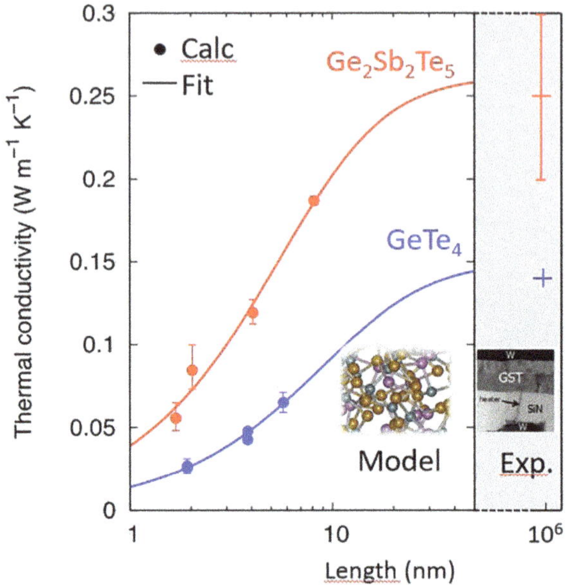

Figure 12.5. Thermal conductivities of amorphous GST calculated by AEMD (red dots). The results obtained in [10] for amorphous $GeTe_4$ using are also reported (blue dots). The lines are fits following an analytical model of size dependence [12]. The experimental points are taken from [13] for $GeTe_4$ and from [14] for GST. Courtesy of E Martin (ICube, Strasbourg).

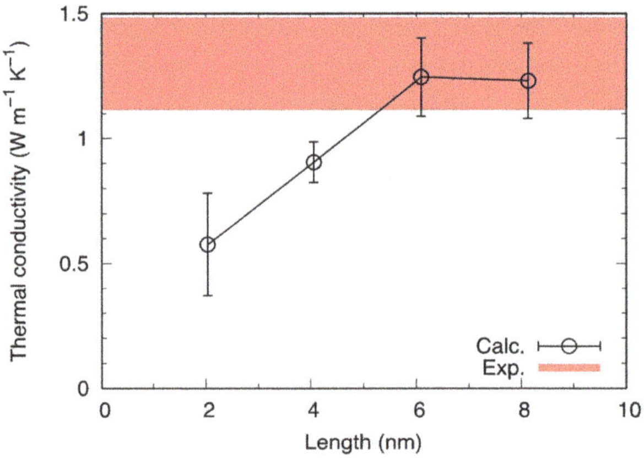

Figure 12.6. Thermal conductivity of amorphous SiO_2 as a function of length in the direction of the heat flux in between 2 and 8 nm. The error bars have been obtained by taking advantage of distinct calculations for each size. The shaded area has been drawn to include different experimental values obtained with various techniques (see [17]). Courtesy of E Martin (ICube, Strasbourg).

shows that thermal conductivity does not change beyond 6 nm, by being asymptotically in agreement with experimental results (figure 12.6). Therefore, there is a common behavior of reduction of the thermal conductivity with size that can be found in disordered $GeTe_4$, $Ge_2Sb_2Te_5$ and SiO_2, despite their different degrees of chemical order. This means that while structural disorder establishes on the scale of a few Å, the vibrational mean free paths can extend over the nanometer range.

To end this section, I would like to mention that working on thermal conductivity appears to be a recurrent event of my scientific path of computational materials scientist using MD. As early as 1984, together with my thesis adviser G Ciccotti, a paper was published on the calculation of thermal conductivity in a Lennard-Jones fluid at the triple point, by exploiting a non-equilibrium approach based on an atomic-scale expression for the heat flux. By subtracting step by step the equilibrium trajectory from the non-equilibrium one, one was able to extract a signal free of noise on temporal lengths significant enough to allow the calculation of the thermal conductivity in a wide range of values for the temperature gradient [18]. This methodology is however impractical in combination with FPMD calculations and, in addition, it does not allow for following temporal trajectories for sufficiently long intervals, since the correlation between non-equilibrium and equilibrium does not last more than a few ps.

12.3 Future: the quest of quantitative predictions goes on, thoughts, recommendations and some very recent results

The future of atomic-scale modeling in the area of disordered, glassy materials is intimately related to the ability of computational materials science to describe, predict and complement experiments when it comes to the properties of interest. Several strategies can be followed and are being followed by myself, my closest colleagues and, of course, many fine scientists around the world. In what follows I will list and detail what I think are the most adapted to achieve the goal of extended predictive power. This is achieved by developing a set of considerations that reflect my point of view on these issues. Examples of studies in progress will be given whenever available, by sticking to the general guideline (employed throughout the book) of highlighting our own realizations for the simple reason of being more reliable and documented on those than on what produced outside of my research environment. *This is just a choice, not implying at all that what is done 'in-house' is any better or more important that what achieved outside.*

1. The improvement of available schemes in terms of comparison with experiments has to remain a goal by itself. There is no point in settling for qualitative models on the grounds that there is a compelling need for some microscopic explanations of complex phenomena. Theory cannot be taken as an academic exercise open to anybody willing to get some 'non-experimental' input and available for any request. We have seen in this book several examples of structural properties calculated at the highest available level of theory (in terms of theoretical developments) and yet still far from reproducing experiments. For instance, let's take the case of the partial pair correlation function $g_{GeGe}(r)$ of chalcogenides as those reported in chapters 6 and 8. While the agreement with experiments is somewhat satisfactory and

in any case much better than any other obtained with empirical models, there is still room for improvements. This is even more compelling by keeping in mind that certain intermediate range properties, related mostly to Ge–Ge correlations, are far from being realistic (as the concentration–concentration partial structure factor $S_{CC}(k)$ of binary $GeSe_2$ systems). Does this mean that the search of the best description of chemical bonding and the study of bonding itself should be a consistent part of any research effort in the area of computational materials science and, in particular, in the case of disordered systems? My answer is an unconditional yes. An instructive example has been provided in [19] for the case of PCMs, for which a rational accounting for the specific bonding features of these systems have been proposed. Interestingly, the issues treated are much more general and applicable to almost any material.

2. In some laboratories, there is a lot of pressure on computational materials scientists (and, in general, to any scientist willing to look for explanations by using modeling tools) to get results as quickly as possible to complement experiments and complete papers with some theory flavor. This can occur at the expense of accuracy and overall reliability of the results. In that specific context, waiting a few months or even a year is not considered acceptable, promoting the choice of empirical methods (as inadequate interatomic potentials) requiring less computational power and producing data much faster. I am strongly against this trend and I would like to encourage anybody getting involved in atomic-scale calculations to put the stakes very high in terms of quantitative predictions. A good strategy consists in planifying calculations on the basis of a reasonable period of time (at least a year) by looking for available resources, without shying away from writing proposals or doing it in connection with more experienced users. Should this not be possible, because of several reasons (complexity of the problem and conceptual issue unresolved, computational time beyond what is feasible to ask and obtain) there is no point in taking such a burden, the best choice being not to take any engagement more than doing low quality research of limited impact.
3. I have been working on disordered chalcogenides for many years and part of the production contains calculations on the same systems with close FPMD schemes (different exchange–correlation functionals (see section 6.7), dispersion forces (see chapter 10), cell dimensions (see chapter 9)). What is the use of being so repetitive on the choice of the target system? The idea is to have a well-defined benchmark choice for testing and improving the level of theory, by acquiring experience and consolidating results. This has to remain as a background project, not taking too much time and resources at the expenses of more innovative areas (see below).
4. Having established our strategic priorities: (1): be quantitative as much as possible by relying on the knowledge of chemical bonding, (2): devote the needed time and resources to reach this goal without falling into the trap of

suffering from the pressure of producing results that might be as 'quick' as 'dirty' and (3): rely on a family of systems available to work intensively on methodological aspects as an ideal sparring partner, it is time to move forward and introduce some ideas on research lines in progress or to be followed.

5. This book has shown applications increasingly closer to real experimental issues, for binary and ternary systems. Focusing on chalcogenides, results were first oriented toward the understanding of specific structural features (as in chapter 6) and became later motivated by the existence of systems having a strong technological impact, as phase change materials (chapter 11). Finally (section 12.2), we have combined structural and thermal properties in the search of specific behaviors on the nanoscale. This evolution in the topics addressed marks exactly the direction worth pursuing in the years to come: FPMD is the unavoidable atomic-scale, theoretical counterpart of applications in the area of energy management and optimization, via the conception and control of new, disordered materials.
6. An example of what can be targeted when aiming at amorphous materials having strong potential for applications is provided by a recent project on energy storage systems (batteries) currently in progress within our research team and in collaboration with the IRCER (Institut de Recherches sur les Céramiques) in Limoges (France). The focus is on alkali ion batteries (AIBs) such as lithium ion batteries (LIBs) and sodium ion batteries (NIBs) for which one would like to improve the electrochemical performances by selecting a cathode materials with enhanced performances. In principle, crystalline compounds with formula $A_xTm_yMO_z$ or $A_xTm_yOMO_z$ (A = Li or Na; Tm = transition metal (V, Fe); M = P or Si) have to be preferred to simpler transition metal oxide systems [20]. However, it occurred that some drawbacks inherent in the crystalline host materials can be circumvented by relying on the glass phase counterparts and, in particular, on glass–ceramics materials (crystallites embedded in a glassy matrix) [21]. In view of the lack of precise understanding of the glassy and glass–ceramics atomic-scale structures and their interface chemistry, an approach based on atomic-scale modeling of the kind detailed in this book appears to be highly valuable (see figure 12.7).
7. A final word concerns Machine Learning approaches (for a strictly non exhaustive list of examples, see [22–25]), considered and almost worshiped by many colleagues as the holy grail of materials modeling for the future. This approach has been briefly touched upon in chapters 3 and 4 as an intelligent and quantitative tool to generate interatomic potentials devoid of any empirical ingredient and based on a large collection of first-principles data. The ultimate goal of ML efforts (when combined to first-principles calculations) is to increase the system sizes of models with essentially no limits and no sacrifice in terms of chemical accuracy. While there is no point in arguing about the interest and the tremendous impact of ML potentials, I have to

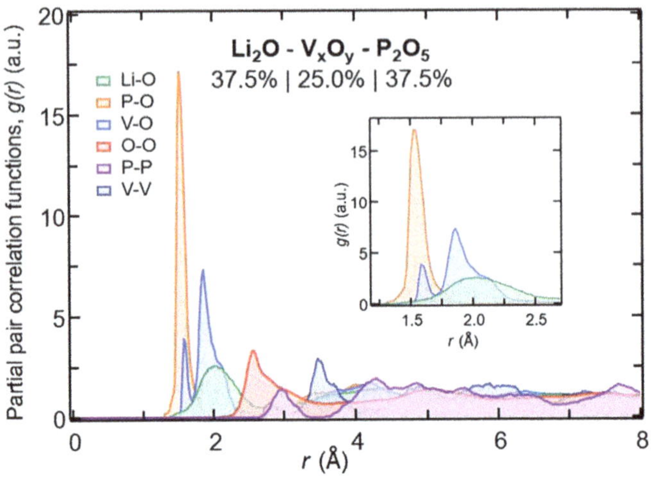

Figure 12.7. Partial pair correlation functions for a LiVP oxide glass. Calculations have been performed with the Born–Oppenheimer methodology (employed, for instance, in section 4.5) for a system of $N_{at} = 215$ atoms. Courtesy of Guido Ori (IPCMS).

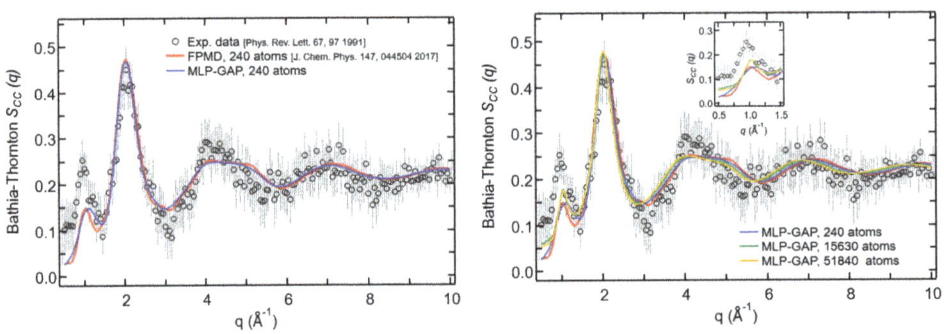

Figure 12.8. Concentration–concentration partial structure factor $S_{CC}(k)$ of liquid $GeSe_2$. On the left: comparison between ML potential (MLP), FPMD data and experimental data. On the right, ML results for three different system sizes. Courtesy of Guido Ori (IPCMS).

admit not having enough time to devote to such developments, by preferring to lay on the side of those collecting the best possible FPMD data out of the calculations described in this book. However, my coworker Guido Ori together with Assil Bouzid (IRCER, Limoges) have rapidly become experts in this area, by producing data on a set of systems like those invoked above (complex transition metal glasses) and chalcogenides. The prototypical case of liquid $GeSe_2$ (treated extensively in chapter 6) is shown in figure 12.8 by considering the concentration–concentration structure factors $S_{CC}(k)$. In view of the elusive nature of intermediate range order properties in this system, it appeared worthwhile to promote the creation of a large simulation cell evolving in time with an affordable atomic-scale tool. The prescription followed is the kernel-based Gaussian approximation potential (GAP) scheme [26]. The FPMD database was built from 300 configurations of 120–480 atoms models sampled at different temperatures in between $T = 300$ K and $T = 1100$ K. Two considerations are in order. First, the GAP scheme is able to reproduce the first-principles data with a very high degree of accuracy. Second, the largest system features a moderate and yet encouraging increase in the intensity of the FSDP in $S_{CC}(k)$, by stimulating further calculations of this kind for even larger systems.

References

[1] Giacomazzi L, Massobrio C and Pasquarello A 2007 *Phys. Rev.* B **75** 174207
[2] Vashishta P, Kalia R K, Antonio G A and Ebbsjö I 1989 *Phys. Rev. Lett.* **62** 1651–4
[3] Vashishta P, Kalia R K and Ebbsjö I 1989 *Phys. Rev.* B **39** 6034–47
[4] Iyetomi H, Vashishta P and Kalia R K 1991 *Phys. Rev.* B **43** 1726–34
[5] Umari P and Pasquarello A 2005 *Diam. Relat. Mater.* **14** 1255–61
[6] Umari P and Pasquarello A 2007 *Phys. Rev. Lett.* **98** 176402
[7] Umari P and Pasquarello A 2003 *J. Phys.: Condens. Matter* **15** S1547–S1552
[8] Lampin E, Palla P L, Francioso P-A and Cleri F 2013 *J. Appl. Phys.* **114** 033525
[9] Zaoui H, Palla P L, Cleri F and Lampin E 2016 *Phys. Rev.* B **94** 054304
[10] Duong T-Q, Massobrio C, Ori G, Boero M and Martin E 2019 *Phys. Rev. Mater.* **3** 105401
[11] Bouzid A, Zaoui H, Palla P L, Ori G, Boero M, Massobrio C, Cleri F and Lampin E 2017 *Phys. Chem. Chem. Phys.* **19** 9729–32
[12] Alvarez F X and Jou D 2007 *Appl. Phys. Lett.* **90** 083109
[13] Zhang S-N, He J, Zhu T-J, Zhao X-B and Tritt T M 2009 *J. Non-Cryst. Solids* **355** 79–83
[14] Lee J, Bozorg-Grayeli E, Kim S B, Asheghi M, Philip-Wong H-S and Goodson K E 2013 *Appl. Phys. Lett.* **102** 191911
[15] Duong T-Q, Bouzid A, Massobrio C, Ori G, Boero M and Martin E 2021 *RSC Adv.* **11** 10747–52
[16] Martin E, Ori G, Duong T-Q, Boero M and Massobrio C 2022 *J. Non-Cryst. Solids* **581** 121434
[17] Anufriev R, Tachikawa S, Gluchko S, Nakayama Y, Kawamura T, Jalabert L and Nomura M 2020 *Appl. Phys. Lett.* **117** 093103
[18] Massobrio C and Ciccotti G 1984 *Phys. Rev.* A **30** 3191–7
[19] Lee T H and Elliott S R 2020 *Adv. Mater.* **32** 2000340
[20] Stanley Whittingham M, Siu C and Ding J 2018 *Acc. Chem. Res.* **51** 258–64
[21] Nagamine K, Honma T and Komatsu T 2011 *J. Power Sources* **196** 9618–24
[22] Behler J and Parrinello M 2007 *Phys. Rev. Lett.* **98** 146401
[23] Noé F, Tkatchenko A, Muller K-R and Clementi C 2020 *Ann. Rev. Phys. Chem.* **71** 361–90
[24] Bartók A P, Kermode J, Bernstein N and Csanyi G 2018 *Phys. Rev.* X **8** 041048
[25] Zhang L, Han J, Wang H, Car R and Weinan E 2018 *Phys. Rev. Lett.* **120** 143001
[26] Bartók A P, Payne M C, Kondor R and Csányi G 2010 *Phys. Rev. Lett.* **104** 136403

www.ingramcontent.com/pod-product-compliance
Ingram Content Group UK Ltd.
Pitfield, Milton Keynes, MK11 3LW, UK
UKHW051902100926
8556IPUK00005B/736

9 780750 324342